Dahlem Workshop Reports
Life Sciences Research Report 27
Minorities: Community and Identity

The goal of this Dahlem Workshop is:
to describe and illuminate the humane options
for the relations between minorities and the larger society

Life Sciences Research Reports
Editor: Silke Bernhard

Held and published on behalf of the
Stifterverband für die Deutsche Wissenschaft

Sponsored by:
Deutsche Forschungsgemeinschaft
Senat der Stadt Berlin
Stifterverband für die Deutsche Wissenschaft

Minorities: Community and Identity

C. Fried, Editor

Report of the Dahlem Workshop on
Minorities: Community and Identity
Berlin 1982, Nov. 28 – Dec. 3

Rapporteurs:
A. M. A. Dummett · A. D. Murray · V. Saifullah Khan · H. Shue

Program Advisory Committee:
C. Fried, Chairperson · C. Adler · C. F. Graumann
A. D. Murray · O. Patterson · M. Walzer

Springer-Verlag
Berlin Heidelberg New York Tokyo 1983

Copy Editors: M. A. Cervantes-Waldmann, M. Grunwald, K. McWhirter
Photographs: E. P. Thonke

With 4 photographs, 4 figures, and 14 tables

ISBN-13: 987-3-642-69313-7 e-ISBN-13: 987-3-642-69311-3
DOI: 10.1007/987-3-642-69311-3

CIP-Kurztitelaufnahme der Deutschen Bibliothek:

Minorities: community and identity: report of the Dahlem Workshop on Minorities: community and identity, Berlin 1982, Nov. 28 - Dec. 3/ C. Fried, ed. Rapporteurs: A. M. A. Dummett ... [Held and publ. on behalf of the Stifterverb. für d. Dt. Wiss., Sponsored by: Dt. Forschungsgemeinschaft ...]. – Berlin; Heidelberg; New York; Tokyo: Springer, 1983.

(Life sciences research report; 27)
(Dahlem Workshop reports)

NE: Fried, Charles [Hrsg:]; Dummett, Ann M. A. [Mitverf.]; Workshop on Minorities, Community and Identity <1982, Berlin, West>; 1. CT

Softcover reprint of the hardcover 1st edition 1983

EDP and Photocomposition: Satz-Rechen-Zentrum, D-1000 Berlin 30

2131/3014-5 4 3 2 1 0

TABLE OF CONTENTS

THE DAHLEM KONFERENZEN

Founders

Recognizing the need for more effective communication between scientists, especially in the natural sciences, the Stifterverband für die Deutsche Wissenschaft*, in cooperation with the Deutsche Forschungsgemeinschaft**, founded Dahlem Konferenzen in 1974. The project is financed by the founders and the Senate of the City of Berlin.

Name

Dahlem Konferenzen was named after the district of Berlin called "Dahlem", which has a long-standing tradition and reputation in the arts and sciences.

Aim

The task of Dahlem Konferenzen is to promote international, interdisciplinary exchange of scientific information and ideas, to stimulate international cooperation in research, and to develop and test new models conducive to more effective communication between scientists.

Dahlem Workshop Model

Dahlem Konferenzen organizes four workshops per year, each with a limited number of participants. Since no type of scientific meeting proved effective enough, Dahlem Konferenzen had to create its own concept. This concept has been tested and varied over the years, and has evolved into its present form which is known as the *Dahlem Workshop Model*. This model provides the

---

* *The Donors Association for the Promotion of Sciences and Humanities*

** *German Science Foundation*

framework for the utmost possible interdisciplinary communication and cooperation between scientists in a given time period.

The main work of the Dahlem Workshops is done in four interdisciplinary discussion groups. Lectures are not given. Instead, selected participants write background papers providing a review of the field rather than a report on individual work. These are circulated to all participants before the meeting to provide a basis for discussion. During the workshop, the members of the four groups prepare reports reflecting their discussions and providing suggestions for future research needs.

Topics
The topics are chosen from the fields of the Life Sciences and the Physical, Chemical, and Earth Sciences. They are of contemporary international interest, interdisciplinary in nature, and problem-oriented. Once a year, topic suggestions are submitted to a scientific board for approval.

Participants
For each workshop participants are selected exclusively by special Program Advisory Committees. Selection is based on international scientific reputation alone, although a balance between European and American scientists is attempted. Exception is made for younger German scientists.

Publications
The results of the workshops are the Dahlem Workshop Reports, reviewed by selected participants and carefully edited by the editor of each volume. The reports are multidisciplinary surveys by the most internationally distinguished scientists and are based on discussions of new data, experiments, advanced new concepts, techniques, and models. Each report also reviews areas of priority interests and indicates directions for future research on a given topic.

The Dahlem Workshop Reports are published in two series:
1) Life Sciences Research Reports (LS), and
2) Physical, Chemical, and Earth Sciences Research Reports (PC).

Director
Silke Bernhard, M.D.

Address
Dahlem Konferenzen
Wallotstrasse 19
1000 Berlin 33, F.R. Germany

Minorities: Community and Identity, ed. C. Fried, pp. 1–7
Dahlem Konferenzen 1983. Berlin, Heidelberg, New York, Tokyo: Springer-Verlag.

INTRODUCTION

C. Fried
Harvard Law School
Cambridge, MA 02138, USA

Structuring relations between minorities and the larger society of which they are a part poses a major challenge to both goodwill and human ingenuity. Some political problems are engendered by need and ambition in situations of scarcity. In such cases the contestants have no difficulty recognizing and acknowledging the validity of their competitors' demands: for they are not different from their own. Individual "guestworkers" in West Germany desire better housing and the same right as German nationals to make a permanent home there. Individual French Canadians seek the same opportunities for economic advancement or for political and social status as English speakers. Individual Pakistanis in Great Britain desire security from personal violence, and West Indians a sense that they are treated fairly by the police. But our concerns have additional poignancy because so often they arise where no contestant wishes to harm the *individual* members of an opposing group, where all contestants want only what is best for each other. For instance, opponents of bilingual education are often motivated by a genuine desire to enhance the life chances of children of minority linguistic communities.

The United Nations' Universal Declaration of Human Rights, for instance [General Assembly resolution, 10 December 1948], recognizes that "everyone has the right of peaceful assembly and association" [section 20-1] and that "no one may be compelled to belong to an association" [section 20-2]. These principles are immediately followed by the statement that "everyone has the right to take part in the government of his country" [section 21-1]. These unexceptionable principles gain ready acceptance by men and women of goodwill in the "humane societies" mentioned in our workshop goal. Unfortunately, these same principles may and often do come into what many consider deep conflict. Where persons feel strong ties to a linguistic, cultural, or ethnic minority and this minority is either too small, too weak, or too dispersed to attain representation in a community's governing bodies, the members of that minority may feel with some plausibility that they have indeed been "compelled to belong to an association," i.e., to whatever coalition they must make with outsiders to make their "right to take part in the government of their country" effective.

Similarly, while the Declaration states "that everyone has a right to education" [section 26-1] and acknowledges that "parents have a right to choose the kind of education that shall be given to their children" [section 26-3], and that "everyone has a right freely to participate in the cultural life of the community..." [section 27-1], yet if education is compulsory, resources are limited, and a minority too weak to organize alternative school systems, then the guarantee of such rights seems hollow. In the same way, participation in the "cultural life of the community" depends on a conception of community which minority and majority may not share. Moreover, even the conception of participation, if it is to be participation in the larger community, is controversial. A linguistic or ethnic minority may believe that its members may best participate in the larger community by contributing the richness and diversity of their particular culture as a group. But such a contribution may be impeded through lack of resources or through lack of institutional channels to receive contributions aggregated in this particular way.

Indeed the perplexities and ambiguities of the subject are nicely captured by section 29-1 of the Universal Declaration which holds that "everyone has duties to the community in which alone the free and full development of his personality is possible." Is the "which" clause a clause of apposition or of modification, i.e., are these primary duties owed to, and does the possibility of fulfillment come first of all from the larger community, the nation, or are an individual's duties to a community relative to the degree to which that individual may find his "free and full development" in just that community?

Both the intellectual fascination and the practical urgency of this topic derive from these ambiguities and perplexities. These ambiguities and perplexities are concrete expressions of the perennial theoretical conflict about the moral, political, and ontological claims of individuals and the community to which they belong. Do communities have rights, indeed even an existence, which are not merely the hypostasis of the individual rights and existences collected in them?

This conflict is then more striking as it was a conscious decision of the organizers of the workshop to focus attention on what might broadly be called liberal democracies: those societies which share a commitment to the principles of democratic participation, to the right of equal concern and respect of all members of the community, and to the basic liberties of association, expression, and thought. Ours was not the smug premise, however, that every society which proclaims these principles is sufficiently or even truly devoted to them. But we did assume that we would have enough to do if we explored the implications of these widely shared ideals for the topic of linguistic, ethnic, and national minorities as these problems arise in societies where an appeal to them is not an empty gesture. The nations from which our participants were drawn are societies in which appeal to these principles has some point. They are all societies in which the efforts of politicians and the intelligence of scholars need not be devoted exclusively to the tactical issues of winning some modicum of respect for basic human rights from unwilling regimes. And yet all these societies have experienced significant difficulty in determining what the concrete meaning in actual situations of these general principles might be.

The concrete fields of struggle have reappeared in all of these societies:

Political representation: What is the role of minority communities in mediating the interests of their members in representative bodies? Universalistic claims which deny any role to units mediating between individuals and the state, coupled with political representation strictly drawn on geographic lines without regard to realities of group allegiance proclaim the membership and equality of the citizen vis-a-vis the state, but often only at the price of depriving the larger community of the opportunity to hear in an effective way the voice of constituent groups. Claims for guaranteed representation of minorities are often greeted with opposition, on the other hand, for such claims are thought to undermine the very principles of the universalistic state. And yet rarely if ever are political or representational subdivisions drawn without some consideration of the identity of interests of those within them.

Language: Although no liberal democracy formally penalizes the use of minority languages, language — like culture — requires not only a critical mass but opportunities for continuing and pervasive involvement in daily life for its survival. On the other hand, nothing is as likely so palpably to both encapsulate and therefore to isolate a group as linguistic barriers. It is no accident that the slavic word for those not speaking slavic languages meant those who are dumb, while the Greek word meant those who make senseless noises. Where linguistic communities are roughly equal in size, multilingualism is not only a possibility but a likelihood. For linguistic minorities, however, this accommodation is not always available and the dilemmas can be extreme.

Education: The conflict becomes acute in this domain because our focus is not on higher education, where (as in the cultural domain generally) free choice is available to mediate the divergent pulls of parochial claims and the universalistic claims of the national society. The real concern is with the education of children and adolescents where someone must make choices for those to be educated. Since the French Revolution one democratic strain has claimed that here above all it is the nation-state, the general society which must determine the content of education and thus mold the minds of the young. But such universalistic claims go along with the most certain means for destroying diversity and undermining the identity of minority cultures. The rough homogenizing power of the state may, however, be replaced by the petty tyrannies of village tyrants, imposing limited visions and blocking off at an early age the individual's surest avenue of escape from a parochialism he may not wish to share. An emphasis on parental choice has seemed to some nothing but an unprincipled compromise leading to a random resolution of a conflict which should be resolved rather on reason and general principles.

Employment and economic rights: The claim of minorities to represent their members and indeed to be counted as distinct units in the productive and distributive processes recaptures in the economic realm the preceding dilemmas. National markets, national and even international aggregations of capital, and large-scale organizations of workers into national unions appear to disregard the distinct claims of members of minority groups to distinct treatment taking into account the groups' special history, culture, and collective interests. Members are not, it is said, only workers to be employed and compensated according to their potential contribution to a national economy as evaluated by national markets, but, say, black workers with a distinct history, distinct disabilities, distinct potentials, and a distinct claim in justice to a share of economic opportunity as a group. Analogously there is the claim that industries should be developed and helped to thrive not only as their products are valued in an impersonal national or international market, but as they represent an opportunity for the communal release of group productive energies.

Cultural phenomena: Just as the modern state tends to produce a more or less homogenized market for economic goods, the scale and technologies of the media for cultural expression and dissemination are sometimes seen to exert a homogenizing influence, and to deprecate and finally render unavailable the contribution of distinct groups within the society. Yet with only small changes these same technologies — e.g., cable television — might provide facilities for group cohesion and expression.

These concrete areas of conflict are not invoked uniquely in respect to racial, linguistic, or national minorities. Wherever intermediate units claim to stand between the individual and the state these conceptual and practical problems arise: If man is fulfilled only within society, how are the units of that society to be defined, and how are they to interact? In a state where mediating units are universal — as, for instance, in any federated state — the conflict may be moderated, but it does not disappear altogether, and the kinds of minorities which have been the focus of this workshop indicate the potential for the conflict even in such national forms. In a federated republic in a sense everyone belongs to a minority. The problems of the minorities we consider in this volume are especially acute and especially poignant because the group identities they represent are especially and acutely distinct; the members of the minorities are *more* of a minority than are the citizens of a federated region in a federal republic. They are a distinct group in whatever group they claim as their home. Moreover, because membership in these minorities is rarely voluntary — outsiders cannot easily choose to belong, and insiders are often prevented from ceasing to belong by the surrounding society — these minorities pose an especial challenge to the rationalistic and universalistic claims of the modern nation-state. They remind us of the indigestible contingencies of human history, human diversity, and human destiny. Perhaps we cannot control entirely who we are (or who we are not), because we cannot control who we have been. And perhaps it would be unhealthy and even monstrous were we to try.

These problems of accommodation to the special claims of minorities, which are a part of the larger society, were the concern of the fourth working group dealing with minority rights. The issue of a nation's claim to maintain a particular identity and character are raised again with a different force in the work of the third group, dealing with problems of national boundaries and their permeability. The arbitrariness and historicity of all modern states mocks any pretensions to an excessive rationalism and universalism about state forms and responses to the problems of minorities. The few fortunate which can proclaim that for *them* at least no conflict exists, may be e such smugness only by virtue of a peculiar history which drew tight g(cal boundaries around their homogenous society. And as the case o such as Sweden may show, changed circumstances bring these knocking at the door of the neatest and tidiest edifices. It is at least that one of the effective ways in which universalistic dogmas may b(easy dominion is by a stringent and exclusionist immigration policy. a complementary irony that those societies which at one time defi borders and populations in the most confident universalistic terms (] France and Great Britain, who in this respect repeated the univ claims of all great empires), today are faced with minority challenges they scarcely know how to respond.

These are problems both of political theory and of practical policy. / feature of the workshop which this volume records is a recognition theoretical and policy concerns must not be allowed to proceed wi fullest attention to the way the world really is, the way people really first of the working groups and its background papers considered th(logical phenomena on which this problem rests: When do persons identify themselves as and with minority groups; how does such iden come about, and what concretely does it mean? Without such a soli(psychological fact, we are in danger of solving problems which do not commodating sensibilities which are not really felt, while paying a real desires thwarted, real opportunities missed. Needless to say, n theory of minority identification can be hoped for. Psychology will r(within itself — as Plato was the first to remark — the antinomies we ha at the social level. The universalizing vision at the level of statecraft is by a vision of infinite plasticity and infinite rationality at the level of i al psychology. And, of course, the proponents of group rights an individual solidarity at the psychological level will emphasize the cor and historicity of the self's constitution.

The second working group undertook a parallel inquiry into socia They looked into the pragmatic and empirical occasions as well as the cal foundations for conflict between minority groups and the large Here, too, the theory of conflict and its occasions reproduces the gen ceptual problem. Some view the group as an opportunistic vehicle, a constitutive potential of groups is overlooked, if the intrinsic interest in group membership and group realization is not taken into account. In its most radical form, this view does not just put group interests on a par with individual interests to be furthered by opportunistic coalitions, but denies the very conceptual sense of individual interests, insisting instead that it is groups which shape and define those interests. Neither this workshop nor any empirical or theoretical sociological work is likely to lay these theoretical disputes at rest. Nevertheless, the survey of what empiricists have observed provides a necessary anchor in reality for theoretical and policy speculation.

In the end what emerges are a series of reasoned but pragmatic compromises of a dilemma which may be theoretically insoluble, and which in any event is so controverted that practical proposals cannot await theoretical consensus. These pragmatic perspectives are not, however, unprincipled and I believe they are far from trivial. Underlying our divergent theoretical and practical perspectives are shared commitments to democratic principles, to humane and tolerant treatment of persons, and to the vision of a diverse humanity, which in its very diversity unfolds the richness of what man may be. The societies from which we come and the passions which led us to explore these questions share at least a commitment to maintaining the conditions under which inquiry may go forward freely, and human development might prosper. At the very least history has chastened our dogmatism. What we offer here should at every point be useful and in some points even inspiring.

This volume contains two sorts of documents: (a) background papers submitted by some of the participants prior to the workshop and (b) group reports which are the product of the discussions at the meeting. The background papers do not and are not intended to offer a comprehensive survey of data, issues, and perspectives. They are rather the views of some of the participants and provided some of the stimuli for the intensive, week-long deliberations of the working groups. The group reports, on the other hand, do aim at such definitiveness and comprehensiveness as the subject allows.

Minorities: Community and Identity, ed. C. Fried, pp. 9–23
Dahlem Konferenzen 1983. Berlin, Heidelberg, New York, Tokyo: Springer-Verlag.

TOWARDS THE DEVELOPMENT OF A TYPOLOGY OF MINORITIES

F. Heckmann
Hochschule für Wirtschaft und Politik
2000 Hamburg 13, F.R. Germany

INTRODUCTION

In speaking about minorities, one is astounded by the heterogeneity of population groups that fall under this heading: the disabled; guestworkers; convicts and ex-convicts; the mentally ill; "colored persons" (i.e., the racially discriminated); national, ethnic, regional, and religious groups; drug addicts; the homeless; vagrants; gypsies; prostitutes; and homosexuals. This in no way complete list is so unsystematic as to indicate the problematic nature of such a categorization (cf. (4) for a review). It is vastly similar to the concept of marginal groups (cf. (22), p. 204) in that both exhibit an extremely vague spectrum of meaning and both are used in very diverse contexts. This paper will concentrate on this inadequacy and suggest a means of dealing with it by historically and genetically reconstructing a sociostructural typology of ethnic minorities, thus enabling us to clarify what type of minority is involved in regard to the guestworkers in the F.R. Germany. However, an analysis of the general concept of a "minority" will be presented first, and an elaboration and criticism of its faults will show us the need for my proposal of a more precise differentiation.

The following definitions of "minority" are intended to establish the prevalent conceptions. They have all been selected from representative sources:

"Minorities are particular racial, cultural, religious or national groups, who, although living among other groups, do not fully share in the culture that they are a part of" ((2), keyword "minority").

Minority is the "designation of groups that are segments of an embracing group or society, who differ from the majority (in the sense of the dominant group) by certain characteristics (e.g., racial, linguistic, confessional) which the dominant group holds to be of less value than its own relevant characteristics" ((8), keyword "minority").

Minorities are "those nondominant groups in a population which possess and wish to preserve stable ethnic, religious, or linguistic traditions or characteristics markedly different from the rest of the population" ((21) as cited by (19)).

"1) Minorities are subordinate segments of complex state societies; 2) minorities have special physical or cultural traits which are held in low esteem by the dominant segments of the society; 3) minorities are self-conscious units...; 4) membership in a minority is transmitted by a rule of descent which is capable of affiliating succeeding generations even in the absence of readily apparent special cultural or physical traits; 5) minority peoples, by choice or necessity, tend to marry within the group" (23).

The concept of a minority depicted here denotes the common characteristics of diverse population groups, mainly referring to the experience of discrimination, oppression, and certain forms of "deviation" and "being different" as well as to forms of awareness and consciousness of the same. Not included in this concept are differences between these groups in other dimensions of the social situation and economic and social position, as well as differences in behavior. These definitions proceed nominalistically , indexing descriptive characteristics and summarizing certain evident common denominators without taking into account the origin of the situation or the concomitant factors of the sociostructural position and consciousness. This deficiency is certainly not without consequences. The subsumption of very heterogeneous population groups under one heading implies an a priori decision that we are working with a uniform area of investigation of homogeneous population groups whose social situation, behavior, and consciousness can be explained by a common theoretical starting point. It implies beginning with identities, common interests, and similarities of such groups as oppressed Indians, guest-workers, the disabled, national minorities, or the homeless. This being the object of discrimination and prejudice (i.e., "object status") leads to the conclusion that other common features exist.

The following analysis will be limited to dealing with the conceptual differentiation of "ethnic minorities," i.e., various population groups differing socioculturally and in terms of consciousness from the "majority."

THE DEVELOPMENT OF A TYPOLOGY OF ETHNIC MINORITIES

When concepts such as "minority" are used in a multitude of different contexts, thus exhibiting a broad semantic scope with relatively vague boundaries there are two ways in which to attempt to define these meanings more precisely: a) the nominalistic procedure, by which a meaning can be determined and defined; and b) the historicosystematic procedure, by which meanings can be reconstructed historically and genetically, i.e., from the context of their origin and of an analysis of their specific subject area.*

The historicosystematic typology of ethnic minorities presented here will be developed within the context of an analysis of three large-scale historical processes: a) the foundation of the modern nation-state (national and regional minorities); b) the great internal and international migrations as a result of the dissolution of the feudal-agrarian mode of production and social structure (immigrant minorities); and c) the outcome of colonialism and the subsequent foundation of "young" nation-states (minority peoples, new national minorities).

* It is not possible to exclude a measure of arbitrariness by using the nominalistic procedure, in that the criteria for the determination of the premises and empirical components which go into the definition must be selected and fixed. "If one wants to exclude the element of arbitrariness from the definition. . .one must relate to the concept insofar as it excludes arbitrariness, i.e., relate to the history of the concept" ((20), p. 5). The historicosystematic procedure for analyzing and determining concepts, on the other hand, cannot stop at mere reconstruction of differences in meaning; it makes possible the identification of different objects, areas, and frames of reference of meanings within the context of the history of ideas, of science and of history proper, thus allowing precision and differentiation oriented on the traditional historical meaning. In addition, consideration of traditional meanings makes an alternative usage of the concept possible by the "meaningful" transference of these traditional meanings to new contents. Furthermore, creations of new meanings can arise, which start with the his

NATION-STATE AND MINORITIES

National Minorities

The ideas of "minority" and "majority" came to Germany as a concept of parliamentarism along with other ideas of the French Revolution and was germanized to "Minderheit." It referred to those who were defeated in an election or ballot and to those excluded from political control (cf. (6), p. 206). In German, according to the dictionary of Jacob and Wilhelm Grimm (cf. (18), p. 9)*, "Minderheit" also means "inferior in dignity" and "inferior in strength." Ever since the revolutionary process in France, a further meaning began to emerge in Europe: the ideas of a nation and nation-state spread simultaneously with the proclamation of universal human and civil rights by the French Revolution. The concept was thus endowed with a constitutional and international sense describing population groups with a particular national identity, whose domain was located within the political organization of another nation — the majority — but who were excluded from equal participation in the political process and prevented from exercising civil rights. Societal processes underlie this conceptual development which can be briefly described as the changeover from a territorial state to a bourgeois nation-state. Whereas the territorial state, whether feudal or late feudal (i.e., absolutist), tended to be indifferent towards the national identity of its population (its unity being ideologically secured by the institution of the monarch), the bourgeois state strove for the unity of political organization and national identity of its inhabitants. The term "state people" (Staatsvolk) originated as an expression of these developments, and the groups of people with a different national identity living on "foreign" state territory became minorities.

Here is a definitional summary of the historically related dimensions of the concept of a national minority: national minorities are sociostructurally heterogeneous groups who live within a state territory foreign to their national identity, culture, and history as a consequence of the constitution of the bourgeois nation-state which is based on historic settlement structures or changes in state boundaries resulting from agreements or conflicts among nation-states. As citizens they are deprived of certain civil rights or the exercise thereof, and they are typically subjected to strong pressure to assimilate. Their political aim is the accession to the nation-state to which they belong in terms of their historicocultural identity and which furthers their economic interests.

* These elements of meaning are still included in the general concept of a minority.

Regional Minorities

It has become fashionable to speak of "the Indians of western Europe" (cf., e.g., (14)), referring to the Bretons, Corsicans, Alsatians, Occitanians, Sardinians, Catalans, Basques, Welsh, Scots, and other minorities. A more exact conceptual analysis of these minorities is still unresolved and will not be done here, since the function of this paper is primarily one of differentiation. The term regional minorities refers to political and cultural movements in certain regions of France, Great Britain, Spain, and Italy, which have rapidly become influential in the last five years.

The comparison of the above-mentioned politicocultural movements with the American Indians is a superficial one and essentially false: The minority situation of the Indians and other aboriginal peoples is a result of the establishment of colonies and colonial states, of the dispossession, physical annihilation, and forcing of the natives into reservations, thus allowing the settlement of colonists on the stolen territory. In a later section I suggest describing this minority situation with the term "minority people." However, Bretons, Welsh, or Sardinians are not minority peoples in this sense. Their minority situation is a result of the establishment of the modern nation-state.

The establishment of nation-states resulted in the formation of hitherto relatively independent, political, territorial, and cultural units into new political and social structures that aspired to political, economic, and cultural homogeneity. Territories that had once possessed their own identity and history were reduced to the status of mere regions by the use of force, by administrative measures, and above all by the leveling influence of the industrial mode of production. Languages degenerated to mere dialects.

This recourse to a pre-nation-state politicocultural identity which was lost or repressed during the course of the establishment and consolidation of the modern nation-state has become the ideological impetus for the "regional minorities" movements. To my knowledge, the factors that gave rise to these movements have not yet been investigated,* however, it would appear that disparate economic and social development and discrimination are important triggers. "Jean" ((14), p. 11) points out that the underdeveloped regions of the European Economic Community (EEC) — those with the lowest gross

* Two rather recent German publications are available: "Jean" (14) and Gustafson (9). These works are very impressionistic and scientifically not convincing, but they do contain a lot of important information.

national product — are by and large identical with the territories of the regional minorities.* I intentionally referred to *disparate* development above because not only underdevelopment but also comparatively fast development without being able to fully participate in the advantages of this development because of national redistribution, appear to provide a strong impetus for the rise of regional movements. An example of this is Catalonia in Spain (cf. (11)). In Scotland, the stagnation of "old" industries and the prospect of a fast economic recovery based on the discovery of offshore oil reserves seem to have converged to make the Scottish movement a force to be reckoned with in political calculations.

Both national and regional minorities are similar in their heterogeneous social structures. However, regional minorities differ from national minorities in their political perspective: in general they strive towards cultural and political autonomy,† orienting themselves towards the pre-nation-state. They are, however, not liberation movements; and they do not orient themselves towards uniting with the nation-state to which they belong historically and culturally, as do the national minorities. To summarize by way of definition: regional minorities are economically disadvantaged and culturally repressed populations of certain regions with heterogeneous sociostructural compositions, who orient themselves towards political and cultural traditions of the pre-nation-state in order to regain a political and cultural identity and attain economic improvement. These aims lead to the demand for political and cultural autonomy within the nation-state.

* As in the case of the Corsicans, e.g.: "Whereas the Corsicans are obliged to emigrate to the mainland, increasing numbers of French and foreigners are settling on the island. Frenchmen are favoured for open positions. The land owned by non-Corsicans is, on the average, 10 times greater than that owned by Corsicans.... The average income of a Corsican is only 50% that of the general income of the French. Food is 20-50% more expensive on Corsica than on the mainland" ((14), pp. 26-27).

† "Brittany, Alsace, Lorraine, Occitania, and Corsica suffer under the centralist policies of the French State and thus demand cultural and political autonomy for their regions" ((14), p. 11). Catalonia and Sardinia strive for the same (ibid.).
In Scotland and Wales, however, there are political forces that strive for complete political sovereignty for their regions (cf. (14), p. 11).

Immigrant Minorities

The immigration from Europe to the United States historically is the source of the second usage of the term "minority": it describes the social situation of the immigrant groups.*

During most of the 18th century the most important factor in American economic development was the acquisition, opening-up, and agricultural utilization of vast new territories. The turning point in this development occurred around 1890 when industry took over as the most important component of societal production (cf. (5), p. 267). For this rapidly expanding American industry, the immigrants were the most important source of labor. After the closing of the "frontier" in 1890, industrial work became almost the sole means of livelihood for the immigrants, thus marking the point in time when immigration became equated with worker immigration (cf. (12), p. 12).

It is this immigration of workers which provides such a fruitful heuristic reference point for a comparative analysis of the migration of guestworkers in western Europe. I have given a detailed account elsewhere (cf. (10), Part III) of the multitude of economic, social, and political factors existing which demonstrate that the guestworkers are not temporary migrant workers but immigrant populations. Historical comparisons with the immigration of workers to the United States underline these findings,† so that the guestworker population in the Federal Republic of Germany must also be understood as an immigrant minority and *not* as a national minority, marginal group, or subproletariat, as it is often designated in the literature or the mass media.

With respect to the immigrant farmers of the 19th century, one can only speak of a minority status in those cases where relatively closed territorial settlement took place and where the development of a nationally specific system of associations and institutions came about for the purpose of preserving social and cultural identity. The German immigrants in Texas are an example of such closed settlements; but these generally did not play a dominant role. Another

* In this sense the term was first introduced by Young (24).

† For example, a constituent characteristic of the immigration to the United States was the high rate of return to the country of origin. Also the majority of immigrant workers did not go to the United States with the intention of staying, but of returning to the homeland "economically restored" after a particular period of time. That the majority did, in fact, stay demonstrates only that immigration is a complex process and not a one-time decision made by the immigrant in his/her home country.

example of closed settlements, which is very interesting but numerically insignificant, is that of religious communities who "imported" their social system, as it were; they immigrated together and continued the system of their social relations on new territory. Francis ((6), p. 23) has traced the migration of the Mennonites from Russia to Canada, where they have resettled and reestablished an almost identical settlement and social system. They tend to isolate and shield themselves from assimilatory influences, indeed from any changing influences. Here we can also speak of a minority status which is reinforced by specific religious identity.

Thus a fundamental differentiation can be made between two types of immigrant minorities*: immigrant farmers who sociostructurally can be classed as belonging to the agrarian middle class, and immigrant workers who have the status of wage laborers. Immigrant minorities of the agrarian middle-class type, who are numerically not very significant, can be subdivided into those who possess a sociocultural system which was constituted by the interplay of immigration conditions and original background ("immigrant colonies"), and minorities with an "imported" social system, who tend towards isolationism. Immigrant workers are groups of predominantly rural origin, who usually work in the receiving country as the lowest class of the industrial work force and who live under discriminatory conditions. They develop a sociocultural system, the immigrant colony, which is the result of a combination of factors: background in the country of origin and living and working conditions in the receiving country. They are in a process of collective assimilation.

MODERN COLONIALISM AND THE GENESIS OF MINORITIES

In the course, and as a result, of the spread of modern colonialism, specific social structures arose in the colonial territories and later nation-states. In the literature, these structures are also described by the concept of minority, but they have to be differentiated from the hitherto investigated minority situations because of their different genesis and structure. The social status and situation of the blacks in North and South America; the Chinese and East Indian populations who "emigrated" to various parts of Southeast Asia and Africa, mainly under the system of "indentured service" (a system similar to that of slavery); the situation of the American Indians and other indigenous peoples; and also the ethnic heterogeneity of the young nation-states and its

* Our analysis does not include the emigration of members of the intelligentsia.

resultant problem; all these must be considered as important minority situations whose genesis and structure are only comprehensible within the context of modern colonialism. To characteristically classify these different social situations under the one heading of "minority" is superficial, blurring sociostructural differences instead of illuminating them.* Because of limited space, I shall confine my discussion to the minority situations of so-called indigenous peoples and ethnic minorities in the young nation-states, and not to the situation of the Afro-Americans† and those ethnic minorities who were deported under the system of indentured service.°

* This understanding of the concept is the basis of most of the English publications on the sociology of minorities. In textbooks, for example, the term "minority problem" is used to refer to the situation of the American Indians, blacks, but also of the Gypsies, Jews, and other groups (cf., e.g., Rose and Rose (11) or Barron (1) or Rose (17)).

† The situation of the majority of Afro-Americans in the United States — in spite of all the changes since World War II — can most adequately be described, in my opinion, as that of a lower class of quasi internal immigrant workers who are racially discriminated against. However, certain processes of differentiation have taken place: a) a so-called "Black Bourgeoisie" has arisen, which on closer inspection turns out to be a myth, i.e., it is found instead to be a "petite bourgeoisie" (cf. (7)); b) since World War II, the black population has been given more opportunities to hold civil service jobs, thereby increasing their chances of upward social mobility. At the same time, a new group of immigrant workers from Mexico — the Chicanos — have moved in, mostly to jobs at the lowest end of the wage-labor market.

° See Davie ((3), p. 305) for more details. For the present situation of especially Indian and Chinese minorities, what is relevant is that they have differentiated themselves sociostructurally from their original homogeneity as immigrant workers and today form a significant part of the "middle class," to some degree even of the "bourgeoisie," of the countries concerned. It will also not be possible to consider here the minority situation of groups from former colonial territories, e.g., the Indonesians or Moluccans in Holland, who collaborated with the colonial powers and emigrated to the "Motherland" after the establishment of an independent state in their homeland.

Colonial Settlement, Expropriation, Expulsion, and Repression of Indigenous Peoples: The Genesis of Minority Peoples

"...the vast, uninhabited territories of America are fertile and suitable for colonization for they are devoid of any civilized population; there are only savage and barbarous beings here that wander hither and thither — scarcely different from the wild animals of the country!" These words of William Bradford (cited according to Jacobs et al. (13), p. 3) illustrate the attitude of Europeans who began to settle in the colonies. The land of the indigenous population was regarded as No-Man's-Land, just waiting to be inhabited by the European colonists. The original inhabitants who were at the tribal society stage of development were robbed of their land and other possessions, killed, driven out, taken prisoner, decimated by disease, demoralized by alcohol, their culture destroyed, and finally, forced to live on so-called reservations. Examples of such tribal populations (25) are: the Indians of North America, the South American plains, and the West Indies; the aborigines of Australia; the pygmies and bushmen of Africa; and, in part, the Eskimos. The repression of these peoples usually resulted in forms of collective apathy and is seen in the extremely high child mortality, low life expectancy, and the high suicide rate. However, today there are isolated signs of growing political consciousness and resistance, e.g., among the North American Indians.

In most of the relevant literature, the social situation of the descendants of these indigenous peoples is also superficially described by the concept of minority. Included in this description are certain partial similarities to other minorities, but not the specifics of the historical origin of these groups, which is necessary for the understanding of their contemporary situation and its developmental tendencies. As these groups are really the descendants of independent peoples who have preserved certain aspects of their historic culture or are bringing these to life again, it seems that the adequate category here would be the concept of a "minority people" as used by Petrova-Averkiyeva (15). A definitional summary of these points would be: minority peoples are descendants of the indigenous populations of territories that were conquered and colonized, whose traditional economic means of existence was taken from them in the process of expropriation, liquidation, and expulsion, and whose social structure and culture were vastly destroyed. They are mostly in an unresisting state of economic, psychosocial, and physical deterioration and are almost totally excluded from the labor process.

New National Minorities

The young nation-states of the Third World were constituted as political units almost exclusively on the basis of the colonial division of territories and their borders. These territorial divisions resulted from the economic, strategic, and political calculations of the colonial powers. "Divide et impera" ("Divide and Rule") was the political guiding principle of the colonial governments, and it meant the splitting up of existing societies and populations and their reformation into ethnically heterogeneous populations under colonial administration. The ethnic and cultural differences of these populations were often reinforced by disparate levels of economic and social development. Minority situations, i.e., the suppression of or discrimination against certain ethnic groups in the young nation-states, result mainly from the struggle of the ruling "elites" of the various ethnic and regional groups for hegemony, i.e., result from the attempt of one ethnic group to establish itself as the dominant "state people" (Staatsvolk), and from the interweaving of these processes through the system of international relations. The Nigeria-Biafra conflict serves as an illustration of this. Politically and economically oppressed ethnic groups within the young nation-states, who are prevented from exercising their civil rights and from developing their own culture, shall be described as "new national minorities." This concept can only be applied to those states who are actually in the process of developing into (or who orient themselves towards) nation-states. An alternative model of political and cultural organization to that of the nation-state, the pluralistic "nationalities state," with cultural autonomy, would perhaps be in a position to prevent the creation of new national minorities and to establish a relation of equality and balance between ethnic groups.*

* Francis ((6), p. 179) argues along the same lines: "...Nationalism as a principle of political organization has not only led to systems that are oriented towards the model of the nation-state, but it has also brought forth the idea of the nationalities state as a real alternative to the nation-state. The historical fate of certain nationalities states (e.g., the old Habsburg monarchy whose violent demise would seem to imply that it was only a transitional stage on the way towards becoming a fully-developed nation-state) must not obscure the fact that other political systems (such as in Switzerland or Canada) correspond more to the type of system of the nationalities state than of the nation-state and have proved to possess a high degree of stability and strength even during the period of nationalism." In the case of Canada, it seems this assertion is highly questionable.

The problems of the "new national minorities" are characterized by the merging of certain aspects of the situation of the "classic" national as well as of the regional minorities: of national minorities in those cases where the possibility and the interest exist to unite with a nation-state (or just-forming nation-state) to which it belongs by virtue of its politicocultural history and identity; of regional minorities, where the establishment of the principles of a nation-state leads to oppression or repression of traditional culture and identity, without the particular minority orienting itself towards a nation-state to which it "belongs." The interweaving of these elements reflects the incompleteness of the processes of "becoming a nation" for the young nation-states themselves, and is not the result of an unclear conceptualization.

In summary: new national minorities are groups within the young nation-states, sociostructurally heterogeneous, distinct in their ethnicity and history, and discriminated against with respect to their economic possibilities, political rights, and cultural development. Their situation is the result of the formation of political and administrative units out of culturally and historically distinct populations under colonialism. These units became the territories and populational base for the foundation of the young nation-states. In their situations, the problems and orientations of both the "classic" national and regional minorities intermingle.

Table 1 summarizes our discussion of the concept of a minority.

TABLE 1 — Typology of ethnic minorities.

TYPE OF MINORITY	GENESIS OF THE MINORITY SITUATION	CHARACTERIZATION OF THE STATUS AND SITUATION OF THE MINORITY	POLITICAL ORIENTATION OF THE MINORITY
National Minority	Foundation of the modern nation-state	Sociostructural heterogeneity; economically, politically, and culturally disadvantaged or discriminated against; pressure to assimilate	Unification with the pertinent nation-state
Regional Minority	Foundation of the modern nation-state	Sociostructural heterogeneity; subject to economic and cultural, also often political and legal discrimination; highly assimilated into the system of the nation-state	Autonomy; recourse to a pre-nation-state politicocultural identity
Immigrant Minorities; Immigrant Farmers	Emigration as a result of the breakup of the feudal-agrarian mode of production and social structure	Immigrant Farmers: agrarian middle class (with transitional phases of wage labor and great uncertainty); as immigrant farmers in closed, national settlements — the development of a rural "colony"; processes of assimilation;	Representation of interests in associations; loyalty to the immigration country
		as immigrant farmers with an "imported" social system — reconstruction of the socioeconomic and cultural system on the basis of religious norms; isolation, warding-off of assimilating influences.	Political abstinence
Immigrant Workers		Immigrant Workers: sociostructural homogeneity as lowest class of (industrial) wage labor, discriminatory living conditions; development of colonies; collective process of assimilation	Representation of interests in associations and trade unions; loyalty to the immigration country
Minority Peoples	Modern colonialism	Progressive economic, psychosocial, and physical deterioration; outside the politicoeconomic production and work process	Political apathy; sporadic struggles against discrimination and recourse to the past
New National Minorities	Modern colonialism, foundation of the "young" nation-state	Sociostructural heterogeneity; discriminated against with respect to economic possibilities, political rights, and cultural development; pressure to assimilate	Unification with the pertinent nation-state, autonomy or independence

Acknowledgements. I thank G. Custance and M. A. Cervantes-Waldmann for their assistance in translation.

REFERENCES

(1) Barron, M.L. 1967. American Minorities. A Textbook of Readings in Intergroup Relations. New York: Alfred A. Knopf.

(2) Bernsdorf, W., ed. 1969. Wörterbuch der Soziologie. Stuttgart: Enke Verlag.

(3) Davie, M.R. 1949. World Immigration. With Special Reference to the United States. New York: MacMillan Co.

(4) Doerdelmann, B., ed. 1969. Minderheiten in der Bundesrepublik. Munich: Delp.

(5) Fite, G.D., and Reese, J.E. 1973. An Economic History of the United States. Boston: Houghton Mifflin.

(6) Francis, E.K. 1965. Ethnos and Demos. Berlin: Duncker und Humblot.

(7) Frazier, E.F. 1965. Black Bourgeoisie. The Rise of a New Middle Class. New York: The Free Press.

(8) Fuchs, W., et al., eds. 1973. Lexikon zur Soziologie. Opladen: Westdeutscher Verlag.

(9) Gustafsson, L., ed. 1976. Tintenfisch 10. Thema: Regionalismus. Berlin: Wagenbach.

(10) Heckmann, F. 1981. Die Bundesrepublik: Ein Einwanderungsland? Zur Soziologie der Gastarbeiterbevölkerung als Einwandererminorität. Stuttgart: Klett-Cotta.

(11) Hösle, J. 1976. Katalanien — 27 Jahre danach. **In** Tintenfisch 10. Thema: Regionalismus, ed. L. Gustafsson, pp. 100–107. Berlin: Wagenbach.

(12) Hvidt, K. 1975. Flight to America. The Social Background of 300,000 Danish Emigrants. New York, San Francisco, London: Academic Press.

(13) Jacobs, P.; Landau, S.; and Pell, E. 1975. Brüder, sollen wir uns unterwerfen? Die verleugnete Geschichte Amerikas. Munich: Deutscher Taschenbuch Verlag.

(14) "Jean." 1976. Elsaß: Kolonie in Europa. Mit einem Vorwort über Occitanien, Korsika, Wales und Jura. Berlin: Wagenbach.

(15) Petrova-Averkiyeva, J.P. 1974. The Fate of Minority Peoples in the Age of the Scientific and Technological Revolution in Conditions of Socialism and Capitalism. Lecture held at the 8th World Congress for Sociology, Toronto.

(16) Rose, A.M., and Rose, C.B., eds. 1965. Minority Problems. A Textbook of Readings in Intergroup Relations. New York: Harper and Row.

(17) Rose, P.J. 1974. They and We. Racial and Ethnic Relations in the United States. New York: Random House.

(18) Schumacher, H. 1969. Entwicklung eines Begriffs. **In** Minderheiten in der Bundesrepublik, ed. B. Doerdelmann, pp. 9–20. Munich: Delp.

(19) Simpson, G.E., and Yinger, J.M. 1965. Racial and Cultural Minorities. An Analysis of Prejudice and Discrimination. New York, Evanston, London: Harper and Row.

(20) Sünkel, W. 1975. Emanzipatorische Erziehung? Lecture manuscript, University of Erlangen-Nurnberg.

(21) United Nations. 1949. Subcommission on Prevention of Discrimination and Protection of Minorities. The Main Types and Causes of Discrimination, Lake Success, NY.

(22) Vaskovicz, L. 1977. Probleme der sozialwissenschaftlichen Randgruppenforschung. **In** Sozialwissenschaftliche Forschung — Entwicklung und Praxisorientierungen. Festschrift for Gerhard Wurzbacher, eds. D. Blaschke, H.P. Frey, F. Heckmann, und U. Schlottmann. Nürnberg: Verlag der Nürnberger Forschungsvereinigung e.V.

(23) Wagley, C., and Harris, M. 1958. Minorities in the New World. New York: Columbia.

(24) Young, D. 1932. American Minority Peoples. New York: Harper.

(25) Zülch, T., ed. 1975. Von denen Keiner spricht. Reinbeck: Rowohlt.

Minorities: Community and Identity, ed. C. Fried, pp. 25–50
Dahlem Konferenzen 1983. Berlin, Heidelberg, New York, Tokyo: Springer-Verlag.

THE NATURE, CAUSES, AND IMPLICATIONS OF ETHNIC IDENTIFICATION

O. Patterson
Dept. of Sociology, Harvard University
Cambridge, MA 02138, USA

Abstract. Ethnicity is defined as the process resulting from the interplay of two kinds of social concourses: an inner, intersubjective concourse between the individual and a we-group, and an outer structural concourse involving the individual and we-group on the one hand and, on the other, the wider world or out-group. The inner concourse begins with consciousness of a shared crisis of alienation resolved through a commitment to the group sharing the crisis, one symbolically validated as a large endogamous kindred with a common memory. The outer concourse takes account of the sources of the crisis, its context and timing, the resources of the potential group, the arithmetic of its social relations with the wider society, and the policy of its leadership. Specific configurations of these two sets of factors determine the kind of ethnic movement and resulting ethnic groups. A consideration of the dangers of ethnic mobilization is followed by a concluding section on the sociology of ethnic knowledge.

INTRODUCTION

What is ethnicity? What is distinctive about ethnic attitudes and behavior? Why do people behave ethnically? What are the conditions under which ethnic allegiance becomes salient? These are the questions I propose to address in this communication.

What Ethnicity Is Not

Ethnicity is an emotional issue. It belongs centrally to that area of experience which Weber designated the "non-rational." Sociologists have great difficulty dealing with such issues. Their ideal is the rational understanding of all human behavior, including the nonrational. Confronted with the nonrational, sociologists search for underlying factors accounting for it. If, for example, they are positivists determined to exploit the statistical and elementary deductive skills in which they have heavily invested, they tend to reclassify the subject into some broader or smaller category of imagined experience which can be modelled, operationalized, and tentatively explained. A subject such as ethnicity easily ends up being defined in terms of related but really distinct social phenomena.

Two of the most common current misconceptions may be noted here. Many sociologists confuse ethnicity with cultural behavior or cultural awareness. This amounts to reducing the understanding of ethnicity to the study of cultural anthropology. In America expecially, ethnicity, it turns out, is the distinctive traditions of a people. Irish ethnicity refers to Catholics with a peculiar penchant for machine politics and political patronage; Jewish ethnicity is an account of people who go to synagogues, or grew up doing so, vote liberal, are highly educated, and are concerned about the state of Israel (2). I exaggerate, but not too much. The point is that all human beings have some kind of culture. Merely to show that a given group has a distinctive culture or subculture, that this culture is meaningful to them, and that they are aware of it tells us little of which generations of cultural anthropologists have not already informed us and nothing about ethnicity.

A second, more recent view is that ethnic behavior constitutes a form of collective action. This, however, amounts not to an understanding of the subject but to its elimination as a meaningful social process by means of categorical subsumption. Several persons who work in this tradition are aware of, and sometimes even acknowledge the fact that their discussion of collective ethnic action has nothing to do with ethnicity per se. As Hechter, Friedman, and Appelbaum candidly acknowledge (see **Special Note**, Murray et al., this volume), "there is nothing to distinguish the causes of ethnic collective action from the causes of any other kind." Such candor is refreshing, but one is left to wonder just what became of the thing-in-itself we call ethnicity. Is it a phantom? Is it perhaps something that appears and then disappears on different levels of behavior — meaningful, e.g., in primary group activity but irrelevant in the "big event," such as social movements? If so, why do such social movements become designated ethnic movements?

THE INNER CONCOURSE

Ethnicity is a distinctive mode of human experience resulting from the interplay of two kinds of concourses. There is an inner concourse involving self and membership, and an outer concourse which relates this inner concourse to the wider socioeconomic context as well as structural and ideological constraints and opportunities.

Three domains of experience are constitutive of ethnicity on the level of the inner concourse: a distinctive form of consciousness, grouping, and symbolizing.

The Domain of Consciousness

Ethnicity is the conscious awareness of a threatened or real crisis and the need which this crisis both stimulates and by which it is resolved. The crisis is the exilic anxiety, the fear of separation, of isolation, of being cut off. The need which this fear, this crisis of being — which existentially always expresses itself as a crisis of being cut off, being in isolation — generates is the need to belong, to be involved, intimately involved with a wider group, and to be inalienably committed both to the involvement and to the object — the group — of that involvement.

Now, all human beings of any sensibility sooner or later experience a crisis or a threatened crisis of being. It is part of our humanness to do so. The crisis comes about in various ways and is variously resolved, although, in a general way, a case can possibly be made that the resolution, whatever its nature, entails some form of commitment.

Ethnicity is distinctive in the way the crisis presents itself and in the peculiar resolution of it. With ethnicity, the crisis comes about as something shared with a particular group of people. Ethnicity is peculiar in that the acute sense of isolation is itself not an isolated experience but a collective one.

Herein lies the source of what I referred to above as the inner concourse. Ethnicity, at its very inception and continuously thereafter, comes upon the isolated individual as a shared crisis, a shared threat of a shared loss, a shared cutting off, a shared being-in-isolation. And the resolution is immediately implicit in the crisis. There is an intersubjective concrescence between the ego in isolation and all the other egos in similar isolation. Together they become we — we who share the crisis, we who resolve it together.

How is this done? The concrescence occurs in four ways. First and foremost, there is a declaration. I declare to all who share the crisis with me that I am one with them — I am Jew, black American, Irish-American, West Indian-British, Turkish-German, whatever. Then I declare it to the world, but especially to the most important contradistinctive others. Invariably, the most significant of these others in the world are those who have been most responsible for my isolation. It may be the whole white race, as is true of black Americans, or only a specific group of historical oppressors, as, for example, the attitude of the Negritude School to French civilization; it may be members of a particular religion (Armenians and Turks) or a particular class, or both (Irish Catholics and Protestants in Northern Ireland).

The declaration, it must be emphasized, need have no sociological substance, no objective sociocultural reality. A professional outsider with his questionnaire schedule may well conclude that there is no basis for these declarers to claim oneness with each other since they have little or nothing in common. This would be specious.

Second, declaration is reinforced by commitment. In the inner concourse, the ethnic profoundly commits himself to those sharing the crisis. Not only is a commitment made, but the ethnic is committed to the belief that this is his most basic and profound commitment. It may not be his most important ongoing commitment (that may be to his career, for example), but it is always his most profound, the one which, if and when the crunch comes, he is least likely to abandon.

The third element of the inner concourse involves expression. The ethnic becomes aware, declares and commits himself in a certain way, with a certain style. The style usually takes its cues from the language of kinship. We are all brothers, all kith and kin. In more specific terms, the ethnic concourse either draws upon specific areas of shared cultural experience, or where these do not exist, invents distinctive modes of verbal and nonverbal expressions for communicating intimacy and empathy to the exclusion of others, especially significant others. Style is more important than content in such expressions. But content is not unimportant; such expressions refer, however, to the symbolic mode of becoming and being ethnic, as we shall see later.

Finally, the inner concourse involves a very paradoxical kind of choice. The ethnic chooses to belong. It is imperative that he or she always believes that a choice is being made; otherwise, the commitment is meaningless. We cannot commit ourselves if we have no choice in the matter. And yet, at the same time, it is a choice predicated on the strongly held, intensely conceived belief that the individual has absolutely no choice but to belong to that specific group. This may seem mystifying, but there is another area of experience in which we choose, believing that we have no choice but to choose a particular commitment: it happens to all of us when we fall in love. Need I elaborate?

These, then, are the main and distinctive elements of ethnic consciousness. There is, however, another dimension of ethnic consciousness which must be pointed out. This is the consciousness which manifests itself as an ideological commitment to the idea and the necessity of ethnicity. The ethnic is not only committed to his own particular ethnicity; he strongly believes that all human beings are inherently ethnic, whether they know it or not, whether they like it or not. The world is an ethnic world, sometimes a benignly pluralistic ethnic world, as in the ethnic philosophy of Kallen (4), sometimes a malignantly hostile universe of other ethnics toward whom one requires constant watchfulness and readiness, as in the ethnic philosophy of Elijah Mahammad of the Black Moslems or Menachim Begin.

This is philosophical dogma, of course, but understandable — if not acceptable — in the light of our opening observation that the ethnic experience is essentially nonrational. What is most distressing, however, is the tendency of many so-called social scientists colluding with such dogma and claiming to "prove" from their " objective" studies that we indeed live in an inalienably ethnic world.

Consider, for example, the following statement by the sociolinguist Jessel in a recent work presented as a serious academic treatise:

> The ethnic process may be compared to an anatomical system where the simple group principle is the bare skeleton and the ethnic group is a corporal pool inclusive of integument, physiology, and biochemistry. In relationships between the ethnic group and its individual members, countless interactions take place mentally, linguistically, and societally. If we are to assume that under evolutionary conditions this might indeed resemble the operating behavior of an ethnic complex with a resultant effect of an ethnic society, then a non-ethnic group in an ethnic world must be regarded as an anomaly. It can be conceived of only as a transitory social phenomenon. Either it had once belonged to an ethnic system and had been squeezed out for reasons presently unknown or it would ultimately find itself as an integral part of an ethnic system in the future (3).

This is dangerous nonsense, and it points to one of the major problems in the sociological study of ethnicity, namely, that the social scientist is in constant danger of becoming implicated — indeed, hopelessly involved — in the inner (and outer) concourses which he purports to examine. This does not mean, let me hasten to add, that I hold to any naive objectivism. The social scientist is always morally involved with his subject, and rarely is this more the case than in the study of ethnicity. The danger lies in unacknowledged bias, the pretense of scientific objectivity in the face of active incitement and enhancement of the subject studied. Like his subject, the social scientist is morally required to declare himself.

The Domain of Groupness

The second mode of ethnic experience concerns the conception of the group with which the individual engages. The degree to which the ethnic individual actively participates in groups varies considerably from one individual to another, and the frequency and intensity of such group activities differs from one ethnic group to the next.

It cannot be maintained, however, that the strength of ethnic allegiance is always a function of participation in ethnic group activities. The ancient Greeks of the classical period were a strongly ethnic people yet shared few organizations as Hellenes. The Chinese in Jamaica are similarly highly ethnic yet tend to avoid ethnic Chinese organizations. In this regard, it is important that we do not confuse groups which recruit their membership mainly or evenly entirely from a given ethnic group and the specifically ethnic organizations of such groups. Not all, or even most events involving only Jews, Greeks, and blacks are ethnic Jewish, Greek, or black events. This point must be emphasized because there is a near universal tendency for sociologists and journalists to make this error. Greeks of the Periclean age participated in many organizations which were exclusively confined to Greeks, but which were not ethnically Greek events. This is true of all the many activities in which Greeks engaged in their capacities as citizens of their city-states. It was necessary to have been Greek to be a member of one of these state organizations; but not only was this not a sufficient basis for participation — citizenship by birth being more important — but Greeks from other city-states were just as rigidly excluded as ex-slaves and free barbarians who together made up the metic class.

Similarly, there are many exclusively black American events which are, or were until recently, in no way ethnic. This was true, for example, of the black church. The vast majority of churches with all-black membership and black fundamentalist worship are in no way ethnically black. Some of them, we know, came to be very important in the rise of black ethnicity, but all that we can say of most of them is that they are culturally black events.

What is the difference between a non-ethnic group, all of whose members belong to an ethnic group, and a genuinely ethnic group? The difference points to what is most critical about an ethnic group. A group is ethnic only to the degree that, on the one hand, it actualizes through social intercourse ethnic consciousness, and on the other hand, to the extent that its activities ritualize validating myths and symbols (these constituting the third ethnic mode to be discussed shortly). The Olympic Games among the ancient Greeks and the cult of Apollo at Delphi were such group events. The Southern Christian Baptist Church under Martin Luther King became such an ethnic group after not having been so for most of its history. *Commentary Magazine* and the group of intellectuals who write for and produce it, was for many years a non-ethnic, though largely Jewish, organization. It is now a highly ethnic Jewish organization.

The ethnic group, then, mediates between consciousness and validating myth as well as other forms of symbolic expression. It always has a cultic quality — it involves a sharing of rites that strengthens group identity and enhances we-ness. Its activities reinforce, too, the specialness, even chosen-ness, of the group and its exclusiveness.

Such ethnic group events need not take place with any great frequency. Indeed, they may rarely ever occur, for there is one other aspect of the ethnic group which makes it peculiar and which can more than compensate for infrequent group events. This is the fact that all ethnic groups either are or aspire to be endogamous. They may, in fact, not be very endogamous, but they always are in principle. The endogamous principle reinforces the conception of the ethnic group as a vast kindred.

The Domain of Symbolizing

And this brings us to the third domain of the ethnic experience. This is the myth of blood, the deeply held belief that the entire group has a common ancestry, a common history, and sometimes a common fate. It is this myth — often having little or no basis in fact — that the specifically ethnic group event ritualizes. The ethnic group is, then, not so much a moral community as is the religious group, but in Daniel Bell's fine phrase, "a community of memory."

All ethnic groups have this validating myth. In addition, however, each ethnic group has a set of dominant symbols which are peculiar to the group. The content of these symbols are taken from significant events in the group's past, or they may be invented episodes and ideas. They are consciously elaborated and manipulated to further the end of the enhancement of the peculiar consciousness discussed above.

We see, then, that the three domains of the ethnic mode of experience support or reinforce each other. Consciousness, however, is the dominant domain of the ethnic concrescence. Ethnicity is, quintessentially, a way of being.

THE OUTER CONCOURSE

The ethnic experience not only does not exist in isolation; it requires a wider context. That wider context may or may not involve other ethnic groups. We have already seen that ethnic ideology invariably defines this wider context as an ethnic universe. This can sometimes have odd results. Thus, in the United States, White Anglo-Saxon Protestants (WASPs) are invariably designated as an ethnic group, though few such persons consider themselves to be members of any such group.

For specific ethnic groups, there is usually, as indicated earlier, a significant counter-distinctive other group, singled out from all the out-groups, which may be called the them-group. Without the them-group, many ethnic groups would simply have no raison d'etre. A good part of the ethnic consciousness of black Americans, for example, comes either from the conscious negation of the dominant white group's conception of them or from the direct embracing of the negative stereotype, itself a form of negation through symbolic co-optation. This, for example, happened with the use of the term "black" by the group to designate itself. There were alternatives — "Afro-American" was for long a strong candidate — but blacks deliberately chose the term black not simply in spite of, but because of the negative associations in the white mind of this term. A more extreme form of symbolic co-optation involves the use of the term "nigger" as a privileged term of familiarity among blacks, even though a white person using the term would still be condemned as a racist, and perhaps rightly so. Persons familiar with Jewish ethnicity can immediately think of parallels.

The outer concourse involves not simply the group's relationship with other groups, but the group's involvement, sometimes its most intimate involvement, with itself. It is in this wider context that ethnicities emerge, are shaped, decline, and are revitalized. The very crisis that forms the seed of the ethnic consciousness is invariably created by the "them-group." The inner concourse, although it is existentially perceived as prior to the outer, is historically often a product of the outer concourse. Them-ness is not only historically but structurally prior to we-ness.

An analysis of the outer concourse, then, really amounts to an examination of the origins and types of ethnicities and of ethnic movements.

Five sets of factors determine the different kinds of ethnic movements and groups:

1. The Sources of the Crisis

The exilic crisis may come about in different ways. It is most frequently the result of migration. Clearly, the more involuntary the migration, the more acute the crisis. But one should be careful not to draw too rigid a line between voluntary and involuntary migration. The Irish who fled the potato famine and English oppression for America were hardly less forced to migrate to America than the Africans who were dragged there in chains: for both, the alternative was death, possibly a more painful one for the Irish if one accepts that death by starvation is a good deal more painful than death by the sword.

External migration is traditionally associated with the crisis which engenders ethnicity, but here again it must be emphasized that many forms of internal migration can be as acute and crisis-laden as external migration. In Asia and Africa, such migration involves not only the loss of anchorage in distinctive tribal and regional cultures, but the added turmoil of movement from agricultural to urban social ecologies. Recent work on black Americans strongly suggest that the mass migration of the group from the rural South to the urban North was almost as acute and disintegrating an experience as the middle passage across the Atlantic.

A second source of the crisis is the experience of, and/or the sense of exclusion from, a wider social entity. This may be the result of defeat in a civil or international war (for example, the Ibos of Nigeria, and on a national scale, the Germans after World War I) or simply a sense of having lost out economically or socioculturally to other groups (the best current example being the French Canadians).

A third source is an acute sense of loss, a feeling of being cut off from all roots, and of powerlessness, as a result of the atomization and apparent heartlessness of modern mass society. There is a vast literature on the subject of alienation and modern industrial society, so the point need not be labored. Revolutionary activity is one response to this sense of alienation — or so hoped the early

Marx. A more likely response is the conservative retreat to the ethnic haven (6).

I do not wish to over-schematize. An element of all three sources may be found in the inception of most ethnicities, although there is a tendency for one or the other source to be more pronounced.

2. The Context and Timing of the Crisis

The second set of factors determining the ethnic outcome concerns the wider socioeconomic context of the crisis and its timing. Most societies differ considerably in the socioeconomic opportunities they provide migrants or the possibilities of disalienation they offer the alienated. An intangible, though critical, factor is the exclusiveness of the dominant society. Ancient Greece and Rome are the classic contrast here. The former was highly inclusive yet hostile to other ethnicities within the body politic — a heterogeneous metic class was about all it would permit during the classic period. And even during the Hellenistic age when imperial self-interest prompted greater flexibility, the Greek ruling elites only grudgingly permitted the hellenization of the native elites. Rome, on the other hand, was a remarkably open society with a high tolerance for ethnic mobilization. Clearly, in this context, the tendency of forced and free migrant alike to resolve the exilic crisis in ethnic terms was much greater than in Greece.

Modern counterparts abound. America and Germany or Sweden are cases in point, the former highly inclusive and tolerant of ethnic movements, the latter two highly homogeneous, exclusive, and themselves rather self-consciously ethnic polities.

The timing of the arrival or crisis of the potential ethnic members is another critical factor. Indeed, this factor alone can explain a good deal of the variance in the subsequent successes or failures of ethnic groups. If a group arrives during a period of economic buoyancy, its propensity to mobilize ethnically may be higher or lower, depending on the socioeconomic resources it brings with it.

Clearly, the explanatory power of context and timing variables are greatly enhanced by their interaction with the third set of factors to be considered.

3. The Social Resources of the Incipient Ethnics

Nearly all incipient ethnic groups have some social resources which they can draw on in their struggle with the wider society, especially the dominant them-group. But the type or range of these resources vary a great deal. The resources referred to are the traditional set of skills, including educational background and orientation, experience of urban life, language facility, and — to a much lesser degree — attitudes and values, all of which a group brings to the new context. The appropriateness of the resources depend upon the context. Thus, migrants who came from the urban parts of Europe to urban-industrial America at the turn of the century were clearly at an advantage over those who came from rural areas, whether from the north or south of Europe. A skilled blacksmith or shoemaker from Europe had little to offer in the new urban context of turn-of-the-century New York. Skilled tailors, on the other hand, were in great demand in the booming garment industry. As recent studies have shown, to the extent that Jews came from the more urban areas of Europe and tended to concentrate in sewing and other urban skills, they were in a much better position to adapt both as individuals and in ethnic terms to the new situation.

I will mention here cultural factors such as traditional attitudes and values as well as child-rearing patterns as possibly contributing factors. But I mention them largely to emphasize the dangers of overemphasizing them, which has been and still is a major error in ethnic studies. I wish to make two negative statements in this regard. One is that there is no foundation whatever to the common assumption that because a group in crisis has a highly cohesive, rich, and advanced culture, it is likely to become highly ethnic on the basis of this cultural distinctiveness. In my study of the Chinese in the Caribbean (7), I have tried to show how the same group of people in different social contexts responded in very different terms — one ethnically, the other non-ethnically. In the United States, the experience of the Irish and their Celtic cousins, the Welsh, is instructive. The Welsh had and still have as distinctive and cohesive a culture as the Irish. Indeed, the Welsh language is still a living tongue in North Wales, whereas Irish is an artificially revived official language. And yet in the United States, Irish ethnicity has thrived while the Welsh have all but disappeared as a distinct group and have never mobilized on ethnic terms.

The second negative observation is that cultural factors play only a minor role in the relative success of ethnic groups measured by the economic status of their members. I do not want to take an extremely structuralist and anti-cultural position on this matter, although after reading Sowell's recent work (11), one is inclined to pardon anyone who does so. I am prepared to allow that Jewish and Japanese mothers do something to their children in the course of their upbringing which may in some way account for the higher frequency of college graduates and successful businessmen in the group. I suppose, too, that the way German fathers brought up their sons in the early part of this century may have had something to do with the rise of German national ethnicity and its initial militaristic success. But I do not know what these things are, and I am not altogether persuaded that such knowledge would be of much use, either in understanding the nature and fate of German, Jewish, and Japanese ethnicities, or as Dr. Spock-type guidelines and prohibitions for the parents of other ethnic groups.

4. The Arithmetic of Social Relations

The fourth set of factors explaining both the propensity to ethnicity and the success of resulting ethnic groups is closely related to the last two. What my colleague Rytina and his associate Morgan (9) call "the arithmetic of social relations" has been largely neglected in the study of ethnicity, although these factors may well be among the most critical. Rytina and Morgan, drawing on the axiomatic theory of Blau and on network theory, have persuasively shown how the relative and absolute size of an aggregate, the degree of segregation of persons, and the frequency of their contact have profound, interrelated consequences for the salience (the degree to which ties are restricted within or without the group), cohesion or density (the probability that two randomly selected members share a tie), and most important, the capacity of the group to control its environment and to achieve success with respect to the out-group.

While the authors themselves do not relate their findings to the problem of ethnicity, the implications are clear, and in what follows I propose to draw them out. If a minority group is small and has many in-group contacts, the resulting high sociometric density will curtail privacy, increase mutual visibility and access to information about others, and apply informal sanctions against those who violate shared norms. What I find most suggestive is the view that the "threat of relative isolation" becomes an important component of peer pressure in groups of certain sizes and proportions. What this suggests is that some groups, by virtue purely of their relative and absolute size, contact frequency, and density, are able to sustain the original crisis of isolation which forced the individual into the ethnic mode. Like the fundamentalist Christian whose faith is kept alive by the constant threat of hell and damnation, the ethnic's commitment is kept alive by the arithmetic of social relations producing the constant threat of being cut off. Aggregates which are so constituted that fear of isolation is both a prehistorical and a continuously historical fact, a precondition and a terrifying condition of existence, will be highly inclined to become and remain ethnic groups.

Equally fascinating are the implications of sheer size for minority groups. Below a certain critical size, a group may become too small to sustain an ethnic movement. It may have high density and cohesiveness, but unless it is rigidly segregated, it may not be able to withstand the absorptive or disintegrative pressure of the out-group. At the same time, a minority group may be too large to either facilitate ethnic development or to succeed in socioeconomic terms as a group. There is an ideal absolute and relative size of groups which, combined with the frequency of in-choice contacts, largely explains both the strength and persistence of their ethnicity and much of their relative socioeconomic success.

Thus, Rytina and Morgan argue that relatively small minorities can combine considerable out-group contact with an in-group density that approaches the point of saturation. However, "much larger minorities, such as residents in ghettos and urban villagers, can be substantially segregated yet lack sufficiently extensive contacts to unify the group" (9). The experience in America of Jews and Japanese, on the one hand, and blacks and Chicanos on the other hand, are immediately suggested by this analysis.

The individual and aggregate isolation of blacks for a long time resulted only in their segregation. This and the large size of the group resulted in an aggregate with low in-group density or cohesion. The much smaller size of the Jewish group and the traditional skill of internal networking based on centuries of experience which they brought with them to America meant that the performance of in-group members could remain highly visible even in situations where out-groupers figured prominently. Such groups are highly flexible and can even enforce innovations in the face of new out-group challenges. The group's size and density also makes it easier to deal with the "free rider" problem, since the sanction of shame is more effectively enforced.

On the other hand, a large segregated aggregate such as the black group finds it difficult to establish effective networks of cohesion aimed at the attainment of individual and social goals, is confined to traditional responses in the face of new external challenges and threats, and has a chronic free rider problem.

Most suggestive of all is Rytina and Morgan's numerical simulation of the role of a highly educated, Ivy League "old boy" minority of 1,000 in a government bureaucracy of 10,000 employees. With a few not unreasonable assumptions, they deduced from purely numerical differences, constraints, and configurations that:

> A major implication of these differences is that the quality and quantity of information available to "old boys" about each other is vastly greater than that available to the others. A further implication is that old boys are in a far better position to make arrangements with each other on the basis of mutual trust and to enforce obligations and reciprocity within their circle. Access to the collective history makes it easier to anticipate reliability and to avoid the untrustworthy. As a result, such a densely interconnected group is a suitable arena for the secure extension of social credit and, resource differences aside, this greater liquidity makes old boys better partners in bureaucratic barter. More extensive knowledge of opportunities and intentions would facilitate the formation of larger, more effective coalitions (9).

And further that:

> They are highly connected and mutually visible, and yet each has more contact with the majority group than majority group members have with each other. Thus the minority combines greater cohesion and positive selectivity with an abundance of out-group ties. The majority has negligible cohesion and negative selectivity; only their out-group ties connect them to an arena of mutual visibility. And only the old boys, with their twin advantages of contact frequency and a small group, can maintain such an arena (9).

This, it seems to me, is a perfect model of the sociometric conditions which engender and sustain one important kind of ethnic identity and group, what I have called elsewhere the "symbiotic ethnic group" (8).

This attempt to tease out the implications for the understanding of ethnic processes from the arithmetic of social relations is mainly speculative. But it is not without empirical support. Prior to and quite independent of Rytina and Morgan's highly formalized conceptual scheme and supporting simulation models, Lieberson had arrived by empirical means at strikingly similar conclusions. In what is certainly one of the best studies on the subject in recent years, Lieberson (5) shows how the segregation pattern of blacks in relation to whites

> suggest that race and ethnic relations operating in the United States had a latent structure such that some of the changes in such relations over time are really shifts along an existing structure due to compositional change rather than changes in the dispositions of the groups toward one another.

Another crucial finding of Lieberson further supports our earlier argument. He has found that many ethnic groups tend to find occupational niches in the new environment and to concentrate disproportionately in these jobs. Such concentrations reflect preexisting skills and social propensities as well as the structure of opportunities in the host society at the time of arrival. The size of the group, however, becomes absolutely critical in its "ability to develop and exploit these special niches." The Jews, Chinese, Japanese, and some European migrants were small enough and grew at a moderate enough rate to use internal networking strategies in furthering their hold on these niches. Ethnic awareness and group formation both enhanced and was enhanced by such niche monopolization and networking (5).

On the other hand, the large and rapidly growing population of blacks migrating to the northern cities in the early part of this century was a positive handicap to the group. There were no niches large or important enough to absorb significant proportions of the group. Their size and rapid growth made cohesion difficult (although not impossible) and at the same time increased negative out-group reactions. Ironically, hostility to the large influx of black southern migrants diverted hostility away from the newly arrived white migrants, providing them with a more favorable environment within which to develop and exploit their own ethnicities. "Ethnic ties and allegiances float and shift in accordance with the threats and alternatives that exist. The presence of blacks made it harder to discriminate against the new Europeans because the alternative was viewed even less favorably" (5).

Numerical factors similarly apply when comparing the ethnic development and relative successes of blacks and the newer Asian migrants. There were 22.5 million blacks in the United States in 1970, a population larger than most member nations of the United Nations. At that time, there were only 591,000 Japanese and 435,000 Chinese. As Lieberson tellingly comments: "Imagine more than 22 million Japanese Americans trying to carve out initial niches through truck farming!"

The arithmetic of social relations alone suggests the dangers of making glib comparisons of ethnic groups. Sheer differences in scale eventually turn into major differences in quality.

5. Ethnic Policy and Leadership

The final set of factors determining the rise and nature of ethnic groups concerns the character of the leadership which promotes such movements and the nature of the policies they pursue.

Given a particular set of favorable circumstances, the emergence or failure to emerge of an ethnic group or its subsequent failures and successes must be explained by the decisions which those who lead it make. This can be illustrated with one dramatic example taken from black American history. Many have wondered why black ethnicity and its product, the civil rights movement, took so long to develop. The answer is partly, perhaps largely, to be found in the structural factors I have discussed earlier. But they cannot be sufficient. For the forces operating against such a movement in the late fifties and sixties were really not all that different from those existing in the twenties. Furthermore, a closer examination of the twenties reveals two things. One is that the crisis of separation, the anxiety of being cut off which is the nucleus of ethnic mobilization, was greater then than it was in the sixties. Second, there was indeed a mass ethnic movement of urban blacks in the twenties — this is the Marcus Garvey movement, involving hundreds of thousands of blacks (1). Why did this movement fail so miserably and so quickly? And why, after its failure when it was clear to all that the time was ripe for mass mobilization, did not alternate movements take its place?

Others may find explanatory comfort in structural forces and in the usual recourse to white oppression. For me, the answer is brutally simple. This missed opportunity was a massive failure of black leadership, one for which the black masses were to pay the price of another forty years of civil degradation. Blaming Garvey would be too easy. Garvey was doing what he did best — giving spell-binding speeches, being fanatical, egomaniacal, pompous, prophetic, overreaching, mesmeric, messianic, absurd, and thoroughly antic — in short, he was nothing more or less than your classic charisma. His grand vision — the return of the entire black diaspora to Africa which would be reclaimed from the European colonial powers in a cataclysmic racial war — was an absurdity. But rationality was never the hallmark of charismata, as every student of social movements knows. And how far-fetched, anyway, was the idea? Consider the idea of a Zionist state of Israel in the context of the late nineteenth or even early twentieth century or the response to Herzl's *Der Judenstaat* when it was first published.

The movement failed not so much because of Garvey's millenarian vision or persecution by the FBI, but because of the failure of black leadership at the time to participate in it, to move in and both routinize Garvey's charisma as well as provide the essential organizational support. This was still possible even after his arrest and deportation. For years after his death, there was literally a vast mass of alienated black urban proletarians searching for a leadership. It did not come. Acephalic and abandoned, the movement petered out. A revolutionary moment had passed, had been quite consciously spurned and squandered by what was already a sizeable black bourgeois leadership dominated by lawyers, insurance and funeral parlor magnates, and a Harvard educated patrician with a French surname espousing an elitist theory of the "talented tenth."

THE DANGERS OF ETHNICITY

Nothing I wrote in the last section should be taken to mean that I favor ethnic movements. To the contrary, I am inclined to be critical of most such movements and groups. From what I have written above, it is clear that many different kinds of ethnic movements and groups are possible, depending on the configuration of the five sets of factors discussed. I do not have the space here to go into an analysis of the classification of such groups. I will only note that there are two basic types of such groups — the symbiotic and the assimilative. Symbiotic ethnic groups are the result of social aggregates which are able to establish one or more niches in the wider society and in whose long-term economic and social interests an ethnic strategy works. Such groups eschew complete assimilation, develop strong commitment to the survival of the group, and, over time, their cultural patterns may well adapt to the exigencies of an advantageous minority status. The Jews are the classic symbiotic ethnic group; so are the Chinese in Southeast Asia and Jamaica, the Japanese in the United States and Brazil, and the Indians of East and West Africa.

The history of these groups indicates that ethnicity is in the best material and social interests of its members, although of course, members interpret their commitment in radically different terms. History also shows the heavy price which members of symbiotic ethnic groups periodically pay for their commitment. Such groups are chronically subject to vicious out-group pogroms from a jealous and prejudiced majority. In times of economic and social stress, it has proven all too easy for fanatical out-group leaders to use the symbiotic ethnic group as a scapegoat for social ills. The European out-group treatment of Jews is in no way exceptional. The fate of the Chinese in Indonesia, the Indians in Africa, and the Armenians in Turkey are only a few of the more horrendous examples of an out group's propensity to scapegoat, terrorize, and commit mass murder against wholly innocent, symbiotic ethnic groups.

Whether it will ever be possible to create a world in which such risks are completely eliminated for symbiotic ethnic groups is an open question. I have my doubts, but I would very happily have them shown to be unfounded.

The assimilative ethnic group is, on the other hand, a transitional and adaptive strategy for the attainment of full citizenship and equality in the wider society. Sometimes their goals are mainly social and economic, as was the case of the European migrants in early twentieth century America. At other times, they are mainly political, as in the case of the blacks in the U.S. They may be moderate, even conservative, or they may be radical, depending on the severity of the exclusion by the out-group and the policies of the ethnic leadership. The ultimate aim of all such movements, however, is full membership and inclusion into the wider society. Members draw inward in order eventually to move more effectively outward. Ethnicity becomes a basis of cohesion and a technique of mobilization.

My own view is that if alternative bases of solidarity and mobilization exist, these should always be explored, because of the dangers inherent in such a strategy. Sometimes, however, the nature of a group's oppression and its peculiar patterns of mystification and internal fragmentation make it inevitable that it must first discover and disalienate itself by means of ethnic consciousness before it can join forces with more universalist movements in its ongoing search for justice and equality. Such certainly was the case with black Americans up to the end of the sixties. It is rare, however, to find so compelling a case for ethnic mobilization.

There are always several dangers inherent in the use of an ethnic strategy. One is the possibility of the means becoming an end in itself. An ethnic leadership might find the ethnic group a convenient turf to achieve and sustain some power and influence in the wider society to the detriment of the group it leads. It will then prolong the ethnic movement beyond the point of usefulness, and to do so, it invariably encourages a descent into ethnic chauvinism.

A major error which an ethnic leadership is likely to make in defending the institutionalization of ethnicity is to point to the success of existing symbiotic ethnic groups as exemplars of what is possible. But as we have shown above, a successful symbiotic ethnic group requires certain conditions both internal to the group and external to it, such as a given size, which may simply not exist. It was, for example, a sociologically tragic error on the part of black American leadership to imagine that it could replicate the experience of Jewish Americans. This tragedy was compounded by well-meaning and supportive Jewish leaders, many of them social scientists who should have known better.

Most of the other dangers of an ethnic mobilization spring from this first. One is the tendency for the ethnic strategy of one group to spark imitative or reactive strategies on the part of other groups who may develop their own ethnic movements either because the principle has become legitimated by the dominant out-group's acceptance of the strategy by the first group, or because they feel their interests threatened in what they see as a zero-sum world. The result is likely to be the balkanization of a polity into warring ethnic groups. American society during the seventies was a depressing example of this. When I first came to America in 1970, I could walk freely in any part of Boston. Now I enter many regions such as South Boston at the risk of losing my limb, if not my life.

Without doubt, the most frightening possible danger of an ethnic strategy on the part of one or more minority groups is that of the politically dominant majority out-group developing its own ethnic strategy in response. When this happens, we are well on the road to fascism. It would be alarmist to say that this is happening in the United States. However, it would be excessively optimistic and unrealistic to say that it could not happen. The southern states of America, as a region, have all the preconditions for a successful ethnic strategy. There has already been a resurgence of fascist groups such as the KKK and the American Nazi party all over the region. More alarming is the fact that the moral majority movement, allied with the new right, could easily and spontaneously be transformed into a fascist movement. A severe depression combined with a Hitler-type charismatic figure is all that it takes. There is cause for concern, if not yet alarm.

What I have said of the United States holds equally for Europe, expecially with respect to the guestworker minorities and their descendants. My considered opinion, for what it is worth, is that an ethnic strategy on the part of these minorities would run too many risks ever to be justified. Guestworkers who face the crisis of alienation, who experience a profound need for reconciliation and disalienation must either resolve the crisis in non-ethnic ways (through God, existential commitment, art, good works, hedonism, love, and if all else fails, just learn to live with angst like the rest of us) and work with universalist institutions in the host society for a more just world, or else pack their bags and return home. The history of continental Europe suggests that an ethnic resolution would be sheer folly.

FROM CONCOURSE TO DISCOURSE: TOWARD A SOCIOLOGY OF ETHNIC KNOWLEDGE

In recent years ethnicity has emerged as one of the most widely and intensively studied areas of sociology and related disciplines. What, we must now inquire, accounts for this academic interest? How does it relate to the reality of recent ethnic movements? What motivates a sociologist to study such movements? What are the implications of such a choice? And assuming that we understand the implications of our study, what are we, as sociologists, enjoined to do?

I have interpreted ethnicity as the interplay of two kinds of concourses. As a first step toward an understanding of the subject, this was not only a necessary intellectual stance, but one demanded by the reality of the thing itself. By its very nature, there can be no discourse in Habermas's sense, no critical reflection in the self-understanding of ethnics. Ethnicity in this respect is very much like faith which eschews reason as a path to understanding. As the theologian Weigel observes (12), "Faith cannot debate with faith because they have no ultimate common ground. Faiths can only rail at each other." To the ethnic mind, it must always be: "My group first, right or wrong." On this level, there is only concourse, a spontaneous confluence of feelings and commitment moving toward an ever tighter concrescence, a growing together against all others.

But we still live in a thinking world. The sociologist assumes that he can both understand and remain detached from the anti-rationalistic self-understanding of the ethnic. But can he? If there is no apparent room for discourse in the ethnic experience, there certainly must remain room for it in his own, for the sociologist who abandons rationality and the search for a rational society abandons the most fundamental tenet of his profession. To the extent that one's reflection on ethnicity influences and is influenced by that experience, to that extent does discourse become possible. It is this which I propose to argue in this final section.

No student of ethnicity can be ignorant of the controversies surrounding the value to be placed on the reality of ethnic competition in his society. Nor can any scholar claim to be ignorant of the ways in which "objective" studies of ethnic groups can and are being used. So it follows that a knowledge of such matters must have entered into his decision to choose the subject of ethnicity for study. But if it did, he must assume responsibility for the normative uses of his work, not to mention the fact that such foreknowledge must, in many subtle ways, have influenced his work. And he must assume responsibility not only because he knew beforehand the possible uses to which his work might be put, but also because, even if his subjects spurn his work, he will have contributed to the emerging climate of ethnic consciousness.

When we critically examine the entire ethnic revival in America and elsewhere, we soon find that we are dealing with three levels of consciousness, all mutually reinforcing each other. On the most basic or primary level, there is the consciousness of ethnic reality itself. It is this which I have interpreted above. But in the process of interpretation a second level of consciousness is generated, and this inevitably enhances the primary consciousness. In the first place, the act of study promotes the consciousness of the group about itself. Participant observation, the favorite field technique of students of ethnicity, is sheer incitement. I should hope that there is no longer the necessity to defend this assertion. People are intensely interested in being studied and "written up," and no element of a group is more interested than those members who are most involved in raising primary consciousness — its leaders.

Second, a book on an ethnic group can, and often does, bring it some degree of public prominence which directly feeds back on the primary consciousness of the group. Ethnic leaders respond to such public discussion with shame or pride, as the case may be, and are prompted to action. Very often, too, in groups without a literary tradition, it is to academic works on the group that leaders turn to learn what is special about them and to select what will become of its dominant symbols and validating myths. The works of Myrdal and Frazier had just such an impact on black leadership during and after the late forties and fifties. To take a more personal example, it is common knowledge among Jamaican sociologists that the black proletarian nationalist group known as the Rastafarians now use as their major source of information on themselves a study conducted by three social scientists from the University of the West Indies in the early sixties (10). This study not only enhanced the group's self-understanding but was a critical factor in its legitimation in Jamaican society. Middle-class Jamaicans, previously strongly hostile to the group, concluded that if it was worthy of the academic interest of three of the nation's most eminent scholars, it could not have been all that horrible. Legitimation, however, was won at a price for the cultists. From being despised outcasts the Rastafarians rapidly became colorful countercultural cultists, the inspiration for the island's booming popular music industry, and advertisement for its tourist trade.

But the student of ethnicity does something else by his study. For his secondary consciousness stands between the primary consciousness of the specific group he has studied and unwittingly promoted, and a third level of consciousness, that most general level which is awareness not of any specific ethnicity but of the idea of ethnicity, of ethnicity as a generalized condition. Now it is the idea of ethnicity which many intellectuals have recently discovered in America and other parts of the world, and, as often as not, it is usually the idea alone which they can bring themselves to support existentially. One should not, however, underestimate the importance of this generalized commitment to the idea of ethnicity, for it is usually the most powerful source of legitimation of this form of consciousness in society. As the German film, "Mephisto," recently reminded us, it was Goering who declared that whenever he heard the word "culture" he felt like reaching for his revolver. A more historically ironic remark is hard to imagine. For nothing contributed more to the national ethnic movement which culminated into the Nazi terror than the deep, obsessive commitment of the German intellectual tradition starting with Herder and the early romantics, to the idea of nationality rooted in a distinctive national culture.

So powerful is the mere study of the subject as a legitimating force that it is likely to have this effect even if the scholar in question is critical of particular ethnicities and ethnicity in general. One is reminded of the academic truism that it is better to have a negative review of one's work in a prominent journal than to be neglected by it. Consider the tragically ironic fate of Nietzsche who, in spite of his contemptuous dismissal of the "damn folk soul" and his repeated warnings that "we want to be careful about calling something German," nonetheless ended up being used as a major legitimating source among Nazi leaders.*

* I need hardly emphasize that a subject is not always legitimized in this way by its study. Thus a person who studies racism does not necessarily legitimize it, unless he consciously approves of it or does it so insensitively that he creates more harm than good. To explain and demonstrate, whether implicitly or explicitly, the evil of something known to be evil is always as good as demonstrating the goodness of something good. There is, of course, the problem of who determines what is publicly held to be good or evil. The answer is the moral community which, in turn, is shaped by the political community. Less than 50 years ago when black Americans had almost no role in their political community, it was possible for one of the country's leading historians, U. B. Phillips, to write a standard text on slavery "demonstrating" that slavery was good for blacks and to win a chair at one of the nation's most prestigious universities, Yale, for so doing. Since gaining some power blacks have helped to reshape the American moral community which, in turn, redefined the terms and content of the discourse between scholar and audience over slavery. Now slavery is increasingly known to have been evil and to have been bad for blacks.

The three levels of consiousness are linked, then, in a critical way, one which tends toward closure. So far we have seen that consciousness about ethnic consciousness reinforces consciousness about the idea of ethnic consciousness. But consciousness about the idea of ethnic consciousness feeds directly back upon the primary consciousness of the ethnic group. Indeed, it is not from the scholar that the potential or dormant ethnic group receives its strongest stimulus, but from the general social climate infused with the idea and the assumed rightness of ethnicity. In this regard, it is unimportant whether or not those on the primary level of ethnic consciousness understand anything about the intellectual niceties of those who articulate, on the tertiary level, the consciousness of the idea of ethnicity. The Czech American blue-collar worker, if he tried to read Novak's essay, "How American Are You If Your Grandparents Came From Serbia in 1888?", would most likely either not understand a word of it, or, if he did, remain mystified as to the relationship between the title and the contents of the essay. But this hardly matters. For one thing, the primary level of ethnic consciousness is usually informed, second hand, by the mass media. And for another, it is often enough that a distinguished thinker and a member of one's own group is seen to be publicly concerned with his ethnicity or even ethnicity in general. And when the government gets in on the act and is seen to vote millions of dollars for the study of ethnic heritage (often at the instigation of those who are aware at the tertiary level of consciousness), the die of legitimacy is cast. It is often at precisely this point that the primary level of consciousness turns to the works of scholars to support itself. The circle of reinforcing consciousness is then complete.

The student of ethnic groups, therefore, is in the very thick of the ethnic revival. There is no way in which he can escape responsibility for the reality of what he studies, no matter how "objective" he pretends to be. For his consciousness is supported by it, directly promotes the idea of it, is used in turn by it, and indirectly enhances it.

This being the case, discourse becomes essential. The student of ethnicity must engage in three related conversations. The first is a discourse on method aimed mainly at an audience made up of his colleagues. The second is a discourse on strategy aimed mainly at the group or groups he has studied. And the third is a discourse on the practical value of ethnicity in general, directed at the society at large.

The first discourse concerns the analyst's conception of ethnicity and his choice of perspective in studying it. Two are possible, as I have already indicated, the more structuralist approach which I employ and the culturalist approach which interprets ethnic groups as cultural entities. The culturalist approach is not only wrong in its very conception of ethnicity, but in shifting the explanatory focus from the structural and demographic factors to values and tradition, it emphasizes what is most unchangeable and least susceptible to political action. It also clearly implies that exploited groups are responsible for their own exploitation. Thus, if black and Puerto Rican misery is largely the result of their "defective" heritages and values, and if these groups insist on ethnically celebrating and preserving these heritage and values, they have only themselves to blame for their condition. This is an old conservative argument. It is hardly surprising that it is the position of the American New Right. What is painfully ironical, however, is the fact that the work which currently most represents this viewpoint and is being celebrated by the New Right as the masterwork on ethnicity is the recently published *Ethnic America*, authored by the conservative economist Sowell, who is black.

The second conversation concerns the sociologist and the group he studies. An ethnic movement, I have already indicated, is always a potentially dangerous development. There are only a few special circumstances under which it can be justified. A sociologist is enjoined to make this absolutely clear to the group he studies. Sociologists were right in not discouraging black ethnicity in its early stages, for reasons previously mentioned, but they were irresponsible in not pointing out the dangers in an ethnic strategy, especially when it was clear that the movement had shifted to a counterproductive chauvinism. Hopefully, European social analysts will not make such errors in their conversations with the potential leadership of the guestworkers.

Finally, there is the widest dialogue, that on ethnicity itself with the society at large. In a debate I once had at the American Academy of Arts and Sciences, my former colleague Moynihan wittily supported his position that ethnicity is a fact of life which we had better learn to accept by telling the story of Margaret Fuller, who one day finally declared that she had come to accept the universe, to which Thomas Carlysle responded, "Gad, she'd better!" I may have no choice but to accept the physical universe (except the stoic choice of suicide), but there is nothing inevitable about my social universe. The view that our social universe is an essentially ethnic one is, as I have already argued, itself an ethnic view of the world. Ethnicity cannot be complacently accepted, even by those who belong to symbiotic ethnic groups. The history of Europe should make that abundantly clear. Sociologists are required to disclose the irrationalities and inherent dangers of ethnic identification. Weigel (12) went on to say of faiths that they "are human things and we can analyze them. The acceptance of faith is a free act, and we cannot compel any man to believe one vision rather than another. However, we can show up the inadequacies of some faiths for the human situation. With this done, a given faith loses much of its recruiting power." This is a theologian speaking, an enlightened one, but one still committed to a particular faith. I suggest that we can, at least with the secular faith which is ethnicity, go further than that. We can disclose the inadequacies of all forms of ethnicities in the hope that we will thereby undermine their recruiting power. Wherever ethnicity exists, or is a potential force, I consider this an essential first step in any movement toward a rational society.

NOTES

(1) Cronon, E. 1955. Black Moses: Marcus Garvey and the Universal Negro Improvement Association. Madison: University of Wisconsin Press. See also Cruse, H. 1967. The Crisis of the Negro Intellectual. New York: William Morrow and Company.

(2) Greeley, A. 1974. Ethnicity in the United States. New York: Wiley and Sons.

(3) Jessel, L. 1978. The Ethnic Process: An Evolutionary Concept of Languages and Peoples. Hawthorne, NY: Mouton.

(4) Kallen, H. 1924. Culture and Democracy in the United States. New York: Boni and Liveright.

(5) Lieberson, S. 1980. A Piece of the Pie: Blacks and White Immigrants Since 1880. Berkeley: University of California Press.

(6) Novak, M. 1971. The Rise of the Unmeltable Ethnics. New York: Macmillan.

(7) Patterson, O. 1975. Context and choice in ethnic allegiance. **In** N.Glazer and D.P.Moynihan, eds., Ethnicity: Theory and Experience, Ch. 10. Cambridge: Harvard University Press.

(8) Patterson, O. 1977. Ethnic Chauvinism: The Reactionary Impulse. New York: Stein and Day.

(9) Rytina, S. and Morgan, D. 1982. The arithmetic of social relations: the interplay of category and network. Am.J.Sociol. **88 (1)**: 88–113. (For two earlier theoretical formulations which take account of absolute numbers and percentages in minority-majority relations, see:

Blalock, H.M. 1967. Toward a Theory of Minority Group Relations. New York: John Wiley and Sons; Blalock, H.M. 1982. Race and Ethnic Relations. Englewood Cliffs, NJ: Prentice Hall; and Granovetter, M. 1976. Network sampling: some first steps. Am.J.Sociol. **81**: 1287–1303.

(10) Smith, M.G.; Augier, R.; and Nettleford, R. 1960. Report on the Rastafari Movement in Kingston, Jamaica. Kingston: Institute of Social and Economic Research.

(11) Sowell, T. 1981. Ethnic America. New York: Basic Books. (For an excellent critique of this work, see Waldinger, R. 1982. The ethnic as self-made man: a review article. Soc.Pol. **13**: 1.)

(12) Weigel, G. 1959. The Modern God: Faith in a Secular Culture, pp. 15–16. New York: Macmillan.

Minorities: Community and Identity, ed. C. Fried, pp. 51–60
Dahlem Konferenzen 1983. Berlin, Heidelberg, New York, Tokyo: Springer-Verlag.

GROUP IDENTITY AND SOCIAL COMPARISONS

T. F. Pettigrew
Stevenson College, University of California
Santa Cruz, CA 95064, USA

Abstract. Most studies of minority identity have utilized black American children. Drawing on the diverse traditions of this research, four summary conclusions are advanced. Particularly relevant investigations of minority identity conducted on other groups are also reviewed, and they suggest two further conclusions about the generalization of these black American results. Implications are drawn for future research and social policy. To understand better how social policy shapes minority identity, a systematic program of cross-national studies of guestworkers in Western Europe is especially urged.

INTRODUCTION

The importance of human identity is widely recognized, even if its precise definition is elusive. The quest for self-identity is fundamentally a search for answers to such questions as: Who am I? What am I like as a person? How do I fit into the world? These are complex questions for anyone, but they are even more complex for minority group members.

We learn who we are and what we are like largely by comparing ourselves to others and observing how others react to us. But our salient group identifications structure this social comparison process. In turn, the larger societal context conditions this identity-forming process by providing meaning to our group identifications and social comparisons. In particular, to use Henri Tajfel's words, "the social climate of group differentials" (5) in the society molds the way in which even young children come to think of themselves and their group.

This paper focuses on how group identifications and social comparisons shape minority identity. It first reviews the extensive research on black American identity. Next it contrasts these results with those of other groups around the world to test their generality. Finally, the paper considers how state policies influence the minority identity process — a little understood domain that raises practical and scientific questions for future research.

THE MOST STUDIED CASE: BLACK AMERICAN IDENTITY

Black Americans are the most studied minority group in social science. And this research has devoted considerable attention over the past half century to identity issues. In broad strokes, research and theory on black American identity fall into four discernibly different, if overlapping, eras and emphases: (a) the mark of oppression thesis (1932-1968); (b) the trait liability thesis (1962-1970); (c) the proud, strong minority thesis (1968-present); and finally, the differentiated effects thesis (1971-present).

The Mark of Oppression Thesis (1932-1968)

Subjugation, this thesis asserts, has direct, negative consequences for minority personality. Lowered self-esteem, "soiled identity," even "self-hate" are often marks of intergroup oppression. Regarded and treated as inferiors, black Americans came to view themselves as they were viewed. Hence, both the social comparison and group identification determinants of self-identity contributed to feelings of impaired self-worth. Theorists in this tradition pointed out that much of American society — from racist mass media content to low-status jobs and inferior, segregated housing — reinforced this process. Worse yet, having to play "the Negro role" of socially defined inferior in interracial interaction furthered the mark of oppression.

This thesis was a research product of its time and place. The campaign against racial segregation in the southern United States made it a central contention; and the U.S. Supreme Court cited it in its historic 1954 ruling against public school segregation by race. Empirical support for the thesis came largely from two sources. First, psychiatrists, black and white, provided case data of individual blacks that traced personality damage to being black in a racist society dominated by whites. Second, two social psychologists, K. and M. Clark (1), devised in the late 1930s ingenious research procedures using dolls to measure indirectly racial recognition and preference in young chiildren.

The Clark research and later work using similar techniques repeatedly obtained consistent results. Racial recognition in both black and white children appears by the third year and rapidly sharpens each year thereafter. Especially significant is the finding that many black children apparently preferred white skin. They frequently chose white dolls and friends, misidentified themselves as white, or showed a tense reluctance to acknowledge their race. Moreover, young children of both races soon learned to assign poorer houses and less desirable roles to black dolls. Mark of oppression theorists generalized the Clark work widely, and held racial identity to be a major component of the personal identities of black Americans.

The Trait Liability Thesis (1962-1970)
This thesis, exemplified by the concept of "cultural deprivation," focused on the problems of blacks but obscured the critical links between these problems and the larger racial system. The trait liability thesis thus invited "blaming the victim" contentions from those who resisted racial change.

The greatest influence of this thesis occurred in American public education. Why were black children, now going to school with white children in growing numbers, often so unprepared? In its more benign version, it traced the problem to cultural deficiencies that could be supplied by special remedial programs. In its more malignant version, a few psychologists and one electrical engineer (but no geneticists) argued that the apparent trait liabilities were genetic in origin. In both versions, racial differences were assumed to be "black liabilities" that had to be somehow compensated for to meet the presumed higher and necessary "white standards." And black identity issues were assumed to be consequences, not causes, of these trait liabilities. Not surprisingly, this thesis soon produced a counter-thesis.

The Proud, Strong Minority Thesis (1968-Present)
Another product of its time, this thesis contends that black American identity is not scarred, that pride of identification with a strong, advancing minority has erased any effects of past racism. The mark of oppression thesis erred in viewing blacks as only passively reacting to white domination; it neglected the proactive capabilities of minorities to act on their environment. Moreover, argues this thesis, the principal social comparisons for most black Americans are not whites but other blacks. And black adults evaluate black children just as highly as whites evaluate their children. Besides, what is often seen by whites as liabilities are actually positive assets from this perspective (e.g., "passive withdrawal" is really "proud aloofness"). In short, this thesis maintains that earlier perspectives reflected white American ethnocentrism in its purist form.

A spate of new self-esteem studies support this proud minority thesis. They find the general self-esteem of black children to be as positive, sometimes even more positive, than that of white children. This new research typically differs, however, from the Clark doll tradition. It usually relies on direct self-report measures that omit mention of race, uses older subjects, and does not consider within-group variances. This new conceptualization and research also obscure the importance of the variety of situations in which black Americans now find themselves, a variety that is rapidly increasing.

The Differentiated Effects Thesis (1971-Present)

This thesis draws on all three of the earlier contentions for a differentiated perspective. It distinguishes between *group* and *personal* identity and emphasizes intergroup differences across diverse situations. The most extensive replication of the Clarks' doll research, conducted during the late 1960s, provides solid support for these distinctions. Porter (3) found a sharp disjunction among her young black subjects between their racial identities and their more global personal identities. Her middle-class children tended to possess more positive personal identities but less positive racial identities than her working-class children. In addition, the poorest black children in Porter's sample, those on public assistance, tended to evince the least positive identities of all, both racial and personal. These findings, combined with those from similar studies (4), permit four summary conclusions.

1. Oppression and subjugation have negative personal consequences for minority individuals. Many of these negative consequences are reflected in personality traits that in a range of situations act to maintain, rather than challenge, the repressive social system.
2. Not all minority group members are so affected nor are most traits of most minority members so shaped. Strong minority institutions, particularly firm family and religious structures, often shield individuals. Moreover, a sharp disjunction between the "real," personal self and the group self is generally possible.
3. Minority members in different situations are differentially affected. The middle class often finds itself in marginal situations, caught between its own group and the majority group. Such marginality can lead to a rejection of or at least confusion over their in-group identification; yet the security of middle-class status can also provide a positive personal identification. Working-class minority members frequently face the reverse situation. And the poorest of minority members may often suffer negative consequences for both their group and personal identities.

4. The disjunction between personal and group identities makes possible proud, strong minorities in spite of the marks of oppression. Personal strength can be and often is maintained even in the midst of group identity conflict. Historically, this strength becomes increasingly evident as the minority itself effectively challenges the oppressive intergroup societal system.

GENERALIZING BLACK AMERICAN FINDINGS TO OTHER MINORITIES

The black American experience is exceptional — a 360-year-old racially distinctive group with a history of slavery and segregation in the midst of a wealthy, heterogeneous nation. Though possessing distinctive variations, blacks in the United States also differ from many of the world's minorities in their sharing of a language, religion, and culture with the majority. We must, therefore, exercise great caution in generalizing these conclusions to other minorities.

Even in the United States, the second largest minority — Mexican Americans (Chicanos) — reveals differences consistent with its particular history and societal context. "Belongingness" is not a black American issue. Having arrived earlier than all but Native Americans and the first English settlers, blacks are seen as an integral part of the United States. But this is a central issue for Mexican Americans and many other minorities around the globe.

Chicanos find their group and national identities made difficult by an array of factors: (a) their great diversity across region, generation, and social class; (b) the accessibility of the Mexican border that keeps the old country vividly in view and makes it possible for new immigrants to try out America and easily return; (c) the extent and recency of their massive migration to the city; (d) the difficulty inherent in dealing with American authorities after years of bruising encounters with them along the troubled border; (e) their possession of a world language other than English; and (f) the common minority dilemma between making it in the larger society and retaining a distinctive and cherished culture. Exacerbating these marginality factors is the ambivalence toward them of other Americans — sometimes accepting and supportive, often rejecting and discriminatory. Small wonder that Mexican Americans have an unusually slow rate of acquiring citizenship and disagreements over a common group label.

Tajfel (5) investigated these identity issues among a variety of groups. For young children of majority groups in the Netherlands, Belgium, Italy,

England, and Austria, preferences for their own nationals over "foreigners" crystallizes early in life. Indeed, this initial form of ethnocentrism emerges long before the concept of "nation" is grasped even in rudimentary form. This primacy of affective over cognitive components also emerges among minorities but with contrasting results. Sephardic Jewish children in Haifa expressed for Tajfel their "likes" and "dislikes" for photographs of people whose ethnicity they had previously judged. These young children tended to like the photographs of the dominant group (Ashkenazi Jews) better than those of their own group. These results, paralleling those of the Clarks, come from a group that is not a numerical minority in its setting. The imprecise term "minority," then, can refer to groups of any population proportion that exist in the shadow of politically and culturally dominant comparison groups.

The important distinction drawn by Porter between personal and group identities among black American children has found European support in research reported at this workshop. M.-L. Pradelles de Latour presented fascinating data collected in Lorraine. In an area that has alternated between France and Germany over the past century, residents vividly demonstrate the complexity and flexibility of group identification. In general, age cohort and the first language shape group identity, but not always. One elderly citizen, she reported, maintained staunchly throughout his life, "Ich bin Franzose," without ever having learned French! Pradelles de Latour has found multiple group identities a natural phenomenon in Lorraine — "We are first of all German-speaking Lorrains, secondly French" (Pradelles de Latour, this volume). Similarly, S. Wallman has found multiple group identities common in South London among West Indians and other groups. In her paper (this volume), Wallman emphasizes the fluidity and plurality of group identification — emphases fully compatible with black American findings.

Recent work on guestworkers in West Germany has uncovered a comparable phenomenon. Malhotra (2) compared the responses of six-to-twelve-year-old children from six nationalities in the Wuppertal area and found an interesting disjunction between "general adaptation" and "social integration" in school. The former was measured by such variables as preferences for living in West Germany and speaking German and self-reports of dressing as well as and looking like their classmates. The latter was measured by such factors as reports of having many friends, playing with Germans during recess, liking school, and planning to continue formal education after the school-leaving age. With these measures, Italian and Yugoslav children revealed the greatest general adaptation but the least school integration. By contrast, Turkish and

Greek children revealed the least general adaptation but the greatest school integration.

There are probably many specific factors contributing to this pattern — years spent in West Germany, facility with German, similarity of the home culture to that of West Germany, etc. But the generic point is the complexity of minority identification. It would be reasonable to assume that extensive adaptation to the larger culture would lead to a child's perception of greater integration in school. Yet social comparison processes make such reasonable assumptions tenuous. Those most culturally similar, whether Italian and Yugoslav children in West Germany or middle-class black children in the United States, are most likely to have contact and compare themselves with the majority group. And this out-group comparison has profound implications for minority identifications and perceptions.

A further complication involves identity changes over time. Most relevant research is only cross-sectional. But we have noted in conclusion 4 above that positive personality changes often emerge when the minority group itself mobilizes and challenges successfully the repressive societal system. Vaughan (6) provides important longitudinal evidence for this phenomenon among five-to-twelve-year-old Maori children in New Zealand. His data spanned the years 1961-1970 and four diverse regions. He notes marked increases in Maori in-group preferences over these years, especially among older children in the urban areas. While in 1961 the data resembled the Clark data on blacks, by 1970 older, urban Maori children slightly preferred their own group. Vaughan suggests that a Maori "Brown Power" movement in the 1960s, inspired by the "Black Power" movement in the United States, had its greatest influence on this particular segment of his sample. Hence, his New Zealand replication of recent black American results does not constitute an independent test. Both reflect, as do recent studies of West Indians in England, a common, worldwide movement of assertiveness among non-white minorities.

These results suggest two broad conclusions about the generalization of identity findings on black Americans to other minorities.

5. Findings on black American identity concerned with in-group rejection, social comparison phenomena, the disjunction between personal and group identification, and rising in-group acceptance with social change are often broadly replicated by studies conducted with children from other minorities around the world.

6. Nonetheless, generalization of black American findings must be made

cautiously. Most minorities are in societal situations that contrast markedly with that of blacks in the United States — "middleman minorities" caught between the dominant and subjugated groups, minorities with a culture highly differentiated from that of the dominant culture, minorities whose national "belongingness" is severely questioned, and minorities with fuzzy, unclear boundaries that allow easy passing into the majority group. Though "self-hate" and similar reactions have been noted among many of these peoples, relevant research is too limited to allow confident generalizations.

IMPLICATIONS FOR FURTHER RESEARCH AND SOCIAL POLICY

This terse summary exposes the many weaknesses of work on minority identification. (a) There is far too much reliance on investigations conducted with just one minority. Future work must concentrate on minorities in situations that contrast with that of black Americans. Groups in the Third World involved in static as well as fluid intergroup relations are of special interest. (b) Children constitute the subjects of most of the studies. To complete the developmental sequence, older subjects must be studied. (c) Cross-sectional investigations predominate. More longitudinal research is required to understand the process and change of minority identity. (d) Longitudinal research with older subjects would also allow further exploration of an interesting minority identity phenomenon that appears to underlie recent "revivals" of ethnic identity among highly "assimilated" minorities throughout the world. Long after much of the cultural and institutional base for ethnic identity has eroded, strong evidence of "affective ethnicity" often remains. Its persistence without structural support suggests that group identification may be uniquely qualified to supply meaning to the self and its social situs. (e) Finally, systematic research on how national policies influence minority identification remains to be conducted. There is an urgent practical need for such work. In its absence, we close with speculations and questions about the implications of this research for social policy.

Social policy must respond to a bewildering array of concerns, and minority identity seldom holds high priority among them. Yet such issues come dramatically to the fore when racial, tribal, and nationalistic strivings spill over into collective action. Moreover, such publicized events show that prior governmental actions and policies led, unwittingly to be sure, to the minority politicization and mobilization. For example, when a nation's highest court rules for sweeping intergroup changes long sought by a minority, the group's aspirations rise rapidly, and its social comparisons increasingly are made with more privileged citizens. But if these raised aspirations are not met for a decade, it is hardly surprising if the minority mobilizes and vents its frustrations in the streets.

Obvious as this American example may appear, the ways in which policy influences minority identity are undoubtedly complex. The surprising disjunctions uncovered in research (e.g., between personal and group identities, and between general adaptation and perceived adjustment) guarantee that "commonsense" thinking of both the political right and left will miss critical subtleties. Complicating the picture further are highly publicized, international movements of minorities — racial and nationalistic. Black power was more than a slogan of the 1960s. As a rudimentary ideology, it inspired not only young Maoris in New Zealand but non-white minorities in South Africa, the United Kingdom, and elsewhere precisely because it articulated the rising aspirations of oppressed minorities in many situations.

Allowing for these complications, it should be possible to determine in future research the broad links between social policy and minority responses. Western Europe and its wide range of national policies governing its culturally diverse guestworkers offer a natural laboratory. Consider the possible consequences of three contrasting policies. (a) One alternative discourages families from accompanying workers, sets strict time limits on the employment, and provides little or no prospect of citizenship in the host nation. Such an approach should minimize social comparisons with host nationals and maximize original national identifications. It might also encourage intergroup conflict and attract less motivated and ambitious workers. (b) Another policy allows families but is ambiguous about permanent settlement. While new citizenship is possible, the residential and educational patterns for the group are kept largely segregated. Such a policy might well lead to maximum marginality among the guestworkers between their host and original nationalities. It might also cause a maximum degree of ambivalence toward the guestworkers among the nation's citizens, especially in times of unemployment when the belongingness of the "guests" is likely to be called into serious question. (c) Finally, a directly assimilationist policy encourages workers to bring their families and plan permanent residence and new citizenship. This alternative might well attract the more ambitious and motivated workers as well as leading in the initial generations to group identification conflicts comparable to those documented in the voluminous research on ethnic immigrants to the United States.

Considerable research on guestworkers has already been conducted. What is needed is a more systematic, cross-national comparison using this previous work together with newly commissioned and strategically designed studies. Such an effort could make important contributions to both social science theory and social policy.

REFERENCES

(1) Clark, K.B. 1963. Prejudices and Your Child. 2nd ed. Boston: Beacon Press.

(2) Malhotra, M.K. 1981. The psychological, social, and educational problems of primary school children of different nationalities in West Germany. Eth. Racial Studies **4**: 486–500.

(3) Porter, J.R. 1971. Black Child, White Child: The Development of Racial Attitudes. Cambridge: Harvard University Press.

(4) Porter, J.R., and Washington, R.E. 1979. Black identity and self-esteem: A review of studies of black self-concept, 1968-1978. **In** Annual Review of Sociology, ed. A. Inkeles, vol. 5, pp. 53–74. Palo Alto: Annual Reviews, Inc.

(5) Tajfel, H. 1981. Human Groups and Social Categories. Cambridge, England: Cambridge University Press.

(6) Vaughan, G.M. 1978. Social change and intergroup preferences in New Zealand. Eur. J. Soc. Psychol. **8**: 297–314.

Minorities: Community and Identity, ed. C. Fried, pp. 61–68
Dahlem Konferenzen 1983. Berlin, Heidelberg, New York, Tokyo: Springer-Verlag.

ETHNIC CONSCIOUSNESS AMONG IMMIGRANTS

J. Brun-Rovet
Ecole des Hautes Etudes en Sciences Sociales
Centre d'Etudes Nord-Américaines
75005 Paris, France

INTRODUCTION

For the historian, to chronicle and analyze ethnic consciousness among immigrants is a task of great difficulty. Let us take "consciousness" first: the historian's insight into people's minds often lacks the expertise of the psychologist, particularly of the social psychologist. Besides, "consciousness" of one kind or another is not always well documented, although this can be remedied in a number of ways; through indirect testimony for periods past (court and administrative records, newspaper articles, books) and oral questioning since the 1930s (and thus oral history has come to exist, extremely useful for anyone interested in the history and psychology of immigration). Dealing with immigrants, one encounters all the difficulties which go with studying minorities: e.g., inadequate documentation and prejudice. The question of where one stands seems essential: will immigrants be studied as part of the society at large, which entails playing down their distinct identity (one avenue has proved very fruitful in this respect: that of studying responses to groups of immigrants, *nativism,* for instance, as studied by Higham (10), or will they be studied as "Italians," "Slovaks," "Jews," etc., at the risk of losing the general problem en route, that of the existential realities of immigration in a given society, at a given point in its history?

All this, actually, is a classical plea for limits in chronology and boundaries in space. The space I am dealing with is that of the United States, with some references to the French experience because it provides to the decentralized case of the United States the counterpoint of French centralization and — until recently — pervasive underlying *Jacobinism*. The time period I refer to covers primarily the 19th and early 20th centuries but also touches on the colonial beginnings.

First, I will question the reality of *ethnic* consciousness among given groups of immigrants to North America through time. Was there anything we can call by that name? Which shape(s) did it assume? Second, I will discuss the making of a pluralistic society, the part played by ethnic consciousness in this development, and its effect on ethnic consciousness. Third, I will try to assess the importance of generations on ethnic consciousness.

THE UNCONSCIOUSNESS OF ETHNIC CONSCIOUSNESS

Most authors who have commented about the interaction between immigrants and American society, notably Gordon (8), date the ethnic question (sometimes under other headlines such as "assimilation," "Americanization") from the great 19th century and early 20th century influx of foreign population in the United States. Forty million people entered the United States between 1820 and 1920. Nevertheless, an early example of ethnic consciousness is given in the first census of 1790. Its authors imposed their group categories on their subject matter and found that the population of the United States was: 60.9% English, 8.3% Scotch, 9.7% Irish, 8.7% German, and 12% Dutch, French, Swedish and other. This, then, would seem to show a very early awareness that a heterogeneous society was in the making. And statements of sensitivity towards non-English people are easy to find, particularly in the writings of Pennsylvanians such as Benjamin Franklin or Benjamin Rush. Germans were particularly singled out, but the Scotch-Irish stood out, too, described either as wild rabble or as particularly skilled frontiersmen. Ethnic characterization, however, had then and for a long time afterwards to compete with other types of characterization. Religious affiliations in the 18th century loomed larger than ethnic ones. When a missionary of the Society for Propagation of the Gospel to South Carolina reported to the authorities in England about Presbyterians, the word "Presbyterian" was used indifferently as a religious label and/or as an ethnic presumption (Presbyterians would be either Scots or Scotch-Irish). One expects Anglican missionaries to think of groups in religious terms, but, in fact, their vocabulary is remarkably similar to that of administrators, military men, and colonists. The missionaries either use only religious categories or mix ethnic and religious categorizations without any consciousness of a slip (Brun-Rovet, in preparation). Religion seems to have been the first source of identity in the colonies. Maybe it was more than that and molded the way in which all differences tended to be perceived and accommodated as well.

This importance of religion is not limited to the colonial period. The first serious outbursts of nativism in the United States, the *Know-Nothing* movement, occurred at the time of the first non-Protestant wave of immigation, with the arrival of the Irish. The *No-Popery* theme was dominant. The fear of Catholicism as an obstacle to freedom to the Republic was paramount. Besides, Catholicism meant double allegiance to many American citizens. It was not the Irishness of the Irish that was questioned, it was their religion. Or rather, what was Irishness, if not Catholicism? The hostility towards the Irish makes one think of a remark of the French philosopher Jankelévitch who noted that the Germans came to hate the Jews not because they were different from them, but because they could no longer be distinguished from them. The Irish, after all, were English-speaking immigrants. At first, this common language did not prove a great asset. Religion, then, I suggest, offset ethnic consciousness until after the Civil War and may have competed with it afterwards. This was true of the majority population, but it is likely that it was also true of the Irish themselves. In fact, Catholicism became a basic component of their ethnic identity which they expressed in building churches and a network of parochial schools.

Other factors worked against a consciousness of ethnicity among immigrants. In some cases, a larger framework of ethnic identification to which immigrants could relate was lacking. The Italian immigrants are a good example. The Italian nation-state was a recent creation for Italians arriving in the United States in the 1900s. The heart of the new Italy was beating in the North; most immigrants were from the *mezzogiorno* or from Sicily and did not think of themselves as Italians. They identified with their village of origin and with their families. Certainly "campanilismo" is a form of ethnic consciousness, but was so fragmented as to destroy group identification rather than reinforce it (1,3,4).

What was often lacking was a will for ethnic consciousness. Anthropological and social differences existed and were easily perceived by immigrants and natives alike. They were not often transmuted by the immigrants into values to be held. One should not set too much store on the positive values of the "American dream," but what drove people to America was not a pure economic law. There was a choice and a hope in the move to America; very often it was individual rather than collective, personal rather than ethnic. The will to assimilate of the sons and daughters of the immigrants is better known than that of the first generation who were so busy making a living. There must have been considerable tension between their desire to transmit their own values to their offspring and their consciousness that these values and habits were *obstacles,* to some measure, of success in American society. If it was ethnic consciousness, then it was an uneasy one.

The relationship in the United States of ethnic consciousness with class consciousness has given rise to an abundant literature (15). In a seminal book Gutman (9) has shown how in many cases class consciousness could be rooted in ethnic solidarity. But this leaves out the numerous cases where ethnic solidarity played against class consciousness. All the recent studies underlying the strength of the family among immigrant groups (2,17) explain not the persistence of ethnic consciousness, but rather its appearance. Class consciousness does not seem to loom large in these studies, i.e., it seems to be less related to the strength of family ties than is ethnic consciousness.

These few remarks are meant as a reminder that ethnic consciousness can be associated with or exist in competition with class consciousness. Long ago, in an article about American nationalism, Potter already noted that men in society tend to have multiple loyalties (14).

It is not a taste for paradox that makes me doubt the reality of ethnic consciousness among immigrants. They clearly differed from the majority population and they knew it. The difference was so obvious as not to require the element of will that we usually associate with consciousness (particularly with the consciousness of class); what immigrants strove to maintain and transmit, and what they apparently were successful in maintaining and transmitting (family patterns, family ties) were perceived by them as *values,* moral or religious, but not necessarily ethnic ones.

How the real cultural differences evolved that set immigrants apart brings one to reflect on the notion of pluralism in the United States.

THE MAKING OF A PLURALIST SOCIETY

> "Tout homme qui choisit sa patrie
> est un Américain." – Talleyrand

The question I am addressing here is a double one. What part did the ethnic realities play in the emergence of a pluralist society in the United States? How did pluralism foster ethnic consciousness?

Any study of pluralism is made more difficult by the mixture of description and prescription in the discourse upon the subject. This is also true, of course, in the discussion of the "melting pot." It is one of Gordon's achievements to have shown that underneath the fusional metaphor of the melting pot two different aims were subsumed: (a)creating a "new man," a mixture of the positive aspects of all culture (what is shed and what is kept in this event is a basic problem), or (b) transmuting Irish, Scandinavians, Slavs, Jews, and Italians into Anglo-Americans. Nothing shows the entanglement of these two views better than Zangwill's play, *The Melting Pot*. The play was given with great success in 1908 all over America and was studied in classrooms. It gave lyrical expression to the "new man" idea and to the mixture of cultures. But it was also dedicated to Theodore Roosevelt who had liked the play and was certainly a good example of "anglo-conformism" (8). In the same period, Boas claimed that immigrants' children's physical characteristics were becoming more "American" and so were their gestures.

In 1911, a 40-volume Congressional Report on immigration was published, commonly called the Dillingham Report. It contained a mine of information as well as of prejudices embodied in a *Dictionary of Races*. In these years Kallen formulated his notion of cultural pluralism derived from William James's pluralist universe. In 1915 Kallen wrote "Democracy versus the Melting Pot" in *The Nation*, in which he described a new Federalism of national cultures.

Kallen could see that a pluralist reality antedated a pluralist consciousness: in early 20th-century America one could encounter Swedish and Norwegian rural areas, German cities, and Irish, Polish, and Jewish "colonies." Group settlement was a basic mode of settlement for immigrants in the United States. Kallen's views also meant an egalitarian development of democracy, demanding equality not only for individuals but for cultures, i.e., for groups. This demand for group equality lies at the root of affirmative action today. It raises fundamental political and constitutional questions that have been widely discussed since the Bakke case and its sequels ((5); also, see Cohen and Kilson, both this volume). Kallen's views may also have rested upon an illusion: in the United States, democracy is not only a political ideology and a legal arrangement, it is in itself a full-fledged culture of Anglo-American origins. Taking part in it means that one is already Americanized. In this respect, France in the early 20th century provides a comparison. There, a supporter of Radicalism (then the dominant ideology) would not distinguish between the Republic and the struggle for lay schools. Also, there was spread through the schools a unifying national political culture by which recently naturalized North African Jews — to take an example — but also Polish or Italian workers would adopt French values.

Pluralism was more than a generous attempt of a group of intellectuals to counter prejudice before the First World War and Americanization campaigns during the war. It was an American response to ethnic fragmentation. Americans have always had to face heterogeneity: in early colonial days, it was religious sectarianism within a Protestant framework. The problem was a real one and was "solved" by intense localism, weak federal and state powers, and the "civic religion" of the Constitution. The United States has given a similar answer to the ethnic problem. Pluralism does not only take into account the existence of different cultures, it fosters their survival under given conditions (12,13,16).

GENERATIONS

Let us take again the comparison between the United States and France in their respective attitudes toward ethnic differences. In spite of the periodic outbursts of intense xenophobia (particularly in the 1930s), there are no hyphenated Frenchmen. Second generation immigrants tend to be thoroughly assimilated. There are no Italian-Frenchmen or Polish-Frenchmen, which, given the number of Italians and Poles entering France in the 1920s, is conceivable. The American way has been different: assimilation seems to have taken place within groups (6,7). Ethnicity has been absorbed through the *sectarian* model, along the patterns long set to accomodate religious divisions. In fact, assimilation has largely taken place so as to preserve ethnic consciousness. It may very well be that ethnic consciousness, then, is a later development, concerning not the immigrants, but their grandsons and granddaughters. Ethnic consciousness and its revival in the 1970s in the United States could be what remains when Americanization has taken place: distinctive family structures, marriage patterns, and friendship networks (11).

There are a number of other factors to ethnic revivalism that we can identify: a reaction of defense against assertive black power and the new Hispanic immigrants, and for the Jews, the Israeli-Arab problems. Besides, the "Roots" phenomenon is not purely American. Whereas in the United States it takes an ethnic form, in Europe it stands on more geographical bases: the Basques, the Bretons, the Corsicans, and the Irish are examples. The old nations of Europe certainly never thought of themselves as pluralist nations. They may have to learn.

Ethnic consciousness then would be, strictly speaking, a reactive phenomenon: reacting to other ethnic groups and to prejudice from the majority population, but also, three generations after immigration, to the loss of most distinctive cultural traits. When there was a flourishing foreign language press in the United States, its main function seems to have been to help immigrants adapt to their new country. Ethnicity was a fact. The first mentions of immigrants in American history took the form of "contributions." Contributions of the Germans, Poles, Swedes ... to the American heritage ....

Ethnic consciousness seems to imply a weakening of the majority society values and confidence and a fear of loss of identity. It is also symptomatic of a new striving: that of truly accepting cultural and religious differences.

REFERENCES

(1) Banfield, E.C. 1958. The Moral Basis of a Backward Society. Glencoe, IL: Free Press.

(2) Bodnar, J., et al. 1981. Lives of Their Own: Blacks, Poles and Italians in Pittsburgh 1900–1950. Urbana, IL: University of Illinois Press.

(3) Briggs, J.W. 1978. An Italian Passage, Immigrants to Three American Cities 1890–1930. New Haven, CT: Yale University Press.

(4) Gans, H. 1962. The Urban Villagers. New York: Free Press.

(5) Glazer, N. 1975. Affirmative Discrimination, Ethnic Inequality and Public Policy. New York: Basic Books.

(6) Glazer, N., and Moynihan, D.P. 1963. Beyond the Melting Pot. Cambridge, MA: MIT Press.

(7) Glazer, N., and Moynihan, D.P., eds. 1975. Ethnicity, Theory and Experience. Cambridge, MA: Harvard University Press.

(8) Gordon, M. 1964. Assimilation in American Life. New York: Oxford University Press.

(9) Gutman, H. 1976. Work, Culture and Society in Industrializing America: Essays in American Working-Class and Social History. New York: Vintage Books.

(10) Higham, J. 1955. Strangers in the Land. New Brunswick, NJ: Rutgers University Press.

(11) Laumann, E.O. 1973. Bonds of Pluralism, the Form and Substance of Urban Social Networks. New York: J. Wiley.

(12) Mann, A. 1979. The One and the Many. Reflections on the American Identity. Chicago: University of Chicago Press.

(13) Patterson, O. 1978. Ethnic Chauvinism. New York: Stein & Day.

(14) Potter, D. 1962. The historian's use of nationalism and vice-versa. Am. Hist. Rev. **67(4):** 924–950.

(15) Steinberg, S. 1981. The Ethnic Myth: Race, Ethnicity, and Class in America. New York: Atheneum.

(16) Thernstrom, S., ed. 1980. Harvard Encyclopedia of American Ethnic Groups. Cambridge, MA: Harvard University Press.

(17) Yans-McLaughlin, V. 1977. Family and Community: Italian Immigrants in Buffalo. New York: Cornell University Press.

Minorities: Community and Identity, ed. C. Fried, pp. 69–78
Dahlem Konferenzen 1983. Berlin, Heidelberg, New York, Tokyo: Springer-Verlag.

IDENTITY OPTIONS

S. Wallman
London School of Economics
London WC2A 2AE, England

Abstract. Nobody identifies with the same group or in opposition to the same set of "others" all the time; everybody has more than one answer to the question "Who am I?" To understand Minorities: Community and Identity we need to take serious account of the process of identity shift, and to recognize the implications of multiple identity options.

INTRODUCTION

In this era, an unqualified use of the term "minority" invariably refers to people defined by their ethnic or "racial" status. Similarly the "Community and Identity" subtitle of this workshop implies ethnic identity and ethnic community. No doubt this is what we intend. But our discussion will be prejudiced if we focus on the ethnic dimension to the exclusion of every other. No one — not even the members of visible or beleaguered ethnic minority groups — consistently identifies himself or is always identified by others in ethnic terms. Ethnicity is only one identity option, only one way of defining a community. It is structured by and dependent on other things happening. And insofar as the "other things happening" are not fixed, the significance of ethnicity to individual or group identity must be fluid.

It may be that these points are too commonsensical to be the focus of serious academic attention. Or it may be that ethnic identity is now so "hot" a political issue that only single-stranded analyses of it catch the public ear. Either way, the study of ethnic identity continues to be dogged by three quite wrong assumptions. The first is that the way people look or behave or associate is an infallible or at least a reliable indicator of who or what they identify with. The second is that ethnic identity is a fixed and inflexible commitment, maybe (such as instinct) primordial, maybe (such as citizenship) achievable, either way steadfast, once-for-all. The third is that ethnic identity must be singular: in one version it cannot have more than one locus, in the other it is unhealthy or unviable when it has.

This paper argues against these strawman assumptions. The second part sets out an approach to the identity processes which accounts for changes in its focus from one situation, setting, or period to another. The third part underlines the normality of multiple identity with evidence showing that individuals have a more or less extensive repertoire of identity options which they call upon or engage with in different contexts and for different purposes. The fourth section suggests that identity with work and local community are alternatives to ethnic identity in industrial Europe and so may, in some circumstances, override its economic or political significance. The final section begins to draw out the practical and analytic implications of this approach.

IDENTITY PROCESSES

Identity issues arise whenever the question "Who am I?" is asked or answered. Essentially, therefore, identity is a personal and introspective matter of the sort that belongs in the purview of psychology or philosophy. The entry of the social sciences into "Who am I?" debates changes both their scope and their significance. Social scientists, although individually no less identity-anxious than the rest, are professionally concerned with the patterned behavior of groups. And even those among us who, like social anthropologists, do not assume that the meaning of behavior is the same for the actor as it is for the observer can only base their analyses on what people say and do "about identity" — i.e., on external expressions of it — and on the external circumstances in which they say or do it. Nevertheless, our interest in identity begins where "something" in the individual reacts to or on the organizational/affective/ecological options of the environment and "causes" him or her to identify in a particular way. It is at this point that psychological and social perspectives meet; they start at opposite ends of the identity spectrum but converge on common ground in the center (4,5). In this workshop, in spite of our diversity, we were by this token talking about a single set of issues.

The question "Who am I?" necessarily requires me to consider who I am not. Like all identity questions, it is a prelude to differentiation which, in its turn, entails the drawing of some kind of line between things like and things unlike. The place of the line will vary according to the criteria by which it is marked. These are inevitably ad hoc because they are dictated by one classificatory purpose or set of purposes: they cannot possibly be appropriate to every such purpose. It is important to distinguish between the need to differentiate, which is necessary and consistent, and the boundaries of differentiation which are contingent and variable. We cannot make sense of the world without sorting the things and people in it into categories of similarity and difference, but we make nonsense of it if we imagine that the same similarities and differences count in the same way all the time. The point, as Bateson has put it, is "difference which makes a difference"(2).

The similarities and differences which people may use to identify themselves or others as members of one or another ethnic group can be listed quite readily. But it is a great deal harder to pinpoint those they are identifying with in a given context, or to predict those they will identify with on any future occasion. On the one hand, classifying man is like classifying anything else to the extent that no one line of difference will suit every purpose. On the other hand, it is unique because it is mixed up in identity processes, i.e., because the facts of human difference are enormously complicated by man's social or psychological or existential involvement in those differences. For this reason, the relation between the "fact" of difference and the significance of difference is not predictable. Neither, however, is it random. The systematic relation between the two levels can be monitored quite simply.

Ethnic identity is marked by ethnic boundaries. These, like all lines of differentiation, have two sides. But ethnic boundaries are lines of *social* differentiation. Precisely by virtue of their socialness, they not only have two sides, they also have two kinds of meaning.

The first is structural or organizational. A social boundary marks the edge of a social sytem, the interface between that system and one of those contiguous upon it. The notion is borrowed from traffic management: the point at which the flow of traffic changes in speed, direction, or vehicular type is the point of interface between two traffic systems. It is also the point at which confusion and/or collision is most likely to occur — viz., the junction between a busy feeder road and a main highway. By the same logic, a social boundary is the point of interface between two systems of activity, of organization, or of meaning.

The second kind of meaning is subjective to the extent that it inheres in the experience of participants. Because it is social and not simply mechanical, the boundary marking the ending edge of one social system and the beginning edge of another has significance not only for the observer but also and more importantly for the members of those systems. It marks members off from nonmembers or nonmembers from members: the boundary can be read from either side. It is the point at which or the means by which "we," the members, can be identified. By this logic ethnic boundaries must be both an interface line between inside and outside and an identity line between "us" and "them." The interface element marks a change in what goes on. The identity element marks the significance given to that change and expresses the participants' relation to it.

The crude variations on any one boundary theme may be read off a four-part matrix:

	Identity	Interface
INSIDE (us)	We identify "us" in opposition to "them." We use the boundary for our purposes, according to our need(s) at this time/in this context.	The border around the familiar, the normal, the unproblematic.
OUTSIDE (them)	"They" identify themselves by contrast to the rest of us. They use the boundary for their purposes.	The beginning of another system. Performance, appearance, activity, social or symbolic structure is different.

When the four boxes are set up, the boundary process most relevant to intergroup relations of the sort called "ethnic"is shown by linking top left to bottom right. It is important to notice that these relations cannot be read off only one column or only one row: they are a function of the way things are *and* of what we make of them – and it is the step-link between the two that heats them up. Thus, it is genetically significant that two batches of people differ in respect of color, stature, or nose form; structurally significant that they marry according to different rules; and culturally significant that they eat different food and/or with different utensils (all these fitting readily into the Interface column above). But these differences only become "ethnic" when participants on at least one side of any of these boundaries use the difference to identify themselves as a group — to enhance the sense of "us" by distinguishing "them" more narrowly.

MULTIPLE IDENTITY

An inside/outside definition of identity does not solve the problems of limitation (Which of the many things that people do and say may reasonably be interpreted as identity items? Which of the myriad dimensions of context are relevant to identity processes anyway?), but it does serve to emphasize its relational aspect: who I am depends on who I am opposed to — whether symbolically or competitively — and on who is drawing the line of difference between us. In this view, ethnic identity is not a fixed individual quality which can be predicted on the basis of physical characteristics, mother tongue, place of birth, or ethnic origin. But identity markers like these are part of the symbolic currency of identity processes. Like other forms of currency, they have a potential resource value which improves when they are well used or invested, but declines even to the point of liability when "spent" in the wrong setting, rejected by exchange partners or devalued by the structure or the contrivance of the social marketplace. And like other forms of currency they can be held in reserve, a focus of security or identity "put by" for a special transaction, another time, a rainy day....

These stored identities are not visible to the observer. Sometimes their concealment is a deliberate part of impression management or economic strategy. Normally, however, even the actor is not aware of items in his identity repertoire that are not "in play." Although there is some kind of analogy between identities built up in theatrical performance and identities combined and recombined in the course of real life, it is not true that ordinary people change their identities as specifically and deliberately as they change their clothes. Moreover, because real life identities represent facets of a single self, not separate selves, it is unlikely that the real life actor will be conscious of shifting from one identity to another, or of putting together new and different combinations for new and different purposes.

It must be for these reasons that people so readily collude with census questions or research devices that allow only a single answer to the ethnic identity question. Many, indeed, whether in the role of researcher or the role of informant, are pleased when the question is asked at all; certainly getting people to define themselves is an improvement over ascribing them an identity on the basis of the way they look — or the way the observer feels about the way they look. But the evidence of inquiries which assume the likelihood of multiple or conglomerate ethnic identities and allow the informant to report "having" more than one, or "belonging" to more than one ethnic community, presents a different picture. Brief reference to three such inquiries will serve here to demonstrate both the reality of multiple identity and the social and psychological value it can have.

The first is a questionnaire used in a study of ethnic minority arts as community resources (1). The study set out to monitor the extent to which the actual or potential audience for ethnic artistic performance in London is drawn from members of the ethnic group "represented" by that performance, and so to infer something about the ethnic identity function of ethnic art.

Although the project and the survey covered all ethnic groups, the researcher was impressed by the way in which any discussion of identity needs inevitably settled on the needs of West Indians and the children of West Indians. Their special position seemed to rest on the popular belief that Caribbean minorities in Britain, unlike migrants from Asia or the Mediterranean, have brought with them no "deep-rooted traditions" and few cultural resources. Since they are also "alienated from white society" they "lack identity"; and this lack "causes" their delinquency and disadvantage.

The survey brought a different perspective. In the particular matter of ethnic identity, its questionnaire listed ten ethnic identities as open options – Londoner, South Londoner, West Indian, Asian, English, Black English, Welsh, Scottish, British, and Other, and encouraged all informants to "claim more than one if appropriate." The results show that, given the chance to express multiple identities, a good proportion of the members of each ethnic group did so. More specifically, more than half of those who defined themselves as "West Indian" also defined themselves in at least one other way.

These results are not surprising at the level of common sense. It is not difficult to imagine that a child, e.g., born in England of one (white) Scottish and one (black) Jamaican parent now living in south London could, when asked "What are you?", define himself as English, Black English, West Indian, Scots, a Londoner, or a South Londoner in turn, depending on who asked him the question, why asked it, and what else was happening at the time. Equally, it is not hard to imagine the same child "claiming" all those identities, not because he lacks a single identity focus and is (therefore) "in crisis," but because multiple identity is itself a healthy choice. In this respect, the position of children of West Indian migrants is different from that of other "second generation immigrants" only insofar as they have a wider and perhaps also a more open choice of identity options. The reported anomaly of their "black English" identity reflects the perceptions of the majority society, not their own experience (10).

The second study challenges the claim that the children of immigrants must choose between two cultures, two community identities, and that this either/or constraint leads to problems of identity or "identity conflict." On the contrary: there is evidence that minority group children "are *of* two systems, not *between* two systems, and therefore they have access to, and participate in, and have different degrees of identification with both systems. In the case of Britain at least, there is no concept of a hyphenated identity"(8). And it is this conceptual failure of the majority culture that puts so pessimistic a gloss on multiple or marginal identity patterns.

A third insight into multiple identity is brought by a psychological study of the part played by ethnicity in the "personal identity structure" of minority group adolescents (12). Seeking to correct the assumption that identity is both singular and fixed, the study rejects a blanket notion of "identity conflict" and deliberately replaces it with "the concept of a person's conflict in identification with another." In this perspective, "identity" is shown to be dependent on particular circumstances or role frames. The study is detailed and complex, but several points are germane here: First, it shows that individuals — whether or not members of "minority ethnic groups" — are aware that identity shifts, and they can, when asked, visualize the particular facet(s) of their own identity which would be relevant in different hypothetical situations or at different points in time. Thus young informants could distinguish the identity of "me as I am" from "me as I would like to be" and "me as others see me." Second, it demonstrates that, at least in the adolescent stages, there are significant differences in the identity processes and encounters of the boys and girls of any one ethnic group, and that the boy/girl contrast is not consistent — even across two nonwhite minority groups (West Indian and Asian) who might be expected to have similar conflicts in identification with the white majority "other." Third, it indicates that both Asian and West Indian adolescents report high incidences of "identification conflict" with their own ethnic group (measured in terms of evaluation of adults), but they do not generally devalue themselves in comparison with whites. In effect, conflicts of identification with parents and the choice of white friends or role models are not signs of "pathological self-hatred" or ethnic alienation.

ALTERNATIVES TO ETHNIC IDENTITY

The multiple identities set out in the previous section are all of the kind normally regarded as ethnic. The three examples cited confirm that ethnicity is not a singular identity locus because people ordinarily have some number of options within the "ethnic" range. But this is not to say that only ethnic difference makes a difference, or that we can properly understand the organization of minority "communities" and the experience of their members by reference only to ethnic identity processes. The proper inference is twofold: ethnic identity is not fixed, and ethnic identity does not always count.

Having noted the fluidity and multiplicity of ethnic identity as such, it is important also to consider the negative case. What, in effect, are people doing when not identifying ethnically at all?

There is no logical limit to the number of identity options available to ordinary individuals, but *work* and *locality* so often override the identity potential of ethnic origin in industrial settings that they warrant special notice here.

Identification with work and the products or rewards of work is not new, but it becomes more salient in these times of little economic growth and low employment. People without a job are likely to suffer "identity problems," whether or not they suffer the lack of a wage (6). Some of these problems stem from the fact that prevailing definitions of adult responsibility and the right way to live are closely linked to a job in the formal economy, and other kinds of essential "work" still tend to be systematically devalued (9). But the immediate effect of job loss is a loss of the structure of time which in turn may "break down" the identity of the jobless (7). Those without socially valued work are likely to "pull themselves together" around other identity loci; those with ready ethnic identity options may then "become more ethnic." This effect may show itself in expressions of racism and/or patterns of association. The statements of individuals are hard to interpret, but some young black Londoners report that work is a setting in which ethnic identity is irrelevant and that their friendships (and identifications?) with whites diminish when they are unemployed.

In the matter of locality, it is normal for people to identify to some extent with the area in which they live, but the potential for local identity is not everywhere the same. Where origin, culture, and locality are congruent, ethnic and local identity may be indistinguishable (3). In other areas, by contrast, the two are quite distinct. Not all "mixed" localities are alike. In some, the resource value of local identity clearly outweighs the potential of other options, particularly the ethnic option (11). It is as though certain kinds of local structure enhance the value of localism over ethnicity and others have the opposite effect. It will pay us as social analysts and as policymakers to examine the structure of identity options and to take interrelations of this kind very seriously.

IMPLICATIONS

Individuals do not necessarily identify themselves by the way they look, or the way they behave, or the people with whom they associate; nor do they necessarily identify with everyone or anyone they resemble in phenotype or culture.

The significance put upon external signs of similarity by nonmembers "outside" may well differ from their meaning for members "inside." Either way, it is informative. Whatever its relation to the way insiders identify themselves, it says something about the identity needs of the ousiders defining them – whether as private individuals or offical observers.

Individuals are not normally limited to a single locus of ethnic identity, nor does some kind of ethnic identity always count: ethnic origin is not inevitably the one "difference which makes a difference." While there are many new instances of ethnicity persisting among immigrants, reviving in their children, and burgeoning among natives all over Europe, there are also signs of the growth of identities based on other social groupings and defined by other kinds of boundary. Among these, identities based on work and on locality may be the most important non-ethnic options. Certainly they are sometimes strong enough to counterbalance the associative and affective pull of ethnicity.

Identity is a two-way process, and these observations apply to majority as much as minoriity populations. But it is not an equal process. Where the majority is politically as well as numerically dominant, it is majority definitions that order the distribution of political and economic resources. It is therefore peculiarly important that the majority takes account of identity shift and recognizes the positive value of identity options.

REFERENCES

(1) Baker, W.V. 1982. The Arts of Ethnic Minorities. London: Commission for Racial Equality, forthcoming.

(2) Bateson, G. 1979. Mind and Nature: A Necessary Unity. London: Wildwood House.

(3) Cohen, A.P. 1982. Belonging: Identity and Social Organisation in British Rural Cultures. Manchester: Manchester University Press.

(4) Epstein, A.L. 1978. Ethos and Identity: Three Studies in Ethnicity. London: Tavistock.

(5) Erikson, E.H. 1968. Identity: Youth and Crisis. New York: Norton.

(6) Fagin, L. 1979. Views from three other disciplines: psychiatry. **In** Social Anthropology of Work, ed. S. Wallman. Association of Social Anthropologists of Great Britain 19. London: Academic Press.

(7) Jahoda, M., et al. 1972. Marienthal: The Sociography of an Unemployed Community. London: Tavistock.

(8) Saifullah Khan, V. 1981. Some Comments on the Question of Second Generation. Conference Paper. Linguistic Minorities Project, University of London, Institute of Education.

(9) Wadel, C. 1979. The hidden work of everyday life. **In** Social Anthropology of Work, ed. S. Wallman. Association of Social Anthropologists of Great Britain 19. London: Academic Press.

(10) Wallman, S. 1978. The boundaries of "Race": Processes of Ethnicity in England. MAN **13 (2):** 200–217.

(11) Wallman, S., et al. 1982. Living in South London. London: Gower/LSE.

(12) Weinrich, P. 1979. Ethnicity and adolescent identity conflicts: a comparative study. **In** Minority Families in Britain, ed. V. Saifullah Khan. London: Macmillan.

Minorities: Community and Identity, ed. C. Fried, pp. 79–92
Dahlem Konferenzen 1983. Berlin, Heidelberg, New York, Tokyo: Springer-Verlag.

IDENTITY AS A COMPLEX NETWORK

M.-L. Pradelles de Latour
E.R.A. du C.N.R.S. No. 974, Université Louis Pasteur
67000 Strasbourg, France

"Berliners know they're Berliners, Parisians know they're Parisians, but who are *we*?" An apparently astonishing question, unless one is confronted with the history of the Lorraine coal basin. In the last hundred years, it has seen such historical and demographic upheavals that every generation experienced a specific historical situation that never repeated itself in the next:
- language and nationality changes took place five times in a century;
- frontiers were shifted seven times since 1815 (10,11);
- the rural, traditional way of life, changed into an industrial, semi-urban one.

The complexity of this situation, in which French parents became Germans who bore German children who later became French and had German children, etc, is best summed up by a 65-year-old man's self-description: "I come from Lorraine, of Germanic culture, of French nationality, and I think in our provincial dialect." He adds that he prefers to read in standard German, watches German or Luxemburg television, but feels most at ease writing in French. He then goes on to say that he has three children and four grandchildren.

Such a description of his identity is not exhaustive. He could have added that he was a schoolteacher, a member of the local horticulturists' association, etc. Most of all, it implies the existence of not one, but of multiple identity options (7,12). This *network* of identities is based on such diverse elements as belonging to a nation and a region, referring to socioeconomic and linguistic groups, as well as to a kinship system. It clearly shows that it is possible to consider oneself of Germanic culture without being a German citizen, or a French citizen without speaking French. Nevertheless, the choice of terms used to speak of one's identity is never made haphazardly (8). The issue is to discover how people will set up their identity supports, i.e., *identification references* that will allow them to define themselves in terms of collective or personal identity (13,14). A subject's attachment to a kinship system is one of the first manifestations of the insertion in the symbolic universe of language. The early identifications constituting the ego take place within the kinship system. These identifications are sustained by references that develop from social and familiar discourses (3). For this reason, a subject's narration of his genealogy, personal, and family — as well as regional and national — history reveals his identification references, i.e., what allows him to say: "I am such and such..."

INTRODUCTION TO THE FIELD OF RESEARCH

The Lorraine Coal Basin and Its Characteristics (Fig. 1)

Located in German-speaking Moselle, bordered northwards by Belgium, Luxemburg, and Germany, the Lorraine coal basin is the French part of a coal measure extending across the border into the German Saarland.

Although it began before the French Revolution, truly industrial coal-mining was given a real start in the second half of the 19th century, bringing about far-reaching changes in its population's life-styles. First, the mostly rural agricultural inhabitants became so-called "worker/miner peasants"*. Later, successive waves of immigrants poured in ("they came in wagonloads"); Polish and Italian before 1945, Portuguese, Spanish, and North African after 1945.

Already by the beginning of the century, small craftsmen and local industries were giving way to a large monolithic coal industry. After World War I, the coal industry was nationalized and enjoyed a first economic boost (a second one took place in 1973 during the oil crisis). Land was gradually bought by the coal enterprises; housing estates were built around the old village to lodge immigrant workers. These were at first kept sternly at a distance: "My father (Polish) married a woman from an old local family. Her parents kicked her out." Slowly they melted into the local population, although in varying degrees, depending on their community.

* "Paysans ouvriers/mineurs."

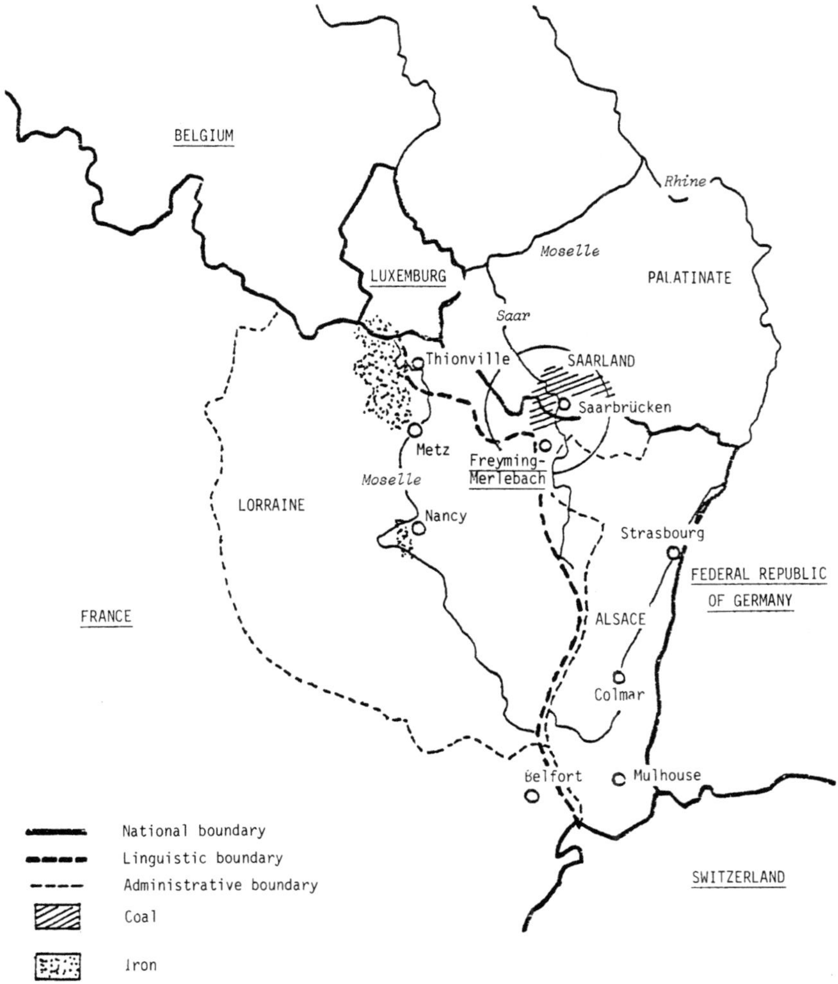

FIG. 1 — Map of the coal-mining regions of the Lorraine Basin.

Linguistic Boundaries and National Frontiers

Lorraine is divided by a linguistic boundary which crosses most of Western Europe without ever coinciding with a national state frontier (9). It cuts Lorraine in two, "thioise" Lorraine in the Northeast, Romanic Lorraine in the Southwest, runs into Belgium, follows the Vosges mountains, and continues into Switzerland. The result of conflicts or ancient contacts between Frank and

Germanic influences on the one hand and Romanic influences on the other, this linguistic boundary has not changed much since the 5th or 6th century, despite numerous territorial vicissitudes through the centuries (2). The area's history was dominated by feudalism until the present frontiers were laid down at the Vienna Congress in 1815. According to Guichonet and Raffestin (5), during the 19th century the tracing of these state frontiers (see Table 1) was at issue for radically different concepts of the meaning of nationality.

1. The German concept of ethnic nationality maintains that language is a "blind fate driving the individual." From this point of view, frontiers must include all those who speak the same language, since according to Fichte "they constitute an entity that Nature alone joined together beforehand through multiple, indivisible ties." The annexation of Alsace and Lorraine was justified with the words: "We Germans know Alsace and France. We know what's good for the Alsations, better than those poor people themselves. We want to give them back their own soul against their will"((5), p. 87).
2. The French concept emphasizes *elective nationality.* According to this point of view, a nation is made up of a group of persons joined together mainly by the "contractual ties of working together as a whole." "What makes a nation," said Renan, "is not speaking the same language or belonging to the same ethnographic group, it's having achieved great things in the past and wishing to do so again in the future"((5), p. 88).

The ups-and-downs brought about by these two concepts are best summed up by a German-speaking Lorrainer's biting sense of humor: "In 1870, my grandfather who was French was taken prisoner by the Germans. In 1918, my father, a German, was taken prisoner by the French. In 1939, I was drafted in the French army. In 1940, I was taken prisoner by the Germans. In 1943, I was drafted in the 'Wehrmacht' as a 'malgré nous'.* In May 1945, I was taken prisoner by the Americans. We always ended up in the prisoner's uniform."

* "In spite of us." This is what men compulsorily drafted in the German Army after 1943 called themselves.

TABLE 1 – Summary notes on changes in the boundary between France and Germany from 1815 to today.

In 1815, at the Congress of Vienna, Saarland is attached to Germany and becomes Prussian.

1815 to 1870

The national boundary runs between Saarland and Lorraine.

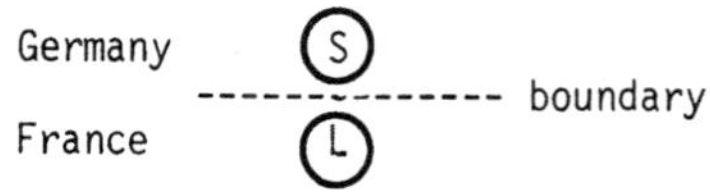

1871 to 1918

Germany annexes Lorraine.

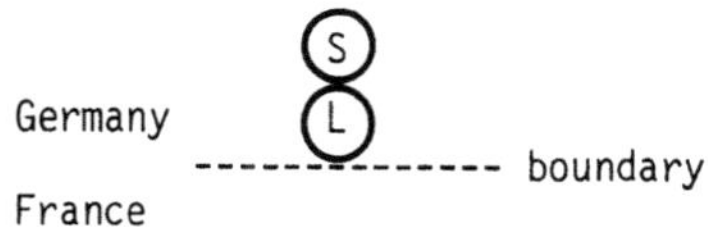

1918 to 1935

Thanks to the Treaty of Versailles, Saarland becomes economically attached to France. The customs line runs between Saarland and Germany.

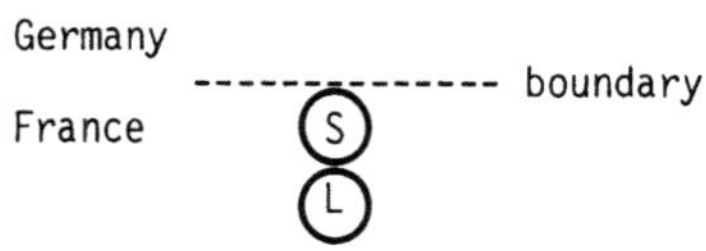

1935 to 1940

Hitler "violates" the military provisions of the Treaty of Versailles and occupies Saarland. Referendum on January 1, 1935.

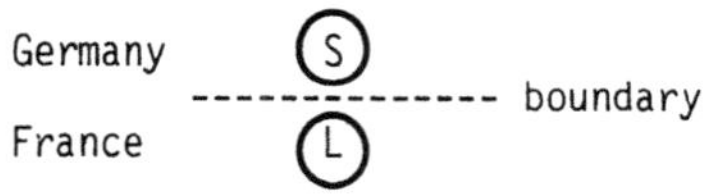

1940 to 1945

Annexation of Lorraine.

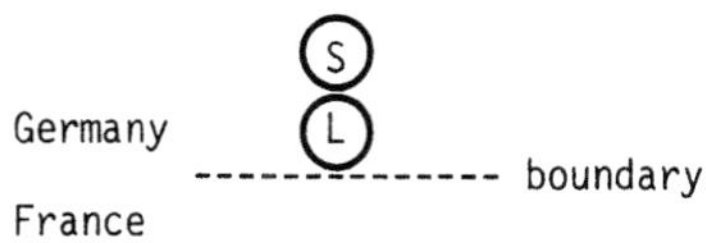

1945 to 1955

Saarland is once again economically attached to France, though politically independent.

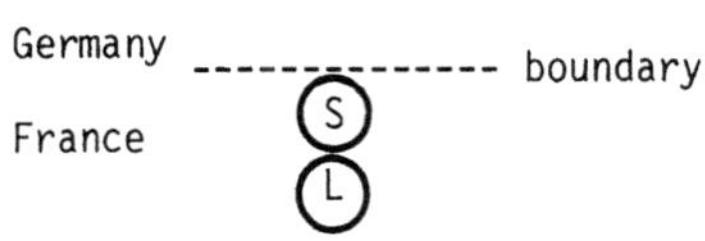

From 1955

Franco-German negotiations. Referendum; Saarland returns to Germany. January 1, 1957; Referendum. 1959: end of economic union with France.

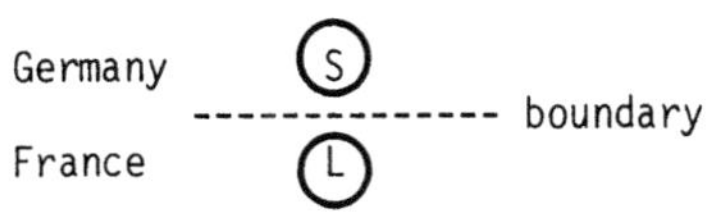

The Village of Hombourg

The village of Hombourg (11,000 inhabitants) was chosen as field of research. Three kilometers from the German border, it consists of the old village, a miners' housing estate dating back to 1946, another so-called "transit" housing estate, and two building plots. The village of Hombourg was never totally given over to farming, although land and land ownership were of utmost importance. In fact, since the 15th century, there have been several small local industries: nailsmiths, weavers, stonehewers, and blacksmiths. In 1856, a metallurgical company previously established in Saarland moved in, bringing along about twenty Saar families. It was followed in 1939 by another steel factory that shut down in 1979. After 1946 mine shafts were sunk, housing estates were built for immigrants, and population grew from 2,400 to 11,000 inhabitants. Worker/miner peasants, though, were only continuing an old "double day" tradition: a day at the ironworks, at the quarry, at the factory, later at the mine, then a day ploughing the fields. This "double day" tradition was only possible through a vast system of mutual aid among families; women played a vital role in this process by exchanging services as well as through daily upkeep of land and livestock.

IDENTITY AND GENEALOGICAL STRUCTURE

Every genealogy is inserted within a time and a space that vary according to several factors, for instance, belonging to different ethnic groups. Depending on whether the grandparents are of local or foreign origin, the subject's statements about genealogy will stress either direct lineage or alliance by marriage. This particular way of structuring the kinship network results in two different ways of establishing one's identity.

Direct Lineage (Fig. 2a)

When grandparents are of local stock, the kinship network is very well-known: an average of 100 persons per genealogy, going back five to twelve generations in time. Interest in the genealogical tree is motivated by the search for one's roots: "How far back does it go?" Reasons put forward to explain this genuine quest always revolve around the same theme: one's origins, the wish to know "where one came from." For this reason, the public records become a reference point and are thoroughly consulted and examined in detail.

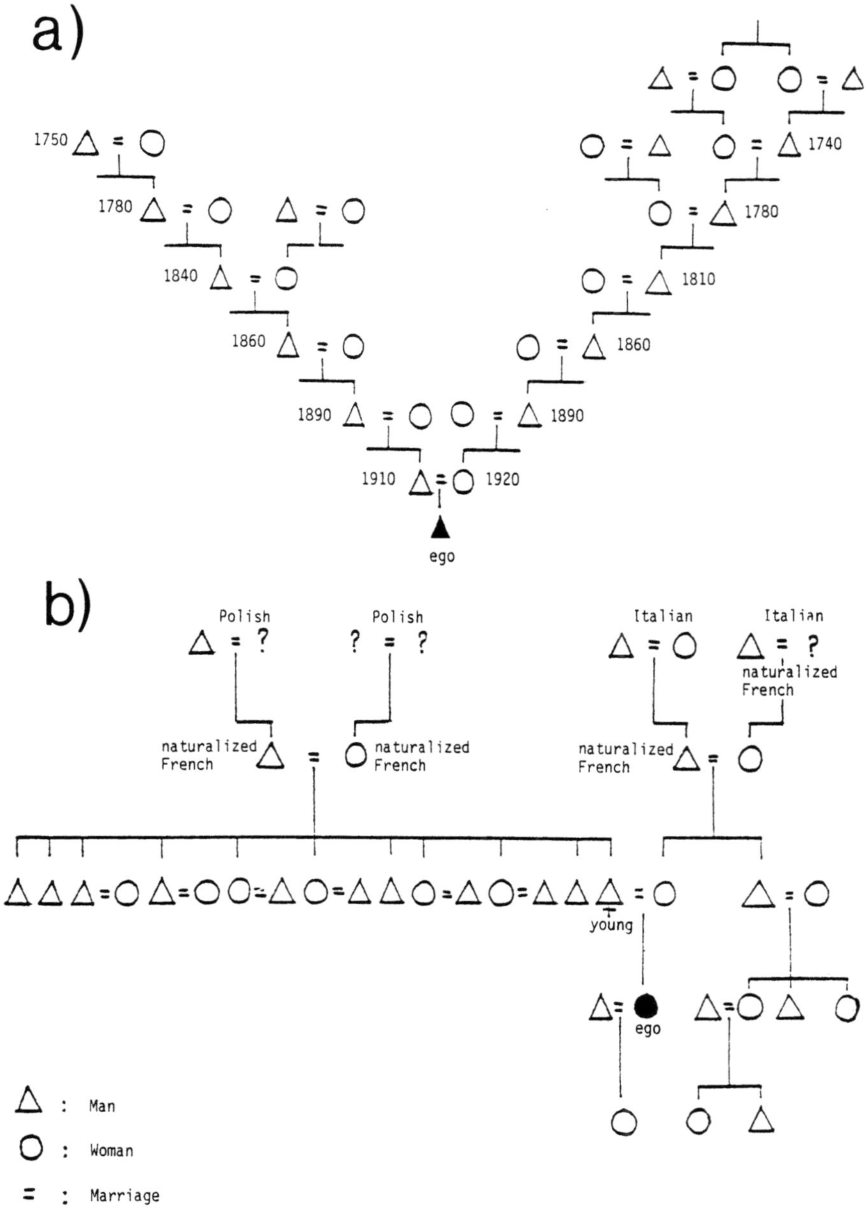

FIG. 2 — Illustration of genealogical structures.

"I have just come from B. Town Hall. I was looking up the rest of my genealogy, I've gone back to 1600." Or: "On my mother's side, I had to stop because records didn't go further than 1600. On my father's side, I've gone back to 1756."

The ancestor's line, that is the *direct line*, is given preference over collaterals. More specifically, the *male* ancestors — bearers of the family name — will be privileged, and research will center on the origin of the family name. When found, it is no wonder that it usually refers to a place name rather than a trade.

This place of origin becomes, in a way, "the hole one came out of"*. "I don't know where we come from," said a man embarrassedly after drawing up his genealogy down to the 9th generation. Indeed, discourse clearly shows that reference to a real or mythical original locality serves as support in affirming one's national identity.

In this concept of genealogy, each generation represents a link in a chain. No link is relevant in itself; it is only there to form the chain. So the period of time determined by genealogy, according to the direct line, is purely symbolic. It marks time, like a clock in reverse: Pierre is Louis's father who is Joseph's father who is Louis's father... in a slightly desperate attempt to find the first stroke, the original stroke. For this reason, genealogy is first and foremost a man's domain, even when his wife is concerned: "I know my wife's genealogy," says Mr. F., "better than she." And a village woman: "On *my* genealogy, you'd better ask my husband."

Alliance (see Fig. 2b)
When parents are of foreign descent, the parentage network is far less well-known: an average of 50 persons per genealogy, going back three to five generations. Grandparents are often not mentioned, great grandparents are seldom referred to at all. They come from somewhere else, Saxony, Poland, Italy, or Ireland, and conversation automatically turns to emigration accounts: "In 1854 or '64, I think, the Irish rose in rebellion against the English Army. One of them fled and finally landed here. Seems that was our great-grandfather... but we're not so sure anymore." The genealogical voyage no longer takes place in time; one doesn't "go back" in time, but in space, one comes "from somewhere." A trajectory is described here, sometimes broken by long stages following the opening of new mines, marked out by births or deaths — these become reference marks in the family's memory: "My

* "Le trou d'où on sort."

grandfather was a miner in Siatkowa. He first went to Germany. I know he stayed there some time because my father and two sisters were born there. Then he went to Pas De Calais in 1919; someone had told him he'd earn more money there. Later on my father went to Merlebach because new mines had just opened."

Although the tree has no roots, it spreads. Relatives and relations by marriage are remembered, or those one meets occasionally on holidays, first communions, and burials. Questions about genealogy are an excellent excuse for visiting family members: "Sunday, I'll go see my mother (my sister, my godmother). I'll ask her then." Sources of information are always women. They are pointed out as representing the family's memory. In comparison to time determined by the direct line, this memory relates to a very different dimension of time. References are no longer names passed on from one generation to the other, but life's outstanding events: births, first communions, marriages, and burials. This family time is no longer symbolic. It has length, depth, and stability. Genealogy becomes a woman's affair: "I know my husband's genealogy," says a Polish woman, "better than he does." The individual is not only a link in a chain but also a ring from which new chains will start. Alliance takes precedence over lineage. *Matrimonial ties,* rather than a search for one's origins, are foremost the quest of immigrants and their families for identity.

IDENTIFICATION REFERENCES

Identification references function in close relationship to the field of kinship. Indeed, both axes of the kinship structure that we just emphasized (marriage ties or direct lineage) (4,6) support more collective identifications such as "being French" or "being a Lorrainer." The first statement stresses direct lineage on the father's side, whereas the second is backed up by direct lineage on the mother's side.

Being French

"My great-grandfather," says a man, "was born in 1866. He didn't speak a single word of French. And yet he never diverged from his belief until his death in 1950. He still would say 'Ich bin Franzose'." This kind of assertion was heard again in several interviews, often through the phrase "being a patriot." "To be French" means to those born before 1945 "to be patriot." This statement about one's identity possesses the following characteristics:

1) In the first place, it clearly shows that national and linguistic identities do not necessarily overlap. It is clear that cultural practices are void of any identification references of a "francophile" nature (15).

2) Although it sometimes results in acts (deserting the German Army), it shows mostly in discourse: "My mother talked a lot about patriotism," says a young man, "she used to say someday there would be a liberation for us and that Lorraine would go back to France."

3) It is closely related to a family name. During an interview, a woman whose maiden name is "Duchamps" (according to family tales, it originated in Southern France) explains how her father, a true patriot, deserted the German Army in 1914 to join the French lines. Her husband breaks in: "With a name like that," he said, "he just couldn't sing out of tune."* What he meant was that equivalence between the name and his particular behavior was normal, regardless of the family's true history. In fact, this family who had settled in Hombourg four generations earlier came from a Saarland village, according to the Records Office. Time and history are telescoped in favor of a mythical or real place of origin suggested by a French-sounding name. This name reminds the subject of the place it comes from and allows him to recognize himself as French.

Although the name is not always the father's (sometimes it is the grandmother's maiden name), it comes nevertheless from the father's side.

Finally, men alone may actually embody this "being French." A woman may transmit a feeling of admiration or respect to her daughters for the one who will incarnate in the family's eyes this ego ideal — father, grandfather, or husband — but she can hardly take his place since she cannot transmit the name. A woman recalls how in the German days her family name "Brun" was germanized to "Braun": "In 1920, we had to pay — and a lot — to get our name back. Father paid for the boys but not for the girls. He said it was no use since we would change names anyway." Sixty years later, she still resented it bitterly.

What counts for the subject is a symbolic identification with a name transmitted in direct paternal line, insofar as it vocally suggests a determinate spot, called "the interior,"† that functions as an unquestionable reference point. The name in question is recognized as French and legitimates the entire discourse by confirming national identity.

* "Il ne pouvait pas chanter faux."

† This is what the rest of France is called in Alsace and Lorraine: "The interior of France" or simply "the interior."

Being a Lorrainer
"On both sides, my mother was a Lorrainer, that gives you a firm footing. If it weren't so, it wouldn't feel the same. One feels of pure Lorraine-stock." "It's a pity the dialect isn't spoken any more because feeling for the homeland gets lost. I'm very fond of Lorraine but I wonder if my children will feel the same love."

Words used to describe one's relationship to the region have strong emotional undertones. Terms such as being fond of, losing, love, or feeling are common. They vividly emphasize a particularly important attachment to land: to be *deeply rooted,* to descend from pure Lorraine-*stock.* The references are punctuated during interviews with words describing chores performed mainly by women: wives, mothers, or grandmothers are here the main protagonists. Earth-bound skills are evoked, not only the soil from which crops are raised and harvested, but also its extension upwards, the house, and downwards, the grave. Throughout the interviews, the father is evoked in terms of social status — railwayman, miner, foreman — whereas the mother is spoken of in terms of abilities, skills, and maintenance. Mother makes preserves and pastry, she works in the fields, does the housework, looks after the animals... All are fragmented, parcelled activities, centered on the land and home. When the mother knows a trade (seamstress, for instance), it is not looked upon as a status symbol, as in the case of men, but above all as a skill: "My mother sewed for herself and the family. You'd give her a ham or a rabbit and she'd sew you a dress."

The produce of her labor is not sold but exchanged for goods produced by another skill. Women are an axis for these elements: the home, the land, and the grave. These serve as a background to help form a regional identity. Women are the *medium* that transmits its, not for the sake of abstract, established rules, but in relation to daily behavior, practices, and customs. These determine how the kinship network functions in daily life. Here are brief summaries of the home, the land, and the grave.

The Home
Belongings, including the house, are handed down in the family according to a truly characteristic rule: not on the father's side, but down the female side, from mother to daughter or grandmother to granddaughter. This rule developed from a norm of conduct I shall call "duty of assistance" of a daughter towards her parents (especially her mother). This duty to assist is very strong: on the parental level, it draws a clear demarcation line between direct kinship

and relatives, especially between daughters and daughters-in-law. A woman explained that when helping her mother-in-law with the housework, she would take care to close the shutters before washing the windowpanes. "Neighbors shouldn't see," she said, "that it's her daughter-in-law who comes to help. It's shameful because it's up to the daughter to do it. I can't do her gardening or shopping either. Just clean indoors and help her wash." Thus when parents get old, one of the daughters stays home to take care of them until death. With her brothers' and sisters' consent, she will inherit the house.

The Land

The alliance system has another most interesting characteristic: repeated marriages between two families. The most frequent cases are of a man marrying two sisters successively or two brothers marrying two sisters. (Someone even mentioned the case of three brothers marrying three sisters or even brothers and sisters of one family marrying brothers and sisters of another family.)

This marriage pattern might be connected to a strong matrilocal tendency (until the mid-20th century, a man would leave his parents after marriage and join his wife at her parent's home). There is a strong link between land transmission and the choice of marriage partners, as in the case of the transmission of the house, it must be grounded on customs and behavior centered on women (this point has yet to be thoroughly studied).

The Grave

Graves are nearly as important an investment as houses. One needs only walk through the Hombourg cemeteries to realize that. As far as the eye can see there are thick black granite slabs overloaded with flowers, sometimes surrounded by a little garden. "Before we used to buy nicely carved stones. Now we prefer granite — it looks neater and it's weather-proof." Graves are more of a status symbol than houses, I was told. People should be able to see what they cost. Graves, like houses, are a woman's affair, and a close relationship between a particular mode of behavior and a rule of transmission is found again here.

Women must take care of graves on Sundays after Mass, of course, but sometimes during the week as well. Taking off wilted flowers, removing rust from artificial wreaths, watering plants, swilling the granite slabs. "It takes her a lot of time," says a man about his wife. Families living too far away pay a village woman to perform this task.

Although an enquiry on how burial sites are assigned is difficult to carry out ("The first one who needs it takes it!"), the choice is left to the wife. In most cases, she will decide to be buried by her parent's place (i.e., by her mother's) and men follow suit.

CONCLUSION

A person has a wide array of identities to chose from. "Everybody has more than one answer to the question 'who am I?' (Wallman, this volume). Nevertheless, these choices are not random. They are completely determined by the subject's personal and familial history, itself inserted in the more global history of his community, region, and nation (1). Thus people affirm that they are French, or its equivalent "a patriot," on the basis of a French-sounding name, supposedly originating from the "Interior" of France and transmitted through the father's line. This assertion may be free of any linguistic or cultural practices. It is essentially based on family stories sometimes several generations old.

Declaring that one is "a Lorrainer," on the other hand, is supported by mostly earth-bound tasks performed mainly by women, and symbolically extended in the way belongings are transmitted down the female line. Although the Germanic dialect and specific cultural behavior are still closely linked to one's declared identity, it may equally well lose its real substance and remain merely as a profession of faith, all the more solid as its rests only on discourse.

Here lies "the danger of ethnic mobilization" mentioned by Patterson (this volume). This danger turns into a "burning" reality when an ethnic claim, detached from the past and all common cultural practices, arises and becomes the support of purely arbitrary pretentions such as, for instance, "racial purity."

REFERENCES

(1) Abou, S. 1981. L'identité culturelle, Relations interethniques et problèmes d'acculturation. Paris: Ad. Anthropos.

(2) Allies, P. 1980. L'invention du territoire. Collection Critique du Droit, 6. Grenoble: Presses Universitaires de Grenoble.

(3) Freud, S. 1970. Essais de Psychanalyse. Paris: Payot.

(4) Goody, J. Comparative Studies in Kinship. 1969. Stanford: Stanford University Press.

(5) Guichonnet, P., and Raffestin, C. 1974. Geographie des frontieres, pp. 86–88. Paris: PUF.

(6) Heritier, F. 1981. L'exercice de la parenté. Paris: Seuil.

(7) Hurstel, F. 1982. Idendité de père et classe ouvière à Montbéliard aujourd'hui. La Pensée **288**: 58–72.

(8) Lacan, J. 1978. Le Moi dans la théorie de Freud et dans la technique de la psychanalyse, Book II. Paris: Seuil.

(9) Revue. 1982. Pluriel Débat. Frontières. Problèmes de frontières dans le Tiers-Monde. Journées d'étude des 20 et 21 Mars 1981. Année 1982, nr. 30. Paris: L'Harmattan Université Paris VII.

(10) Rohr, J. La Lorraine Mosellane 1871–1946, Tiré à 100 exemplaires. Imprimerie Sarregueminoise.

(11) Roth, F. 1976. La Lorraine annexée 1870–1918. Annales de l'Est, Mémoire, nr. 50. l'Université de Nancy II.

(12) Saifullah Khan, V. 1981. La "seconde génération" ou les "nouvelles générations," transl. X. Couillaud. Paris: Centre d'information et d'études sur les migrations.

(13) Tabouret-Keller, A. 1975. Un champ sémantique: les noms d'appartenance raciale au Honduras Britannique. **In** La linguistique, nr. 11, pp. 123–133. Paris: PUF.

(14) Tabouret-Keller, A. 1977. La notion de nation en défaut: le cas de Belize. Equivalences. Langues et nations (special edition) **2-3**: 85–99.

(15) Tabouret-Keller, A. 1982. Identités et Evolution de Situations Linguistiques complexes. **In** l'éducation bilingue dans l'école Valdotaire. Bibl. valdotaire nr. 21. Aoste: Musumeci.

Standing, left to right:

Friedrich Heckmann, Jeanine Brun-Rovet, Sandra Wallman, Carl Graumann, Tom Pettigrew, Lorraine Baric, Werner Schiffauer.

Seated, left to right:

Marie-Lorraine Pradelles de Latour, Lale Yalcin, Verity Saifullah Khan, Francoise Morin, Sheila Patterson.

(Not shown): Serif Mardin, Orlando Patterson.

Minorities: Community and Identity, ed. C. Fried, pp. 95–107
Dahlem Konferenzen 1983. Berlin, Heidelberg, New York, Tokyo: Springer-Verlag.

FORMATION OF CONSCIOUSNESS
Group Report

V. Saifullah Khan, Rapporteur
L.F. Baric, J. Brun-Rovet, C.F. Graumann, F. Heckmann, S.A. Mardin, F. Morin, O. Patterson, S. Patterson, T.F. Pettigrew, M.-L. Pradelles de Latour, W. Schiffauer, S. Wallman, L.A. Yalcin

INTRODUCTION

Discussion focussing of the formation of minority consciousness preceded discussion about mobilization and claims. But in asking under what conditions people become conscious of being a member of a particular ethnic category, there are four major problems.

First, there is the problem of evidence. The only evidence of ethnic identity is what people say or do. What people say may not correspond to the objective situation. As well as what people actually say, there are the cultural assumptions and connotations underlying why they say it and what they mean. And then, of course, there are the concepts others may use to articulate or interpret what has been said.

Second, there is the problem of the different perspectives on ethnic identity. In this case there are clearly different *linguistic* usages and meanings attached to the same phenomena which may cause problems of communication between social scientists with different first languages. The different meanings may reflect a particular *cultural* perspective which has been built up within a specific historical and socioeconomic context. Third, there are different perspectives between, but also within, different *disciplines.* The terms "identity" or "ethnicity," for example, may reflect very different interests and approaches among psychologists, social anthropologists, or political scientists. And, of course, these terms may not be known by, or may have a negative meaning for, the "subjects" of the research.

The fourth problem in discussing the formation of consciousness arises from the need to look at interaction, not "the" group itself. The focus is on consciousness at the boundary which is created by the dynamic interaction between the in-group and the out-group. Despite these difficult problems, however, universals can be sought in this realm if the importance of social context is kept constantly in mind.

UNDERSTANDING CONSCIOUSNESS AND IDENTITY

To understand how people develop an awareness of their identity or membership in an ethnic collectivity, it is necessary to adopt a *social* definition of *identity*. The conception of this social reality is culturally determined and articulated through language. Understanding of the social context is obviously necessary in the following example. A person from the Bassin Houiller Lorrain (in North East France) who died in 1965 declared throughout his life "Ich bin Franzôse," without ever having learned a word of French (Pradelles de Latour, this volume). Clearly there should be no assumption that ethnic and linguistic identities overlap (Tabouret-Keller, in preparation).

The same problem faces social scientists who assume that if they discuss such issues in a common language they will understand each other's terms in the same way. They, too, are constrained by the meanings of their concepts and the particular contexts in which they have worked. Terminology that is acceptable in some countries, such as "ethnic minority," may be unacceptable to social scientists in other countries. "Minority" may be a pejorative term in some countries, or "ethnic" may carry the connotations of conservatism.

Adopting a social definition of identiy should also involve an acknowledgment of *power relations*, i.e., power relations between individuals, between ethnic groups, and between ethnic groups and other minorites. Some of these cases are clearly relations between a dominant culture and a subordinate culture. Over and above this is the relationship between the state and the individual or group.

The other conception needed to understand the formation of consciousness is that of the construction or *creation of social reality* (1). This creation of social reality is contributed to by both the minority and the dominant majority, and a reactive dynamic is created by the lack of fit between the two.

Social scientists and other professionals are also involved in this process and have to be extremely sensitive to the ideological implications of their work. It is not only their concepts or categories which may not be accepted by the people they study, it may also be their interpretation of the social reality of those they study.

An unquestioning acceptance of the term "ethnic minority" for indigenous autochthonous national minorities in France could lead to both an imposition of unacceptable outsider categories on the population but also to an erroneous analysis of the situation.

This example brings us to the fundamental *historical dimension* and the dynamics of superordinate-subordinate relations (13). The national indigenous minorities in France represent themselves as nations or national minorities, but for the state they are cultural groups (5,8). For example, Occitan is claimed as a national language by segments of the Occitan population, whereas the Ministry of National Education considers it a regional language.

The imperial history of England has played a part in the English preoccupation with black or colored minorities as causing the greatest problems. In the case of the West Indian minorities, this perception greatly exaggerated the degree of cultural distinctiveness, as many migrants of West Indian origin came to England with a language and religion common to the majority, despite differences in dialect and religious practices. The South Asian minorities may have represented distinctive cultural traditions (and few migrants came with the intention of cultural assimilation), but they did not perceive themselves as "black."

The common experience of racial prejudice and discrimination based on color experienced by both of these populations has led some members of South Asian populations to share a consciousness of being "black" in English society. This is especially true for the more politicized members of the populations and for an increasing number of British-born youngsters of South Asian origin (10).

A dynamic historical perspective on social identity needs to incorporate change over historical, geographical, and personal time. The mutual interaction between insider and outsider perception of difference is mediated by the nature of power relations in the context in question. The formation of individual and collective consciousness is always interwoven and provides another dynamic underlying the process of ethnic consciousness.

THE NATURE AND TYPES OF MINORITIES

A demographic definition of minority does not necessarily help the understanding of the reactive process in superordinate-subordinate interaction (Heckmann, this volume). For example, to analyze the position of women from a numerical perspective, e.g., is to evade the importance of power in understanding male-female relations — the strategies and institutions which maintain the dominance of men.

The culture of the dominant majority is not only powerful in the way it conceptualizes the position of mnority groups, but also in the dominant institutions, such as the education system, which may reproduce the unequal distribution of economic, social, and linguistic resources. The dominant culture may, of course, constitute a numerical minority, as in South Africa.

The sense of powerlessness which is a characteristic common to most members of minorities may be constrained by the mechanisms of the dominant culture, such as lore concerning universal fair treatment and legal or economic strategies which restrict communication between members. In some cases it is more appropriate to talk about a latent group or category as the common sense of powerlessness has not yet been shared or does not have the channels through which to articulate it. This is why knowledge of the local demographic features is so important.

The distribution of certain minority populations is so scattered that even personal interaction is restricted, whereas in others, high concentrations may facilitate the setting up of institutions or intermediaries to act as spokesmen.

In the context of this workshop, ethnic and national minorities emerge as among the most important minorities. One type of "ethnic" minority has already been mentioned. The autochthonous minorities in France are examples of regional cultures with their own territory. Territory is the main criterion for their sense of common belonging, e.g., Corsican, Basque, Breton. Members of these populations do not accept the "ethnic" minority label, which has very conservative connotations. It is important to analyze their position within the French state in terms of center-periphery relations. In cases where other criteria for solidarity exist, territory may play a lesser role.

The guestworker minorities are a product of labor migrations, and the obvious similarity between both guestworker and immigrant populations is their displacement from their own territory and economic base. Guestworkers are typically perceived by members of the dominant culture as aliens, temporary residents who will return to their homelands. In Germany, e.g., they are expected either to "assimilate" or to return home. Many Turks in Berlin would perceive the former as involving a break with their own background and Islamic culture (Schiffauer, in preparation).

While the legal position of guestworkers and immigrants may be very different, their actual circumstances may be quite similar: the exclusion from certain sectors of the labor market, social discrimination, and a sense of insecurity or "loss."

There are many important questions which arise in any assessment of the relationship between class and ethnic groups and also in analyses of the relationship between ethnic groups, both indigenous and non-indigenous. These cannot be considered here, but a couple of points need to be mentioned. It is dangerous to assume that either the ethnic minority or the dominant culture is homogenous. The dominant majority may also be considered an ethnic category within which there are subordinate and superordinate subcultures. Any analysis of a particular situation involves an assessment of the socioeconomic differentiation of both the majority and the minority.

In some cases the formation of consciousness of a common boundary is produced by the realization of its external imposition by outsiders. The case of Panjabi-Pakistan migrants in Bradford, Northern England, shows how this may interact with other demographic and historical processes. Early male migrant settlers from very different ethnic, linguistic, and religious backgrounds shared accommodation. On the arrival of their families and a larger number of kin and fellow villagers, they chose to mix with the villagers from their own (sub)ethnic group (based on region of origin). They became increasingly conscious of the class differences, while middle-class migrants became aware of the general English classification of "Pakistani" and its negative associations. In those days it was only rarely that Mirpuri villagers identified with a wider Panjabi or "Pakistani" population (9).

One of the reasons why Panjabi-Pakistanis ten years ago in England and Turkish migrants today in Berlin did not develop a strong sense of ethnic salience (based on nationality) is that the beliefs in the likelihood of return and of common heritage or origin were focussed on specific localities and experiences in the homeland. Some migrants in the United States initially hoped to return to their regions of origin; some of those who intended to, did not return, but the existence of the idea of return certainly hindered group consciousness for a period. The belief in return may reduce tension with the wider society; it may also help the psychological and social strains of migration.

Ties of family, marriage, and kin may also support consciousness of more specific ethnic identites. For migrants these ties may span across continents. Among members of indigenous national minorities these ties may provide the main protection from the encroachment of the dominant culture.

MULTIPLE IDENTITIES

The notion of multiple identities is essential to an understanding of ethnic identity as a social process. Ethnic identity is not fixed, unitary, or homogenous: its different components shift over historical time and situational change, affecting both individual and group processes. The idea of identity options has to be conceptualized within this context of constrained choices. The constraints on ethnic identity options come both from within and from without. It cannot be assumed that all the structural constraints are external: class, for example, may operate within the minority group as well as the majority.

People have more than one locus of commitment. It is important always to look at the relationship between ethnic identity(ies) — and other social and cultural identities — on a horizontal level. Identification with a particular neighborhood or employment identities are often salient in many inner-city areas of London (15,16). Furthermore, within any ethnic identity, there are often increasingly specific (sub)ethnic identities nesting vertically one within the other.

The interaction of these two types of compound identities contributes to the process of collective identity formation. The changing salience of each level of ethnic identity (e.g., Pakistani — Panjabi-Mirburi) is not dependent on the nature of interaction within the ethnic population but on the frequency and nature of contact with the English subcultures of the wider population in that context (12).

There is often an assumption that multiple identities will inevitably lead to "identity conflict" or low self-esteem. But to return to the example of the Northe East of France, it is possible for the local inhabitant of Lorraine to say "je suis Lorrain, de culture germanique, de nationalité française et je pense en dialécte"(Pradelles de Latour, this volume).

While there may be conflict between an individual's sense of self and the collective definition of the kin, or ethnic group, one of the main sources of conflicts of identities among ethnic minorities is the denial of their salience or of the legitimacy of these identities.

These conflicts may be produced by circumstance or by the imposition of an alternative classification. Middle-class Haitians in France have a very different sense of identity compared to their middle-class relatives in New York. In France, their emphasis on their ability to speak French and their knowledge of the literature is played up and their color is overlooked. In the United States, the French language is no longer a resource and their skin color makes them invisible in a visible minority. Within the black population they are rejected as French speakers (8).

These are other examples of the changing salience of ethnic identity markers according to context *within* the same country and over time. The constant dynamic between the insider's and the outsider's perception of each other is effected by very different economic and social contexts. The same population in different countries may be perceived and treated in very different ways and so gain very different perceptions of themselves and available resources. This is understandable considering the power of the institutional and conceptual strategies of the dominant culture.

Another restriction or influence on the formation of consciousness comes from the power of the classifications of the dominant culture. In both France and England there is no concept of a hyphenated identity. In France, until recently, there has been an open policy of immigration. Naturalization has gone together with an attempted acculturization through the public school system, where French identity has been defined through French culture. As a result, plural identities have been suppressed and the French national and cultural identity reinforced at the public level. At the private level, national, ethnic, and religious consciousness has developed through its own internal dynamics and reaction to the dominant culture (3). The United States and Canada provide good cases of publicly acknowledged hyphenated identity.

Multiple identities are structurally determined, and the disjunction between an individual's identity formation and the collective identity of the groups can be particularly evident in the cases of many young children of migrants who partly identify with aspects of the different (sub)cultural systems to which they belong. Here it is helpful to distinguish between affective and cognitive identifications. Many indigenous and non-indigenous youngsters are very skilled at interacting in different (sub)cultural systems, but their ease in using the appropriate social and linguistic skills does not necessarily reflect fundamental affective bonds.

STRUCTURAL AND POLITICAL PRECONDITIONS

There is an ongoing interaction between the growth of consciousness and external conditions. These external conditions include some of the political, economic, and social constraints on individual and collective expression already mentioned (e.g., state policies and dominant classifications, economic exclusion, patterns of settlement, personal rejection, etc.). There are many problems and dangers in trying to identify universal conditions which will lead to the formation of consciousness.

Many of the structural and political constraints experienced by ethnic or national minorities produce a sense of powerlessness, and this may also be exacerbated by a sense of loss. The trend toward autonomous movements in late industrialism suggests that the growing power of the state and transnational political and economic developments have led to an increasing sense of powerlessness among many groups and individuals. On the other hand, increasing opportunities for education and communication may create conditions for enhanced group identity. Changes in national political and economic circumstances may also lead to an undermining of the dominant culture's sense of identity, as in the case of Britain. The loss of empire and decrease in economic influence are just two factors contributing to the sense of "loss" experienced by the English. As English identity has been gradually threatened, the presence of ethnic minorities could be manipulated, not only for economic and political gain, but also to raise the consciousness of the majority. The presence of outsiders, be they migrants or aggressors in the South Atlantic, helps to create a firmer boundary around "us" especially when "we" are in need of a sense of collective belonging (14).

Reasons for mobilization can be many. The process of ethnicity may move toward separatism or automy. Ethnic mobilization may be perceived as the only possible way of achieving social, economic, political, and/or cultural claims. Economic competition is often a source of ethnic conflict and a cause of ethnic consciousness and mobilization. Whatever the potential reasons for mobilization, sometimes triggering events set the process in motion. There may be a conjuncture of events, such as the May 1968 movement in France and the Algerian war, which gave a boost to national minority movements such as Occitan, Breton, and Corsican. There are other events, such as the black power movement in the United States, which have an effect far beyond their own territory, helping to raise the consciousness of other oppressed minorities (Pettigrew, this volume).

The focus for identification may be territory, religion, language, and/or mythical charters. There are three main "myths"* which may provide a focus for group consciousness: the myth of common origin (real or assumed), the myth of return (to the homeland), and the myth of the promised land (for those who have no possibility or desire to return). And, as illustrated in earlier sections, a sense of belonging can also be generated by exclusion from without.

The articulation of these (imposed or internally generated) foci for identification is often assumed by ethnic leaders who may themselves provide a further focus of identification or, through national and international minority conferences, build wider ethnic consciousness, e.g., Pan-Indianism in South America (7).

Some leaders are from the elite, others from the working class. Some are appointed from "above" (reflecting the demands of the dominant culture and institutions), and others emerge from below as in the case of indigenous tribal or religious leaders of Kurdish rebellions in Turkey, Iraq, or Iran (2).

If economic or social change faced by indigenous and non-indigenous minorities leads to a sense of loss or powerlessness, it is often the leaders who articulate this sense of malaise. On the other hand, many ethnic leaders may harm the long-term interests of minorities. In addition, the "illuminati" who feel they see the future way clearly may not be followed by their reluctant flock.

The opportunities available for developing a shared focus for identification and reasons to mobilize depend on many factors. Those already mentioned are the critical size and demography of the population, social resources of the group (particularly occupational opportunities), the structure of social networks, and the means of communication, including the nature of the media and the opportunities for dissemination of ideas. The national legal framework, the policies of the dominant institutions such as the education system, and police and health services, to name just a few, jointly determine the structural constraints on the potential for the formation of consciousness.

* By "myth" is meant the anthropological concept of a charter for membership and for action which may or may not be "true" but which gains unquestioning credence.

SOME EFFECTS OF MINORITY GROUP IDENTITY AND THE RESPONSIBILITY OF SOCIAL SCIENTISTS

The effects of minority identity can be discussed, in many cases, in terms of trade-offs and double-binds. There are exceptions where no trade-off is allowed and the minority position is "frozen," e.g., by treaty.* There may be economic, symbolic, or psychological trade-offs in the formation of group identity. The psychological benefit of regaining self-respect and dignity for those who, like the descendants of slaves, suffered the most violent denial of belonging, may have been balanced against the economic and social disadvantages that it involved for some in the short- and long-term.

There are frequently double-binds produced by the combination of ethnic consciousness and state policy. Conservation of a group boundary by exclusive practices such as endogamy can raise the hostility of the outgroup. The present situation for many migrant groups in Berlin was epitomized in a phrase from a radio broadcast heard during the conference: "integrate or go home." Another kind of double-bind occurs when the ethnic boundary is so relaxed that the group loses its identity.

Many discussions about ethnic or national minorities assume a model of rational choice. The preceding examples involving the sense of loss, and the question of dignity, and other examples of group mobilization, which are clearly against the group's own economic interest, suggest an alternative model: that of nonnegotiable attachment. The use of only one of these models is likely to lead to distortion in the interpretation of the formation of consciousness. Similarly there is a need for an analysis which incorporates both structure and culture, organization and experience. An understanding of the structural constraints determining a minority's political position, socioeconomic context, and response to the state and dominant group is essential in an analysis of changing cultural processes.

The choice of concepts and methods not only influences the type of analysis undertaken by social scientists but may also unwittingly reinforce conceptions based in the dominant culture or language and thereby harden boundaries or legitimize stereotypes (Patterson, this volume). This problem of legitimization is also evident in media coverage and in the preoccupations and perspectives often adopted by other professionals such as social workers, teachers, and

* As in the Treaty of Lausanne which recognized Greeks, Jews, and Armenians — but not the Kurds — as minorities in the newly founded Turkish Republic.

doctors (4). Social scientists have a heavy responsibility in avoiding legitimating the processes of domination. They should ensure that the study of the dominated is not altogether overlooked. In France, many social anthropologists do not see the dangers of ignoring the study of national indigenous minorities while focusing their interest on "exotic" cultures overseas or among the more recent migrant populations (6).

One way that social scientists could begin to balance out the impact of some of their work is through encouraging the training of minority researchers to research their own populations and cultures of dominance. Another is the development of strategies of active dissemination. In both cases, the social scientist should be fully aware of the social and political implications of his/her work (Patterson, this volume).

CONCLUSIONS

It seems clear from our discussion that there is a wide and complex set of phenomena subsumed under the terms of consciousness, identity, ethnic group, and minority group. The consequences of an unexamined use of the terms and indicators involved can lead to confusion in any multidisciplinary discussion; conversely, the advantage of illuminating the issues through examples and comparative data are great. We propose a greater effort (perhaps through an international exchange system of work in progress and findings) to keep the scholars concerned with minority studies abreast of one another's work in the field.

A most important conclusion of all our deliberations is that it is impossible to detach minority experience from that of the majority. Both minority and majority (however defined — through demography or dominance) are locked into an interconnected system. It is not just the case that the majority "operates" on a passive minority. In the process of relationships the consciousness of both types of groups are changed.

The comparative context of investigation, interpretation, and dissemination of material is vital. There is no universally accepted model of how minority-majority relations operate, and the experience of different countries must be built into research approaches as well as the resulting conclusions. This is yet another reason for establishing an international network of communicating scholars.

Without a historical dimension, the study of the ethnic or minority experience does not make sense. We operated throughout with a flexible model of the intertwining of individual and group history: there can be a group "career" as well as an individual career, and the relationship between the two can provide enlightening insights into the process.

REFERENCES

(1) Berger, P.L., and Luckmann, T. 1966. The Social Construction of Reality: A Treatise in the Sociology of Knowledge. Garden City, NY: Doubleday.

(2) Bruinessen, M.M. van. 1978. Agha, Shaikh and the State, on the Social and Political Organisation of Kurdistan. London: E.G. Brill, Harraffowitz, and Luzac Distributors.

(3) Brun-Rovet, J. 1980. America! America! Paris: Gallimard.

(4) Grillo, R., ed. 1981. Nation and State in Europe. New York: Academic Press.

(5) Morin, F. 1979. Langue et identité ethnique: le cas occitan. Pluriel **15**: 9–26.

(6) Morin, F. 1982. Indigénisme, Ethnocide, Indianité en Amérique Latine, Gral. Ed. Toulouse: Editions CNRS.

(7) Morin, F. 1983. Migration Haitenne, Ethnicité, Milieu Urbain. Urban Anthr., in press.

(8) Morin, F.; Bastide, R.; and Raveau, F. 1974. Les Haitiens en France. Paris: Mouton.

(9) Saifullah Khan, V. 1976. Perceptions of a population: Pakistanis in Britain. New Community **V**: 3.

(10) Saifullah, Khan, V. 1979. Ethnic identity among South Asians in the U.K. **In** Asie du Sud: Tradions et Changements, eds. M. Gaborieau and A. Thorner. VI European Conference on Modern South Asian Studies. Paris: Editions du Centre National de la Recherche Scientifique.

(11) Saifullah Khan, V. 1982. The dynamics of ethnic relations. Block 3, Units 8–9 of a third-level Open University course, Ethnic Minorities and Community Relations, University of London.

(12) Saifullah Khan, V. 1982. The role of the culture of dominance in structuring the experience of ethnic minorities. **In** Race in Britain,

Continuity and Change, ed. C. Husband. London: Hutchinson Educational.

(13) Tabouret-Keller, A. 1982. Identités et Évolution des Situations linguistiques complexes. **In**: l'Education bilingue dans l'école Valdotaine. Bibl. Valdotaine no. 21, pp. 35–61. Aoste: Mesumeci.

(14) Wallman, S. 1978. The boundaries of "race": processes of ethnicity in England. MAN **13(2)**: 200–217.

(15) Wallman, S. 1982. Living in South London. London: published for the London School of Economics by Gower and Associates.

(16) Wallman, S. 1983. Rethinking Inner London. London Journal, accepted for publication.

Minorities: Community and Identity, ed. C. Fried, pp. 109–118
Dahlem Konferenzen 1983. Berlin, Heidelberg, New York, Tokyo: Springer-Verlag.

THE COMMUNICATIVE BASES OF SOCIAL INEQUALITY

J.J. Gumperz
Dept. of Anthropology, University of California
Berkeley, CA 94720, USA

The experience of the last decade has shown that in spite of extensive legislation and positive efforts on the part of public agencies and industry, attempts to integrate minority groups into the economy and to encourage free and equal participation in public affairs have been far from successful. In most Western industrial societies ethnically or culturally distinct individuals are grossly overrepresented among the poorest and least educated sectors of the population. They tend to be the first to suffer from economic fluctuations and the last to be affected by welfare programs and measures designed to correct inequities in employment and access to public resources. Moreover, when minority groups do enter the public arena, their contributions and abilities are likely to go unrecognized, and their demands for recognition are misunderstood.

The current situation of minority groups thus differs markedly from that of earlier periods where, without government action of any kind and after an initial period of adjustment, immigrants soon seemed to adapt to the basic values of the host society and find their own economic niche, attracting little if any national attention. It appears that, contrary to what accepted social theory might predict, some of the very developments that are most characteristic of our era, developments that have often been hailed as indicators of social progress — the elimination of legal barriers to social mobility, the widening scope and centralization of government services, standardization of educational curricula, and near universal exposure to mass media — are in fact serving to increase public awareness of cultural distinctions and are contributing to the growth of ethnic stereotyping, intergroup conflict, and alienation.

In this paper I argue that such problems of multiculturalism — and this applies to their economic and political, as well as to their attitudinal dimensions — are to a significant extent produced by interactive and evaluative processes that are intrinsic to the bureaucratic institutions which characterize our society. An understanding of the communicative constraints which affect the workings of such institutions is thus not only a matter of abstract theory. It is also essential for both minority group members who must seek to gain basic insights into their own situations in order to be more effective in formulating and pursuing their demands and for the planners and policymakers who must interpret and reconcile the often conflicting claims of interest groups.

To say that interethnic difficulties have their roots in interactive experience is of course not new. The claim is familiar from the work of the Chicago school of urban sociologists of the nineteen thirties and forties. Its theoretical basis is clarified by Berger and Luckmann (1). More recently Goffman (6) has produced detailed case studies to show that basic features of social order, embodied in such concepts as value, norm, role, status, and social identity, cannot be taken as stable dimensions which have given and shared meanings for a community as a whole. His work suggests that people's notions of what these categories signify and of the social distinctions they reflect arise out of their contact with other people and vary with the extent and the quality of these contacts, so that generalizations about groups must build on or at least be informed by empirical observations of everyday behavior.

The argument is carried one step further in the work of Garfinkel (5), Cicourel (2), and their students, who go beyond observation of interaction in general to concentrate on decision-making and evaluative processes in such public agencies as health care institutions, police, judiciary and welfare establishments, all of which play a major part in determining the quality of life in modern urban society. Through empirical analysis of the day-to-day activities by means of which these organizations function, they show that mere description of formal organizational procedures cannot account for how these institutions work and how they affect our lives. The evidence indicates that what happens at any one stage in the actual operation of such establishments is significantly affected by taken-for-granted and for the most part not explicitly stated conventions and background assumptions. Institutional decision-making thus takes on an ad hoc character which regularly leads to outcomes that are quite different from what one might expect on the basis of commonly accepted structural descriptions and organizational goals.

The sociologists' findings are important because they reveal the inherent limitations of attempts to explain interethnic relations in terms of moral and legal principles or attitudinal, macroeconomic, and sociopolitical factors alone. If it is true that minority group demands arise out of everyday experience, and if what happens to individuals in their contacts with others at work and within urban institutions is a major component of this experience, then a basic understanding of the mechanisms through which actual cases are handled and problems are resolved becomes essential. Detailed examinations of the interpretive procedures which channel the day-to-day operations of relevant organizations are needed to discover the hidden sources of systematic bias which, although not necessarily reflected in the actual demands made, may nevertheless lead to serious inequities in access to opportunity and in the allocation of public resources.

Yet exactly what these biases are, how we study them, and how they interact with legal guidelines, basic policies, and situational exigencies to affect judgements is far from clear. My own position is that, since what is involved is basically talk, organizational decision-making can be regarded as made up of verbal exchanges, where orally presented information is recoded and translated into preestablished standardized categories and case records and where the written texts produced in this way are further evaluated and reevaluated through face-to-face negotiation. We should therefore be able to achieve more direct insights into the nature of background assumptions if instead of relying on post hoc analyses of case histories or looking at behavior in general terms — as sociologists tend to do — we concentrate on tape recordings of typical interactive situations, reflecting various stages of the organizational decision-making process, and apply discourse analysis to recover the sociocultural, situational, and linguistic presuppositions which affect these processes.

Let me begin with some background on the view of discourse and interpretation that underlies my argument, and on the perspective on language and social organization processes it suggests. Linguistic theory has developed rapidly over the last years and has profoundly affected many areas of social science. Since the most influential work concentrates on communication of referential content and on the grammatical structure of words and sentences, insofar as these can be explained in terms of ideotypically uniform systems of rules abstracted from actual speech events, it is often taken for granted that the generalizations that build on this view of language can account for everything that is linguistically significant about verbal communication.

There is thus a tendency to assume that language difficulties in face-to-face encounters necessarily involve communication failures occurring in bilingual or bidialectal settings, where participants lack the requisite grammatical or lexical knowledge to understand overt sentence content. When someone whose sentences are grammatically and lexically acceptable nevertheless does not seem to make sense, when participants in a conversation do not respond as expected, or when their remarks appear uncooperative or impolite, we tend to conclude that matters of attitude or technical ability are involved which cannot be studied through analysis of linguistic form.

During the last decade this restricted view of verbal communication, which results from linguists' tendency to confine systematic analysis to the grammatical structure and referential meaning of isolated sentences, has been challenged from a number of directions. Philosophers of language, following the later Wittgenstein and Austin, contend that referential theories of meaning in which meaning is defined as the relation of utterances to objects and ideas in the nonlinguistic world have been unable to account for our ability to interpret such common words as "and" and "but," and for the role that historically given human conventions play in interpretation. Meaning, they hold, ultimately resides in human action, so that the proper unit of linguistic analysis is not the sentence, but the speech act, defined in terms of speakers' communicative intent. They go on to suggest that assessments of communicative intent always involve a process of inference which relies on presuppositions or background assumptions not given in the overt lexical content.

Recent research on text comprehenstion by cognitive psychologists and specialists in artificial intelligence provides empirical justification for the philosophers' arguments. To demonstrate the claim that to understand texts of any kind readers must integrate their schematic knowledge, i.e., their understanding of what the text is about and what is being accomplished at any one point in the event, with their reading of actual words; sentences such as the following are cited: "The notes were sour because the seams split." Note that although the two constituent clauses are quite clear, the utterance as a whole seems incomprehensible. Only when we learn that the utterance occurred as part of a conversation about bagpipe playing can we infer what was intended.

It is the notion of schema or interpretive frame (as Goffman, who has written about the issue from an interactionist perspective, would call it), the degree to which schemata are known, how schematic information is signalled and learned, and to what extent learning is a matter of sociocultural background, that is crucial to an understanding of the communicative dimensions of minority group problems in modern urban society. What schematic knowledge does is to provide the overall perspective which enables us to integrate what we read or hear with what we already know and to fit together individual terms of information into a coherent argument. When schematic knowledge is not shared, what in terms of covert content seems like the same message may be interpreted differently by different individuals. For example, participants in a comprehension experiment who were given the same narrative presented from two different perspectives, and who were later asked to recall what they had read, differed significantly in the factual information they recalled. Information transfer is thus rarely automatic. Even with speakers of the same language, interpretations of what is heard or read tend to be affected by unstated assumptions learned through previous experience.

What does schematic information consist of and how is it conveyed? Discourse analysts working with written texts tend to treat schemata as matters of extralinguistic knowledge, knowledge which one learns to utilize as part of the normal process of language acquisition and which competent speakers can be presumed to possess. But when we turn to oral communication, serious questions arise as to the extent to which such knowledge is shared and as to how it is retrieved and made available in message interpretation.

Conversation of all kinds presupposes active cooperation between producers of messages and listeners who provide feedback, either by means of direct responses or through posture, gesture, or other forms of back channel signals. Such cooperation cannot be taken for granted. To enlist conversational involvement, potential speakers must induce others to participate. That is, they must somehow convey at least some advance information as to what the outcome of the intended exchange may be. Once talk has begun, moreover, initial schemata are subject to constant and often quite subtle change, and such changes have to be negotiated as part of the interaction. Further problems arise with allocation of turns at speaking. Individuals do not automatically have space to present or develop an argument; they must work to retain their turn by enabling others to predict where their responses can fit in.

The communicative difficulties that can arise in the course of such interactive processes have been documented in recent work on doctor-patient interviews by Cicourel (3), who found that physicians' questioning strategies often rely on implicit, taken-for-granted organizational knowledge that assumes familiarity with the goals and workings of the medical establishment, a knowledge which patients cannot be expected to have. The resulting differences in schematic knowledge seriously hamper the physicians' ability to elicit needed information and to evaluate patients' responses. This may have serious consequences for the diagnostic process.

A recently completed long-term study of college counselling by Erickson and Shultz (4) demonstrates that the amount of useful information conveyed in counselling sessions is a direct function of participants' ability to find common conversational topics on which to base their comments and questioning strategies. The relevant themes are generally established indirectly through informal repartee at the outset of the interview. Success in this scene-setting stage of the encounter requires that the participants have at least some common background experience. With individuals of similar ethnic background, this is frequently the case. But when backgrounds are not shared, more work is required to establish productive topics. These findings challenge the commonly held assumption that in public encounters where everyone concerned agrees on what is being accomplished, information transfer presents no problem.

It seems that interpretation of all kinds, even in what are normally seen as task-oriented, instrumental events where the focus is on objective, factual information, depends on participants' use of signalling strategies to establish contexts and create interactive conditions favorable to conversational cooperation and communicative effectiveness. Speakers and audiences go beyond surface content to infer and respond to intentions which, for the most part, are conveyed only through indirect signs. Erickson and Schultz' discussion concentrates on the role of nonverbal cues (eye blinks, limb and torso movement, and rhythm) in this signalling process. But gestures are always closely coordinated with tone of voice, pronunciation, and other verbal mechanisms, so that we can say that contextual information is in part conveyed through linguistic cues.

How are such cues learned, and to what extent can we assume that they are used in the same way by all speakers of a language? Socioliguists concerned with the language usage of specific human populations have long argued that any theory that postulates complete sharing of grammatical rules within a speech community simply has no basis in fact (9). While all speakers of a language must control a basic core of categorical rules, there are other systematic aspects of verbal signalling which are inherently variable, and the resulting diversity is as basic to social systems as the division of labor. The incidence of phonological, syntactic, and lexical variants in representative samples of speech has been shown to be directly related to class, ethnicity, occupation, and similar social categories. Occupational specialization in health care, legal, and other urban institutions is another source of variation. It has furthermore been shown that variants associated with either social or contextual categories, when used in interaction, tend to evoke attitudes and stereotypes associated with these categories and may significantly affect the way speakers are judged by others. Choice of speech forms is thus closely related to rhetorical effectiveness.

Research at Berkeley (7,8) has focussed on the interaction of linguistic knowledge and social background in the signalling of schematic information and in creating the preconditions for effective communication. Our data derive from comparative analyses of audio and video taped materials. These materials are collected in naturally organized institutional events and contrast speakers who have good sentence level command of English, but who show significant differences in sociocultural and communicative background, with speakers who share both language and background. Analysis focuses on conversational inference, that is, the situated or context-bound process of interpretation by which participants assess what is intended at any one time, and on which they base their responses. Relying on conversational analysis and on questioning techniques based on those in cognitive anthropology, we have found evidence to show that conversational inferencing relies on perception of a variety of phonological, syntactic, intonational, and indirect semantic cues, as well as on lexical knowledge. We refer to these features of language as contextualization cues, and to the conventions they reflect as contextualization conventions.

While conversationalists everywhere rely on contextualization cues in relating what is said to previous experiences, in evoking appropriate schemata, and in guiding the inferential process, the linguistic forms that cues take depend on participants' communicative history and on the situation at hand. With bilingual or bidialectal speakers, code switching, or alternations among what sociolinguists have identified as sociolinguistic variables, may serve as contextualization cues. In settings where technical expertise is at issue, the alternation between informal and technical language may have similar functions. Prosodic features of stress or rhythm always play an important role in marking formality or informality, in chunking the stream of talk into idea units, in signalling semantic relationships between such units, and in indicating emotion and emphasis. But the way in which prosody interacts with syntax and lexicon is a matter of historically-based discourse convention.

Contextualization conventions are learned as part of individuals' interactive experience. As with grammar, the relevant strategies, once they are learned, become internalized and continue to be automatically applied without conscious reflection, so that speakers generally have difficulty in providing direct information on their own practices. When contextualization conventions are shared, they serve as a communicative resource. But while internalization of grammatical rules is essentially complete by the time a child enters school, contextualization conventions tend to be acquired throughout adult life. Moreover, the principles determining when and under what circumstances cues are used are context sensitive and so complex that they cannot be transmitted through formal schooling or writing. Face-to-face contact in situations of equality which allow for free and nonthreatening feedback, where misunderstanding can be corrected without loss of face or risk of offense, is necessary. Where ethnic boundaries or other conditioning factors limit opportunities for this type of informal interaction, learning opportunities are also limited. As a consequence, individuals who have no difficulty in producing grammatical sentences may react to different contextualization cues and thus come up with different interpretations of the same messages. This is especially frequent in encounters requiring long explanations or cognitively complex arguments, or where persuasion calls for a great deal of indirectness.

Furthermore, in most everyday situations miscommunication resulting from differences in contextualization conventions is not identified as such. Given the subconscious nature of the signalling mechanisms involved, such differences are in fact extremely difficult to identify. What tends to happen is that miscommunication leads to breakdowns in conversational cooperatin, loss of information, and misevaluation of attitudes or ability. Potential learners thus face a dilemma in that they must establish supportive personal relationships in order to learn, yet they lack the strategies for initiating and maintaining the contacts that might enable them to accomplish this.

Special outside conditions must exist which predispose interlocutors to allow for the possibility of miscommunication and to overlook what on the surface may seem like incoherent talk or infractions of accepted norms, in order for barriers to learning to be overcome. Yet the position of minority populations in urban society is such that these conditions rarely hold. During the last decades, all city dwellers, regardless of background, have become increasingly dependent on publically available goods and services for much of their daily needs. Although in theory access to public facilities is open to all, it is never automatic. Individuals require communicative strategies to present and justify their claims, and their cases are judged by others who, in the absence of direct firsthand knowledge, must base their assessments on verbal information alone. Experiences in these institutional contexts reveal some of the real difficulties that may give rise to minority group demands.

Minority group members do significantly less well than others in most spheres of institutional contact: in classroom interactions with teachers and fellow students at school, in large health care establishments, in relations with police and the courts, in counselling situations, and particularly in employment interviews. Although the interactive conditions involved in these situtations vary in detail, they are similar in that what are ultimately judged are such matters as coherence of arguments, validity of justification, and persuasiveness of requests. Where contextualization conventions differ, it becomes difficult to separate personal or culture-bound assessments from objective determinations of ability and attitude.

The experience we have gained through ethnographic observation, interviews, and conversational analysis indicates that there is a close connection between what happens in encounters of the kind I have described and the stereotyping, negative attitudes, and conflict that characterize interethnic relations. The bureaucratic procedures which govern the conduct of affairs in industry as well as in public institutions depend on the assumption that there exist uniform, objective criteria of evaluation which control access to valued resources. Failure to understand that these criteria are themselves necessarily culture and convention bound, and that the conditions under which we live prevent many individuals from learning what these conventions are, leads to a vicious cycle of miscommunication, stereotyping, and indirect discrimination which is difficult to break.

The problem is not one simply of learning to communicate more efficiently on the part of one or another group. New curricula cannot offer any solutions. What we need is a closer examination of the workings of our urban institutions and of the gatekeeping and adjudication processes by which they are maintained.

REFERENCES

(1) Berger, P.L., and Luckman, T. 1966. The Social Construction of Reality: A Treatise in the Sociology of Knowledge. Garden City, NY: Doubleday and Company, Inc.

(2) Cicourel, A.V. 1973. Cognitive Sociology. New York: Free Press.

(3) Cicourel, A.V. 1981. Language and the Structure of Belief in Medical Communication. Paper presented at the Sixth World Congress of the International Association of Applied Linguistics, Toronto, Canada.

(4) Erickson, F., and Shultz, J. 1981. The Counselor as Gatekeeper: Social Interaction in Interviews. New York: Academic Press, Inc.

(5) Garfinkel, H. 1967. Studies in Ethnomethodology. Englewood Cliffs, NJ: Prentice-Hall, Inc.

(6) Goffman, E. 1974. Frame Analysis: An Essay on the Organization of Experience. New York: Harper and Row, Publishers.

(7) Gumperz, J.J. 1982a. Discourse Strategies. London: Cambridge University Press.

(8) Gumperz, J.J., ed. 1982b. Language and Social Identity. London: Cambridge University Press.

(9) Labov, W. 1972. Sociolinguistic Patterns. Philadelphia: University of Pennsylvania Press.

Minorities: Community and Identity, ed. C. Fried, pp. 119–131
Dahlem Konferenzen 1983. Berlin, Heidelberg, New York, Tokyo: Springer-Verlag.

CONFLICT AND COMMUNITY SURVIVAL

J. Rex
Research Unit on Ethnic Relations
St. Peter's College, University of Aston
Birmingham B8 3TE, England

Abstract. Many urban communities are riven by intercommunal strife raising doubts as to whether the separate ethnic communities or the urban society can survive. This paper discusses recent urban riots in Britain both in relation to their internal urban causes and in relation to larger international issues.

INTRODUCTION

The international context in which this paper is given calls for the problem suggested in the title to be treated in a more than localized way. The suggestion seems to be that across the world, or at least across the urbanized industrial communities of the world, there is the possibility that civilized communities as we know them will not survive because of intergroup conflict. Maybe there is something in this, and I shall begin by covering some of the situations in which international issues are projected into an urban context, so that the very urban fabric is torn apart, let alone its community structure. I suspect, however, that the posing of this question arises from a concern about a narrower range of situations such as those which were represented by the American Long Hot Summer of 1967 and the British riots of 1981. Since I have some experience derived from research about these situations, I shall concentrate upon them. But, finally, I shall bring the two themes together by suggesting that the special Anglo-American problem is far from being unrelated to major world issues.

Diagnosis, of course, could be thought of as leading to proposals for remedy. But the very notion of remedy leads to the question,"Remedy from whose point of view?" It is possible, of course, to evade this question by positing peace or cessation of violence as an end in itself, but such an approach usually ends up on the basis of law and order solutions or of the manipulation of consent so that, in effect, peace is achieved through oppressed groups abandoning their legitimate goals. The other possibilities lie in the creation of a structure within which conflicting interests can be peacefully pursued or in the carrying through of the struggle of oppressed groups by all possible means until they achieve equality of bargaining power and consequent just treatment.

There are very few parts of the capitalist world today which do not face some kind of problem of terrorism. We are all accustomed today to the notion that air-travellers have to be protected from hijackers. There are also many cities in the world in which innocent citizens may be killed in explosions as they go about their peaceful business. Sometimes the "terrorists" represent a minority cause which gains no legitimate representation in the state, however just it may appear from outside. Sometimes they see themselves as peacemakers standing for causes and transformation more radical than those envisaged in the normal political spectrum. In either case, terrorism is the weapon of the weak against the strong. Unable to win or have any chance of winning according to the going rules of the game, the terrorists abandons those rules to shift the struggle to a terrain where, even if he or she (terrorism being unisex) may not win, the oppressors will suffer.

In many parts of the world, however, conflict has gone beyond the terrorist phase. What we are discussing, then, amounts to civil war. Such civil war may lead to the partition of whole countries, or it may be projected into great cities which themselves become divided into armed camps. The parties to such conflicts are quite frequently ethnic groups in an unstable plural society, but they may also be class groups or political groups in a developing situation of revolutionary class struggle. What is at stake in these cases is the unity of the state itself. Urban communities become divided because the state itself is divided, or, to put the matter in Weberian terms, because the state has ceased to exist, having lost its monopoly of the use of legitimate force.

The most terrible examples of community conflict of this kind in recent times are provided by Beirut and Belfast. In the first case, one has an already complex plural society into which have been inserted the problems of a neighboring state. What was a divided capital city has now become a battleground. In

Belfast, on the other hand, the conflict has occurred between two indigenous groups of Northern Irishmen, even though one of them is a settler-descended group and the other has its connections of kinship, culture, and religion with the neighboring Irish Republic. Here what was once a reasonably integrated city has become divided into ethnically segregated armed camps.

Police in England, in responding to the civil disorders in Brixton and other places in 1981, often referred to Belfast and the policing strategies which had been worked out there as providing a model for the Brixton of the future. In so doing, they defined the problem in international and quasimilitary terms. So also did the most militant leaders within the West Indian-descended black community. I shall return to this definition of the situation, but before I do, it is important to look at some of the other roots of the situation which emerged in United States and British cities in the sixties and seventies.

Urban sociologists in Chicago had speculated as long ago as the twenties about the class structure of cities as it was expressed in geographical terms (2). The city was seen as having four residential zones. Three of these were straightforward enough. They included "the zone of the working-men's homes," "the middle-class zone," and "the commuter's zone." The fourth, however, did not fit easily into this class and status pattern. It was a "zone of transition" which basically housed the new ethnic immigrants and native problem people and down-and-outs.

On the whole, the Chicago sociologists conceived of these zones in a static way as forming some sort of status order. They do not seem to have recognized the potential dynamic elements in their own theory such as the fact that, once the first "commuters" had been produced, a suburbanizing process had been set in train which involved inner-city abandonment, or that, if the "zone of transition" was indeed in transition, it might move towards some new state. Nonetheless, Chicago did leave behind a concern with inner-city problems which was to have considerable significance as cities in America and elsewhere underwent dramatic transformations in the postwar world.

A really dramatic change occured in American thinking, however, amongst laymen as well as among sociologists writing about cities between 1950 and 1970. In the 1950s the great cities were thought of as great cosmopolitan centers, in which even the slums were thought of as interesting because of their immigrant and ethnic culture. That kind of respect for cities was quite compatible with a developing suburban trend. By the mid-sixties, however, all this had changed. The cities in their centers had become, above all, places of fear.

Large parts of them had actually been abandoned and looked like bomb sites created without benefit of bombing, and even more important than this, with the exception of very special cultural enclaves of the Bohemian sort, the central city area had been abandoned to the poor and the blacks. Rates of unemployment and rates of street crime there had become terrifying high. The drug traffic flourished and the police and the young blacks were virtually at war with one another.

There is a world of difference between this world and the rather folksy, sentimentally described "zone of transition" in the work of Park and Burgess. Their zone of transition was seen as part of a functioning capitalist system. What one has now is a situation in which the ghettoized blacks are not part of such a system, not even really as a "reserve army." They have been rejected by and left out of the system and are basically regarded as a problem of control. When eventually in the mid-sixties their conflict with the police exploded, America was challenged by open riot, burning, and looting. The question of what to do about the inner city now came to occupy a central place on the agenda of American capitalism. On the one hand, programs of affirmative action were initiated which could give some hope to at least a minority that the system had something to offer them. On the other, those who remained had to be more efficiently policed and controlled.

It would seem from this account that the violent conflicts which occurred on the streets of American cities in 1967 were simply the product of urban capitalist development, because, given the long-term process of modernization in industry and urban life, capitalism was bound to reject both the buildings and the people on whom its initiation had depended. What this does not recognize, however, is that people who were rejected were not simply the poor as such, but were specifically the younger blacks. A society which enabled the European "huddled masses of the poor" to find new hope and opportunity in America had no place within it for those who migrated to its cities from the old plantation economy and society of the South.

The problem, therefore, was not simply an internal problem of capitalist development. It was a problem of the intersection of two social systems, that of metropolitan capitalism and the collapsing colonial world. That, too, was a problem which was likely to recur in other capitalist societies. On the one hand, they resisted the incorporation of colonial workers in their midst. On the other, they had sent settlers to the colonial societies who were determined and well equipped to fight for their existence there. It should be noted also that in the American case a new "American Dilemma" arose over her other immigrants from colonial and quasi-colonial contexts, namely, her Mexican and her other poor Latin American immigrants.

By comparison with the United States, Britain had been prior to 1945 a comparatively ethnically homogenous country. She had, it is true, absorbed Irish immigrants for a century or more and, in the early years of the twentieth century, she had accepted Jewish refugees from Europe. These immigrants, however, were not sufficient in their relative numbers to disrupt or change the basic sociopolitical system of the country which remained based upon a class struggle within an ethnically united nation. Problems of a new kind emerged, however, in the postwar world.

All of the successful economies of Northwest Europe after 1945 faced a labor shortage. They had lost population by emigration and that population had to be replaced, and, even more important, they had lost the willing work-force necessary to man the less attractive industrial and service jobs because of the rising levels of aspiration amongst an increasingly educated population. To some extent the gap could be filled by turning to neighboring countries in Europe, or, in Germany's case, to refugees from the East, but France, Britain, and the Low Countries turned also increasingly to their former colonial dependencies. In Britain these immigrants were drawn largely from the Indian subcontinent and the Caribbean. It was their presence and their presence alone amongst all other groups of immigrants which made the "immigrant problem" a new and central one on the British political agenda.

Britain's cities in 1945 were undergoing a double process of suburbanization. On the one hand, there was the indivAlually chosen migration of white-collar and better-off skilled workers to semi-detached houses in private suburbs. On the other hand, there was the process of Council housing, whereby large numbers of working-class families were able to move to good rented housing on largely surburban Council estates. Most English families, therefore, came to expect an improvement of their housing condition either through mortgages provided by building societies or through getting on the list for Council housing.

the West Indian and Asian immigrants to Britain did not, of course, automatically share British values in these matters. They were more inclined to look to private purchase and private renting rather than relying on the Council and, like other poor immigrants, they chose to some extent to live close to their kinsmen. But, insofar as they did look to the established systems to provide for them, they found that estate agents and building societies discriminated against them and that, in their main areas of settlement, they were excluded from the Council list. These processes of discrimination ensured that the new immigrant colonies would have their own separate housing system in the worst houses in the worst areas of the city.

My own first empirical work on race relations in Britain (4) dealt with the kinds of community produced by this alternative housing system. My colleague and I suggested that they represented something significantly different from the black American ghetto. They were not nearly as ethnically homogenous as the black American ghettoes were and actually brought together in economic and social relations members of very different groups, both indigenous and immigrant. The whites who remained in the area were the least successful in the competitive urban struggle for existence and were inclined to blame their own lack of success on the presence of immigrants. The different immigrant groups also led their own separate cultural lives and pursued different interests. The situation was one ripe for conflict. Yet conflict was mitigated to a considerable extent by the local Community Association which brought together the various groups in circumstances where they could pursue and negotiate about their different aims and interests without their differences with other groups being defined in a racialistic or violent way. At that stage we felt that there was reason for optimism that, despite inter-ethnic conflicts of interest, the means would be found for coexistence, and that the immigrant groups would gradually come to accept and be accepted within the overall cultural and political value system of the city.

Ten years later another colleague and I made a more far-reaching study of a neighboring part of the same city (5). Out conclusions were far less optimistic. That is to say, far less optimistic about the prospects of integration. What we suggested was that New Commonwealth immigrants were still concentrated in the least desirable jobs where they were inadequately protected by trades unions, more completely segregated in their areas of residence, and increasingly confined to schools with high concentrations of New Commonwealth children. By 1981 there were six wards in the city with above 48% of their population living in households with a New Commonwealth head, and the majority of remaining whites in the area were of retiring age. Not surprisingly, the schools had more than 90% of their children drawn from New Commonwealth households. In these circumstances it was not expected that there would be, nor did we find evidence of, increasing identification of immigrants and their children with the politics of their society of settlement. Far more important in the lives of West Indian-descended people were ideologies of black consciousness which dealt with their situation within the whole of imperial and colonial history. And amongst the Asians, although there was a more widespread attempt to use the unions and Labor Parties on their own behalf, there was a strong tendency for their organizations to be oriented towards the problems of the Indian subcontinent.

In Britain, unlike the U.S.A, what happened to buildings in the inner city was subject to a great deal of public control. Thus, even though there was no new slum clearance and redevelopment program initiated in the mid-sixties when earlier programs came to an end, there was nonetheless a concerted attempt made and much public money was spent to arrest physical decay through home improvement. Policies were initiated to try to save the inner city from decay and to make it a desirable place to live in despite the overall suburban trend.

Whether objectively these processes of environmental improvement are succeeding is as yet hard to say. Those who have worked on them are struck by the overall poverty of the inner city which prevents private resources being available and continually find that the public resources necessary to carry through their programs are cut off in times of overall economic stringency. Thus a more radical and comprehensive policy has now been proposed whereby local government will work with central government and the Department of Environment with the Department of Industry to stimulate investment and job creation in inner-city areas (1).

This drive for inner-city improvement, however, has not necessarily meant that anything will be done to achieve greater integration or greater justice for ethnic minority residents. Indeed, there are strong indications that the opposite is the case. It is the presence of the minorities which is taken to be the main indicator of decline, and what is therefore proposed is the reconquest of the inner city for a more traditional British, albeit working-class British, way of life. The proposed improvement of the inner city is coupled with pressure on the minority residents to move out. The forced segregation to which the original settlers were subject is now seen to lead to problems, so what is proposed is forced dispersal. There is, however, nowhere for the New Commonwealth minorities to go, except in the case of those who have enough money to move to the suburbs, where discrimination by private vendors has been restricted if not abolished by race relations legislation. For the majority, therefore, the pressures of inner-city policy can only add to the general racist pressures to which they are daily subject. Increasingly, the minority communities have to start thinking in terms of self-defense.

The actual story of this development of self-defense strategies differs in the case of the two main minorities. For the young black British boy from a West Indian home, the major problem is that of perceived police harassment. For the young Asian, it is a failure of the police to offer him adequate protection against race attack. Both of these tendencies are greatly exacerbated in times of increasing unemployment.

There are actually very few areas of Britain in which there is a large West Indian minority living entirely apart from Asians. In Handsworth in Birmingham, for example, which is generally thought of as being a West Indian area, nearly half of the 57% of households with a New Commonwealth head are Asian. But there are a few places where the minority are overwhelmingly of West Indian descent and one such is Brixton. In such areas there is a very high rate of black youth unemployment and, not surprisingly, a high rate of certain types of street crime. For many of the police, including those who keep the statistics of crime in Scotland Yard, street crime is a black problem and special tough measures have to be employed against the black population. This means that virtually all young blacks know the experience of being stopped and searched on suspicion, and the whole community has become accustomed to quasi-military methods of policing. It was surely only to be expected that in these circumstances a small incident could lead to the resistance of whole sections of the population to police action. This is what happened in what came to be called the Brixton riots.

Very different events occured in Southall, which is one of the main centers of segregated Asian settlement. There it was unemployed white youths who triggered off the "riot" situation. These young skinheads began to attack not the police, but young Asians and Asian businesses which they had come to use as scapegoats for their own failures. In these circumstances the young Asians, it might be thought, would have turned to the police for protection. In fact, this was not the case, and the recent experience of young Asians made it extremely unlikely that it would be the case. Lacking any confidence in the police, they prepared themselves for self-defense and, if necessary, for self-defense against the police. Thus, although the Southall riots were started by white skinheads, they ended with a confrontation between young Asians and the police (3).

The kind of resistance shown by the young blacks of Brixton and the young Asians of Southall received widespread official public condemnation. But such condemnation was not universally shared either by their own elders or by comparable groups of black and white youth in other parts of the country. Indeed, there was some danger after the spread of rioting to Liverpool and Northern cities that there would be a general rising of the young unemployed. The government, however, was quick to act to deal with the general problem of unemployment. If it could not provide jobs, it could at least provide "schemes" to keep the young unemployed off the streets, and it was prepared to accelerate investment programs in at least some inner-city areas.

What did not follow in the wake of the riots was any serious attempt to deal with the specific causes of the riots in Brixton and Southall. Proposals were made for more sensitive policing and elaborate plans for police-community liaison councils were set up in the wake of report of the Scarman tribunal (6). It is significant and surprising, however, that these rather weak proposals for improvement (which actually dealt only with policing matters) were widely regarded by the white press as a set of proposals for a new deal in race relations. They were certainly not regarded in that light in the black and Asian community. There was very little sign a year after the riots that there was any greater sense in these communities of the legitimacy of dominant white values and institutions.

In similar circumstances, the United States adopted policies of affirmative action and ensured that a potential radical black leadership would be bought off and that there would be sufficient prospects of mobility for others for them to remain quiescent. Nothing like the same thrust of policy has developed in Britain. What has happened is that for a whole number of reasons the Asian communities and particularly the Indian community have found ways of succeeding despite discrimination, just as many of their cousins have done previously in other parts of the British Empire. They have done this through initiatives in business and through encouraging their children's educational success and entry into the professions. Nothing quite like this has happened, however, in the West Indian community. Though there is some evidence that West Indian children in inner city schools actually do better than white children in those same schools (7), they are still systematically discriminated against in the labor market. The sole sign of an emergent West Indian middle class, therefore, is in the existence of the few hundred potential leaders who have been recruited to the community relations, teaching, and social work professions, that is to say, who are not so much an independent middle class with a stake in the society, but rather the very agents of the white control system.

Any sociologist in Britain today is likely to be asked, "Will there be more riots?" The answer to that is certainly, "yes." The measures which have been taken to deal with the causes are so limited that there is no reason to suppose that there will not be another incident which triggers a new explosion. The police are at the moment exercizing some restraint, but there are clear signs that there are some among them at a high level who would like to go in and take out the blacks, and they may well get the upper hand. Finally, the level of unemployment can only be expected to increase so that there will be more and more youths who have little to lose through confrontation.

We are, of course, still a long way from a situation of total community breakdown. The existence of the level of violence so far seen in Britain is probably well below that which has existed in earlier periods of history. There is also probably a whole range of potentially successful strategies of oppression, of ruling without consent, which have yet to be tried. What may be in question is whether there are any serious signs of a policy of creating a just multiracial society in which minority groups can realize their aims and in which they consent to their condition.

There are a number of different scenarios which are envisaged by sociologists as to the way in which this more just multiracial society might be realized. The most popular, but the least convincing, is that which envisages that central and local government will detect disadvantage where it exists and correct it through appropriate forms of expenditure, if necessary involving affirmative action and positive discrimination on behalf of minorities. Such a scenario is unconvincing because it assumes the existence of a political will to achieve such objectives. In fact, what does tend to happen is that the diagnosis of disadvantage involves an assertion that this disadvantage is at most accidental or that it is due to the special characteristics of the minority. It rarely involves any willingness to recognize racial discrimination and oppression by the dominant group as the cause of the problem. Not surprisingly, therefore, when positive discrimination is proposed as a remedy, there is a backlash in that the minorities are seen as being given advantages which they do not deserve. Such a political background to programs for the disadvantaged is much more evident in Britain than in America, because in America the programs were backed by strong political commitment which had arisen out of the civil rights movement, while in Britain they were being proposed by a government which had taken up a strong anti-immigrant posture.

A second scenario for progressive change sees the integration of the minorities as occurring through social, economic, and geographic mobility. To some extent this is bound to happen. Once the cruder forms of discrimination in employment and housing are prevented by race relations legislation, there are bound to be some who find their way to business success, to better employment in skilled work and the professions, and to better suburban housing. The likelihood of some development in this direction is greater because it does not depend upon the beneficence of government. It is something which better-off members of the minority communities will achieve for themselves. Whether it will imply "integration," however, is another matter. For one thing, the numbers involved will be too small to have much influence on the "ghetto" situation of the minority (8). For another, the mobility which is achieved may be largely mobility within a separate pyramid. The crucial question, then, is whether the elements of the third scenario will come into play. These are effective political action by the poorer and more ghettoized sections of the minority community and integration of the economic, social, and cultural life of the successful and socially mobile members of the minority into the mainstream of the "host" society.

We are beginning to see developments of this kind in Britain. The very process of segregation in the inner city ensures that minorities have concentrated voting power and that Labor politicians particularly cannot ignore the votes of minority workers. The riots themselves may be seen as a different kind of political action aiming not at separation but at integration into the political mainstream. And, not the least important factor in times of grave economic crisis, the destruction of the welfare state forces more and more members of the white working class into trade union and political action alongside of minority people. Meanwhile, it does also seem to be the case that more successful members of the minority communities are entering the world of British business and playing a part in trades unions and professional associations.

While one should not underrate these trends, there are, however, also trends of an opposite kind. Unintegrated and ghettoized minorities are bound to become the target for racist attack in times of recession. It does still seem to be the case that whenever there is a specifically racist intervention in politics it wins a large measure of public support and the more orthodox politicians are seen as covering up an issue which should be more openly discussed. Within such a climate it becomes far more difficult to enforce an antiracist program in the police, and those police who see themselves as engaged in quasi-military activity against aliens are likely to get the upper hand. And, finally, racism makes powerful inroads into the culture and politics of the young unemployed. One thing, then, which must be expected in the years that lie ahead is a growth of political racism.

Against this background, it cannot be expected that minority politics will be directed solely towards integration and winning social and economic rights in the society of settlement. The particular deprivations of immigrants to the metropolis will be seen as part of a worldwide struggle against capitalism, colonialism, and racism. Sometimes this struggle will be defined in Marxist terms, but far more frequently it will be seen in terms of the Third World's struggle for liberation. To some extent, too, it will involve a kind of counteractive racism.

Clearly the problem of urban community cannot be isolated at this point from the politics of the larger world. Nor can the politics of a small metropolitan country like Britain. In the postcolonial era the balance of forces is tipped against the colonialist. Abroad white settlerdom has to fight and will fight viciously for its existence, whether in Africa, Ireland, or the Middle East. At home, colonial immigrants now fight for a larger place in the total imperial scheme of things than they had in the old colonies and will be resisted by nationalism and racism.

In this sense the problems of the British and American cities are obviously not to be considered as quite separate from the more violent confrontations of Belfast or Beirut. Nor is it to be expected that there will be solutions which do not take account of larger questions of racism. The question is now whether the metropolitan countries can rid themselves of their colonialist and racist legacy and treat those members of the colonial minorities who have settled in the metropolis as full and equal citizens.

NOTES

(1) Department of the Environment. 1977. Policy for the Inner Cities (Cnd. 6845). London: H.M.S.O.

(2) Park, R., and Burgess, E. 1922. The City. Chicago: University of Chicago Press.

(3) For a detailed discussion of the British Riots of 1981, see Rex, J. 1982. The 1981 urban riots in Britain. In The International Journal of Urban and Regional Research, vol. 6, no. 1, pp. 99–113. London: Edward Arnold.

(4) Rex, J., and Moore, R. 1967. Race, Community and Conflict. London: Oxford University Press.

(5) Rex, J., and Tomlinson, S. 1979. Colonial Immigrants in a British City. London: Routledge and Kegan Paul.

(6) The Brixton Disorders, 10–12 April, 1981. Report of the Scarman Tribunal (Cmd. 8427). London: H.M.S.O.

(7) Troyna, B., and Cashmore, E. 1982. Black Youth in Crisis. London: George Allen and Unwin.

(8) Wilson, W. 1978. The Declining Significance of Race. Chicago: University of Chicago Press.

Minorities: Community and Identity, ed. C. Fried, pp. 133–157
Dahlem Konferenzen 1983. Berlin, Heidelberg, New York, Tokyo: Springer-Verlag.

THE CHILDREN OF ALIENS IN WEST GERMAN SCHOOLS: SITUATION AND PROBLEMS

D. Hopf
Max-Planck-Institut für Bildungsforschung
1000 Berlin 33, F.R.Germany

GENERAL OUTLINE OF PROBLEMS AND QUANTITATIVE SITUATION

At present, there are over 4.6 million aliens living in the FRG and West Berlin; this is equivalent to approximately 7.5% of the total population (1). Roughly three quarters of the foreign nationals originate from so-called recruitment countries, particularly Turkey, Italy, Yugoslavia, Greece, and Spain, from which workers were recruited by the FRG until 1973. It is the children of these foreign nationals who will concern us in this paper (2).

Approximately three quarters of all aliens have been living in the FRG for five years or longer, and over half of them arrived eight or more years ago. The age and family structure has shifted considerably over this period: whereas there were relatively many "guestworkers" who were unmarried or living alone 20 years ago, the proportion between men and women is now more balanced, and there is a growing number of children and young people.

In the future, the numerical proportions between Germans and aliens, particularly between German and foreign children, will shift still further, as illustrated by Fig. 1, showing the different relations between producers and nonproducers as well as the shift in the numerical proportions among children and juveniles (3).

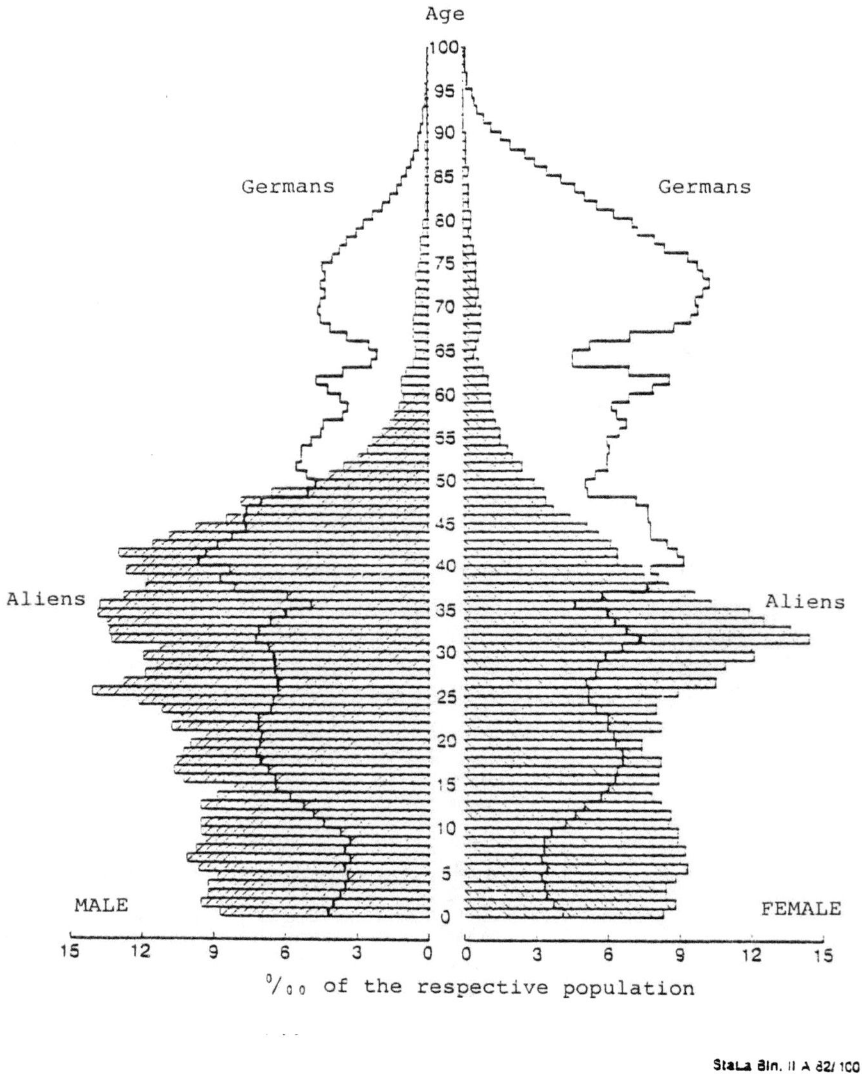

FIG. 1 — Age structure of the population of Berlin (West) as of January 1, 1982 (Germans/aliens).

The following statistics apply for the six-to-ten-year-old children of aliens in the FRG (4).

The prognoses contained in Table 1 are mainly based on the assumed continuation of the known birth rate statistics and, for later years, on an estimation of the different reproductive rates of the German and alien populations.

It is possible that factors other than the birth rate will have greater influence on quantitative developments, for example, legislative changes with respect to aliens, which as of Oct. 1, 1978, changes residence permit conditions, the effect of which will probably be an increase in the number of children and juveniles joining their parents. For almost half of the foreign workers residing in the FRG have close family relations still living in their country of origin. An estimated 1.4 million candidates for immigration (5) (apart from those from EEC countries) could join their families in the FRG in the near future, including 700,000 children and juveniles. Of importance in this connection are government subsidies to families with children which as of January 1, 1979, are only minimal for children living in the country of origin (6).

THE CHILDREN OF FOREIGN NATIONALS IN THE EDUCATION SYSTEM

According to the statistics of the Ministers of Education and Culture Conference (Kulturministerkonferenz) of August 1982, in the school year 1981/82 695,700 foreign children attended schools of general education in the FRG. Eighty-seven percent of these come from the following six recruitment countries: Turkey (51.5%), Italy (11.5%), Yugoslavia (10.4%), Greece (7.5%), Spain (3.5%), and Portugal (2.7%). The proportion of foreign pupils of the total number of schoolchildren varies according to school type. For the school year 1981/82 the percentages were as follows: primary school 13.3%; secondary modern school 11.8%; special schools 9.3%; grammar school 2.1%; secondary technical school 2.5%. This unequal distribution is shown in Table 2.

These statistics show clearly that foreign children compared to Germans are underrepresented at grammar and secondary technical schools and overrepresented at secondary modern schools.

TABLE 1 — Six-to-ten-year-old children of aliens (in% of the age group).

1977	7.99	1981	14.94
1978	9.61	1982	15.76
1979	11.68	1983	15.79

TABLE 2 — Total number of German and foreign school students at general educational schools in the secondary stages I and II, school years 1981/82 (7).

1.	Secondary Modern	Secondary Technical	Grammar	Compre-hensive	Total
2. German Pupils	1,968,058	1,290,909	2,062,866	201,787	5,523,620
3. % of Total	35.6	23.4	37.3	3.7	100
4. Foreign Pupils +	202,230	25,107	22,612	18,465	268,414
5. % of Total of Foreign Children +	75.3	9.4	8.4	6.9	100

+from recruitment countries

These average figures for the whole of the FRG conceal the fact that there are considerable regional differences. For example, the quota of foreign pupils attending special schools (of all pupils in special schools) in the FRG in the school year 1981/82 totalled 9.3%, but the extremes ranged from 18.5% in Baden-Würtemberg to 3.8% in Schleswig-Holstein, whereby the quotas for nationals of different foreign countries varied considerably. For example, only 2.9% of the Greek pupils went to special schools as compared to 7.9% of the Italians. Also, disparities in distributions within the individual Bundesländer must be assumed. The resultant problems for schools and instruction are varied. Many obvious differences — noticeable even by a glance at the statistics cited — have not yet been explained.

An examination of the quantitative relations shows that the end of schooling is a special problem area. Considering the customary close relation in the FRG between career choice opportunities, job training and practice, and school reports and qualifications, it is important that fewer than half of the children of foreign parentage pass their final exams at secondary modern school. For example, in Bavaria in 1978, 50% of the foreign school leavers left without certificates, and in Berlin it was 62% — compared wtih 26% of the German school leavers. The figures would appear even less favorable were they confined to the children of parents from the recruitment countries only.

The available data on the number of foreign students attending vocational schools are unsatisfactory and not very reliable (8). An approximate estimate of the proportion of foreign students at these schools is provided by Table 3.

TABLE 3 — Foreign school students enrolled at vocational schools.

	Total % of Foreign School Students at Vocational Schools (9)	Total % of Foreign Nationals of the Total Population of the Same Age (from 16 to under 19 years old)
1973	2.69	6.08
1974	2.86	5.62
1975	2.76	4.89
1976	2.61	4.89
1977	2.61	4.92
1978	2.68	5.20
1979	2.94	5.78
1980	3.84	7.12

Although the comparison of these age groups is a rough one, these figures show that the percentage quotas in the schools do not correspond to those of the age distribution in the population: the foreign school students are underrepresented at the vocational schools. This tendency is even more apparent in the findings of the Institute for Research into the Labor Market and Occupations of the Federal Institution for Labor (Institut für Arbeitsmarkt- und Berufsforschung der Bundesanstalt für Arbeit) for the school year 1976/77 (10). According to these findings, of the 170,600 foreign juveniles between the ages of 15 and 19 who were no longer attending a school of general education, 111,200 (65%) had no professional training whatsoever; 17,200 (10.1%) received some training but without taking a final examination (part-time trade school or basic training year); 42,100 (24.7%) received full job training (full-time schools 7.1%; dual system 17.6%).

This small selection of data on the quantitative circumstances serves to show that the children and juveniles of aliens are, on one hand, confronted with conditions in the schools and in lessons that do not specifically cater to them and are thus not particularly beneficial. On the other hand, they considerably change the conditions of learning and teaching prevalent in schools. They represent a group which possesses extremely disparate prerequisites for the learning process and which has only scant prospects of successfully finishing school. However, it is not possible for either the schools, teachers, or pupils to accomodate themselves to the situation that has arisen or to expect permanent solutions to the problems that have been identified. For the quantitative situation in this sector changes faster than in other areas of society, the validity of prognoses is questionable, and the effect of measures implemented are in the main uncertain. For example, when the Federal Government ceased recruitment in November 1973, the influx of more foreign workers was hindered and has, in fact, led to a drop in their numbers; but there has not been a corresponding decrease in the total number of aliens because, as before, the guestworkers are bringing their families into the country, and recently this has increased. Due to this factor and the relatively high birth rate, the age structure of aliens in the FRG is changing; not only through an upward drift in the "age pyramid" (cf. Fig. 1), but also — and sometimes markedly — through horizontal entries (family reunification) and other fluctuating phenomena.

The children of aliens present problems for schools and schooling for which there are no traditional solutions and for which the institutions and those affected are not prepared. It is astonishing that it was noticed relatively late that within the school system specific problems would arise both because of and for this group of school students, although a glance across the borders of Germany would have been sufficient to ascertain this fact. For example, the Structural Plan for the Educational System of the German Education Board, 1970, contained plans for up to 1980 but did not deal with the particular school problems of foreign children and juveniles, although in the year it was published there were already nearly three million foreign nationals living in the FRG. Similar cases were the Overall Education Plan of 1973 and the Educational Interim Balance Report of 1976. This is even more surprising considering that, basically, the problem is by no means new for Germany; in 1910, e.g., there were approximately a quarter of a million "Reichsausländer" (settlers from outside the German Empire) — mostly Poles — in Nordrhein-Wesphalia, who in certain districts accounted for over 25% of the inhabitants. At that time, the school administration at least gave the matter of the schooling of children of these "foreign workers" considerable thought (11). Bingemer et al. are perhaps not incorrect in their interpretation of contemporary German hesitancy in dealing with the problems regarding aliens as psychological suppression of the whole problem complex, arising from the experiences with the approximately 5.5 million forced laborers and 1.5 million prisoners of war who were made to work in Germany during World War II (12).

In the meantime, despite the relatively short time span in which research has been carried out in this field, such a tremendous amount of literature has appeared in German that it is almost impossible to command a view of it. Its contribution to solving the existing problems is unsatisfactory on a number of grounds. For one thing, there is a lack of research which concentrates on the central issues taking the specific circumstances in Germany into account. The findings of research studies conducted abroad have only a limited validity for the situation in the FRG. For example, the type of situation in which children of foreign parents grow up in a new culture is of great importance; also whether they belong to the majority or a minority, the future perspectives with which they come, whether their self-image is that of immigrant or "guestworker," and how the residents of the receiving country define their status. On the whole, literature from other countries is concerned with the situation of aliens who explicitly consider themselves immigrants and as a result are in a different situation to that of "guestworkers" from recruitment countries. It was assumed on all sides — including the guestworkers themselves — that their stay in the FRG would be only temporary.

Another problem with many studies published in the FRG is that their authors take up old and trusted themes of research in the social sciences — particularly in sociology and social psychology — and, just as they are, chew them over with regard to the case of the guestworkers. This is a hazardous undertaking, not primarily because of the impending duplication of the social sciences, but because of the curtailment of understanding of the real problem. Thus it is often recommended (without naming the grounds), e.g., that because of the obviously different initial situations of German and foreign children with respect to schooling, we should proceed in a similar way to the old debate over compensatory education for German children from different social classes. High hopes are set on language training programs for small children and school pupils in the dismantling of perceived deficits, for example, or great improvements are expected from merely providing numerous places at the elementary levels. This recommendation overlooks the fact that these children are different from one another in many basic respects and an extremely heterogenous group. Not only do they come from families which, according to the criteria of their country of origin, belong to social classes which cannot be compared to the German lower classes; it has also been suggested that particularly the active and flexible individuals with initiative leave their home countries, which afford them only limited scope for development (13). Insofar as we possess certain knowledge on these matters, their familial constellation, environment, legal position, planning horizons, preconditions and capabilities for learning, etc., are different from those of the group of underprivileged children from German working class families, who are the target group for the development of compensatory educational measures as well as the research that was carried out in this area. In the FRG, the foreign workers and their families often no longer belong to their social class of origin, nor can they be considered as directly belonging to the German lower class (14).

DIFFERENTIATION ACCORDING TO NATIONALITY

The simple differentiation between German children and children of foreign parents, as has been made in numerous studies, in the majority of statements on education policy as well as running through the policy measures affecting schools, is as a rule too coarse a distinction and thus unfruitful. In addition to this differentiation, it is much more necessary both in research and in developing measures to improve the situation of these children to make their specifically national characteristics the central focus. To then apply generalizations to another or more nations would represent an independent step.

The differences existing between nations need no substantiation: Turks and Italians differ more in the central aspects of their way of life and outlook than Germans of different social classes do. The practice of different religions alone implies a different relationship to the way of life in Germany, influences reactions to setbacks and frustrations, to uncertainties regarding the future, or to conflicting norms of education in the family and school.

For example, some of these differences are those of the socialization customs and life style in the home country. Of these, surprisingly little is known. There are but few studies on the family in Turkey, Yugoslavia, and Greece. Apart from some impressionistic observations, there are no studies which focus on children growing up under the normal conditions of their home country, on the one hand, and on the other, the kinds of changes and strains introduced by the conditions in Germany and the effect this invasion and the confrontation with other customs and norms has.

On the other hand, differences between the members of different nations arise in Germany. For example, it is of great importance for foreign juveniles regarding the development of their perspective for the future — or even of life in general — as to which country they come from. For families of non-EEC countries, return to their country of origin is effectively to leave Germany forever, as readmittance is not possible. Thus the "threshold" for returning emigrants varies from country to country. This naturally affects the scope for planning, which changes according to the given legal and/or political situation. The resultant confusion and uncertainty characteristic of the present situation lead to a severe strain within the family, because the dependence of the children and juveniles on their parents' decisions differs from nationality to nationality (15).

Regarding the possible future perspectives of families, the prognosis of change is difficult to make for other reasons. Should they decide to re-emigrate, unemployment will await many who return to their country of origin. However, the situation there and thus the attractiveness of returning can change rapidly; e.g., due to the political discussions of possible membership in the EEC.

Differentiation between nationalities is necessary for other reasons, as in the case of certain behavior characteristics which are not easily detectable but which affect most, if not all, areas of life (16). These range from the national specificity of psychic disturbances among school children (17) to variants of behavioral expectations according to sex in learning style in school, and types of errors when learning German (18) and other subjects, as can be observed in the classroom (19).

These few examples of differences between various nationalities will serve to illustrate that to treat all "foreign children" in the same way can lead to increasing inequalities. The same applies to the productivity of research.

DIFFERENTIATION BETWEEN "GENERATIONS"

In addition to differentiation according to nationality, at least one other is unavoidable, i.e., that of age, or of the very diverse worlds in which the foreign children find themselves or in which they grew up. It has become usual to speak of "generations" in this connection. It should be noted, however, that this occasions a considerable confusion of terms. For although the immigration of foreign workers has taken place over a matter of decades, it really does constitute an essential difference whether a child grows up in the environment and language of its home country and then has to switch over to another language and environment, or whether from a very early age it grows up in a family subjected to many and varied kinds of strain, needs two languages in order at least to get by, and possibly feels a stranger within the family as well as outside of it.

It will perhaps cause the least confusion if the term "generation" is used in its customary sense, defining generations according to their relative ages and not distinguishing a second from a third generation according to the place where they grew up. Thus the children of the guestworkers are the second generation (20), and there are but few who can be considered as belonging to a third generation, as is apparent from the total number of foreign workers over the last decades: in 1957 there were approximately 100,000 foreign workers, in 1961 the number had increased to half a million, in 1965 to a million, and only in 1971 to two million (21). Even if 15% illegal immigrants are added to this (22), one cannot yet speak of a generation of grandchildren.

As mentioned before, the reason for the inconsistent use of the term "generation" was to emphasize the significance of the milieu in which a child grows up. For this reason, I suggest that the second generation be subdivided into a "2nd-H generation" — these the mainly older children and juveniles who were born in the home country, spent their first years there, and then came to Germany to be reunited with their families; and "2nd-D generation" — children that were either born in Germany or came here as babies, who were faced with the problem (and the opportunity) of adapting to the German environment, and who are often not easily identifiable as foreign children.

This differentiation is important both for the selection of a population sample for research purposes as well as for the development of improvement measures, as a visit to any school or kindergarten will demonstrate. The 2nd-D generation differs from the 2nd-H not only in their linguistic ability in both languages, but also in the social forms of their contact with adults and other children, and in their readiness to become involved in the way of life and conditions in the FRG. Thus arrival of a large fresh contingent of 2nd-H generation in Germany represents a considerable burden for all those concerned with foreign children and juveniles, particularly as the situation has become somewhat easier due to the growing up of greater numbers of the 2nd-D generation.

Of undoubted importance to members of different "generations" is the way they are regarded by the host population and which longer-term perspectives for the future they have to show: whether they consider themselves immigrants or guestworkers, or whether the host population intends to "integrate" them. The concept of "integration" covers a number of different models of varying scope and complexity (23). For example, there is the idea of assimilation, or one-sided integration, i.e., in principle, foreign children should become like German children. In contrast, a pluralist concept of integration envisages a) the German and the various foreign cultures as coexisting side by side unmodified, and b) where there are unavoidable points of conflict, tolerance will be exercised. Finally, the concept of "interactionist integration" foresees mutual influence and association on the basis of equality and equal rights on principle. Further, conceptions of integration affect the intercourse of groups of different nationalities with each other — something which has not been discussed hitherto. For the foreigners often have no experience to fall back on when confronted with persons from other cultural backgrounds but have first to develop these forms, sometimes even despite enmities existing between their different nations.

The momentary legal position and the length of stay of over five years of the majority of foreign children would normally have led to a standardization of perspectives over the course of the coming years, had there not been the new influx of family members from abroad. Under these circumstances, the exceedingly great heterogeneity among foreign children and juveniles remains a fact that cannot be ignored by research work or educational policy or practical work in schools. Moreover, this area is influenced in a way that is virtually impossible to predict by the political options and the ideas toward "integration" of those involved – whether it signifies pressure to conform and self-abandonment, or tolerated ghettoization, or whether it means the inclusion of the foreign nationals in society, which is then implicitly recognized as a changing and not a static phenomenon.

SCHOOL AND TEACHING PROBLEMS

From the information given above the following points are of essential importance: a) the high and increasing proportion of children and juveniles of the foreign population as a whole, compared to the German population of the same age; b) the expected increase in the influx of family members, particularly children and juveniles, who grew up in their home countries (2nd-H generation); c) the distribution of young foreigners in the secondary schools, which deviates considerably from that of the German school students; d) the low quota of school certificates; e) the underrepresentation of foreign juveniles in vocational schools; f) the rapid change and thus the difficulty of predicting developments; g) the regional specificity of conditions and their further development.

In order to identify and deal with the school problems of foreign children, the distinction between 2nd-H and 2nd-D generation is of particular importance. When the children are grouped according to age and length of stay in the FRG, the following can be observed: leaving specific national differences aside – Turks particularly are below average with respect to length of stay – in 1978, of children and juveniles up to 20 years old, 30% were 5 and younger, a further 30% were 5-10, and the groups of 10-15- and 15-20-year-olds each represented 20% (24).

Half of the under-fives, in a cautious estimate, can be accounted as 2nd-D generation. Nearly all of this group is within the sphere of familial education, which means that their development is determined by family influences, for example, language learning. On one hand, this represents a relief, for the two cultures in which these children grow up and in which they must operate are not as yet in competition. On the other hand, the opportunity of "functionally" growing into German culture is reduced. The situation in which the foreign children spend their first years is probably of essential importance for their later learning capacity. Their basic problem in learning and development is that of having a double "curriculum" to master. They have to find their way in two different cultures whose facets do not always complement one another and sometimes contradict.

In short, for children of foreign parents living in the FRG there are two alternatives in early childhood (and only during that period): either monocultural socialization and later relearning, or bicultural (if not "multicultural") development. A child who grows up within a narrow family unit has both advantages and disadvantages: the advantage of a relatively undisturbed monocultural socialization (25) with possibly better chances of a differentiated intellectual development (26), but with the disadvantage of not only having to learn everything from the very beginning from the point of contact with German culture, but also having to relativize or even substitute many things that have become self-evident in intercourse with the peer group and in order to meet the expectations of adults, if the child is not to remain isolated. By contrast, a child who is at a crèche and kindergarten from the very beginning learns to absorb parts of both cultures and is able to harmonize them (or at least minimize conflicts). Starting school will not appear as a break in development; on the other hand, the child will probably not feel at home in either culture in the differentiated way that a monoculturally-educated child does.

From this, three consequences can be drawn: in the group of under-fives, there will be many children with the same prerequisites as those of the 2nd-H generation, even some who were born in the FRG, when they start school. Second, the time of arrival and number of years spent in the FRG provide only a very rough indication of the learning prerequisites with which a child starts school. Thus the school must provide alternatives for the children according to their preschool experience. This implies that lessons must be differentiated and individualized to a hitherto unprecedented extent.

Of the 5 – 10-year-old group, an estimated 75% belong to the 2nd-D generation (in the most unfavorable instance, a 10-year-old child would have been in the FRG for four years). Three out of 4 children in this age group are growing up — at least since they reached school age — with the requirement of meeting the demands put upon them by school in Germany and, perhaps even more important, of getting along with German children of their own age every day. Approximately half of the 10–15-year-old group belong to the 2nd-D generation; they have been in the FRG since starting primary school. This probably applies to only about a third of the 15 – 20-year-old foreign juveniles. At least a quarter of this group has been in Gemany only since the age of puberty, so the majority belong to the 2nd-H generation.

Thus, in summary, only a minority of the total of these foreign children can be regarded as belonging to the 2nd-H generation, i.e., confronted with the particular difficulties attendant upon scant or no knowledge of the receiving country. However, this group will be de facto larger, mainly because of the close family orientation of the majority of alien families, which in the preschool period can mean almost total isolation from the German language and environment. Should the estimated 700,000 children and juveniles actually join their relations living in the FRG, as is forecast, the proportions will shift still further, and then the majority will effectively belong to the 2nd-H generation. Here the so-called horizontal entry throughout the whole school year, according to individual arrival time, represents a severe burden.

THE HETEROGENEITY OF SCHOOLCHILDREN

In the following, I shall briefly outline selected problems of schoolchildren. I shall not refer in great detail to the considerable body of literature on this subject (27).

Leaving aside the group of foreign children for a moment, a process of increasing heterogeneity among school children can be asserted. In the primary schools, there are several reasons for this (28). A growing number of Bundesländer, for example, have stopped the practice of requiring some children to repeat school years. Also, fewer children are sent to special schools than formerly or diagnosed as too immature to begin schooling for the time being. Further, there has been an increase in the number of children requiring special attention from the teacher; for example, children from broken homes, only children, or children with certain general learning and behavior disturbances. These and other reasons have led to an increased need for individual attention and, thus, to the teacher requiring more time for each child as well as new forms of teaching and care.

In this increasingly difficult overall situation, which exists even in schools without foreign pupils, those schools with foreign students must experience even greater problems. The foreign children start school with extremely diverse prerequisites, and the heterogeneity thus occasioned is far greater than that existing among German pupils. The significance of the differentiation made above between nationality and "generation" becomes easily comprehensible here: children who differ not only from their German school fellows but also from each other with regard to language, cultural norms, curriculum vitae, and previous school career have to be taught and attended to. School achievements and abilities remain in the background due to language deficits, lack of motivation, or other value orientations; specific gender behavior is far more apparent than amongst German pupils, which obliges the teacher to interpret; in contrast to the stated age, it is often impossible to judge the real age of the pupil, which is of course vital for determining the basis of appropriate demands and aid. Added to this is the problem of children who have just arrived joining classes throughout the school year, which often destroys the stability of groups that have just managed laboriously to achieve a certain measure.

A key problem is the different levels of language competency of the foreign workers' children. The difficulties experienced today are far more severe than the language deficits of the German lower classes, which was discussed extensively in the 1960s and 70s. For heuristic purposes, the foreign children can be roughly divided into three groups: a) children who are linguistically competent, do not attract attention in other areas, and are adjusted to conditions in Germany (about 5–10% of foreign schoolchildren); b) the approximately 40–70% of foreign children in a school class who obviously have difficulty with the language but are able to follow lessons. These pupils try hard for a long time to keep up in class, but they are doomed to failure if special steps are not taken to help them. The members of this group represent the greatest hindrance for the German pupils during lessons, for they (justly) demand consideration, which necessarily leads to a slower pace in the lessons and occasions many questions concerning language comprehension, mistakes, and misunderstandings. c) The third group consists of children whose difficulties with the language are such that they are unable to follow lessons. This group of children, which is expected to increase considerably, naturally comprises those who have only recently arrived in the FRG. It is worthy of note that the children in this group rarely draw attention to themselves by disruptive behavior but try hard to contribute in spite of almost permanent failure or behave quietly or apathetically.

It is self-evident that a school class of pupils which contains members of all three groups of foreign children described above presents great if not insurmountable problems for the usual frontal teaching method on grounds of language competence alone. It must be emphasized here that scarcely fewer problems exist in teaching classes exclusively made up of foreign children, as their language competency differs so greatly in most cases. Even with groups of children of the same nationality, it is likely that the teacher will face more difficulties than if it were a comparable class in the country of origin. Lessons in the children's native language would not solve the problems either, for language competency in the mother tongue cannot be assumed in the case of many children of foreign workers.

In this outline of the special characteristics of foreign children relevant to teaching, reference must be made to their different forms of learning, an area in which no research as yet has been done. At the level of the perception of given signs, symbols, and sounds (for example, when copying from the blackboard, copy drawing, discrimination of sounds) there appear to be processes which are different to those of German pupils first starting school. But a more important observation is that the flexible forms of discourse and the first stages of process-oriented learning that have been used increasingly in German primary schools over the past years, for example, recounting events, such as a birthday party, that happen at home, or projects and open discussions in class, appear strange to many foreign pupils. It appears that they are used to a much more directive approach both at home, in school in their home country (if applicable), in Koran schools, or in specifically national classes. There, more often than in our schools, learning would appear to be guided using straightforward exercises to which there is but *one* correct answer, thorough preparation and learning prescribed texts by heart. Less emphasis appears to be placed on approaching complex problems by the open question method, to which there is more than one correct answer. This collision of disparate learning and teaching practices necessarily leads to dysfunctional behavior of all concerned. Also, this conflict confronts every teacher with the problem of how far the different prerequisites for learning and expectations of the foreign children should be considered and how far the important steps that have been taken in the German schools should be revoked, in spite of the justified claims of these pupils for individual attention.

This example illustrates the importance of a comprehensive concept capable of guiding practical decisions and conduct for the daily contact and coexistence with foreign children and juveniles. For only a clear answer to the question as to how far the various cultures should adapt to one another can make it possible for teachers and pupils to react with self-confidence in specific situations; i.e., in the example given, either formalizing lessons more (also for the German children) or requiring the foreign children to make an extremely difficult adjustment. In practice, the obvious way out of the difficulty would be the strict individualization of teaching, which would allow each pupil to utilize his/her own prerequisites and learning customs, interests, etc. Naturally, this could not be done without appropriate teacher training and the development of new teaching aids. It must be admitted, however, that the question of long-term goal would still be unanswered, for individual teaching could bring about an increasing divergence in the children's development.

With regard to the social behavior of foreign children in school, both informal observations and controlled surveys have found that disruptive behavior in school is expressed differently, namely, less destructively (29). Considering the very high demands on foreign schoolchildren and the unlikelihood of meeting them, refusal or disobedience is seldom. Disturbances in class are freely admitted, if questiond by the teacher; spontaneous cordiality toward teachers and school fellows is often expressed; the equanimity with which excessive demands, frustrations, or even injustice are borne is also striking. Although these are purely observations, one can say that the preconditions for teaching are by no means unfavorable (30) and the teachers are not additionally taxed by having to deal with rebellious and disobedient students. This applies not only to primary school but also to the lower and upper grades of secondary modern school, where it can happen that the foreign students become the most active class group, so that teachers refuse to let them be transferred to other classes (31). In secondary modern schools, foreign students are often outstanding, a fact that is not really surprising since the largest "pool of talent" at present in our schools is probably to be found in the group of foreign pupils (32).

INSTRUCTION

Many of the problems with instruction in class are due to the fact that the students have to master a double curriculum and thus have to meet higher demands than their German counterparts. It is not only a question of the strain imposed by having to cope with school lessons as well as specifically national demands on learning, such as Koran school, etc. The excessive demands extend to such essential areas as the mediation required of the individual between contradictory norms and aims. The education of the ability to criticize, as contained in the preamble to German school legislation and thus a duty of the school, is not compatible, for example, with the traditional decision-making hierarchy in a Turkish family.

With regard to the school curriculum in a narrower sense, the heavy emphasis on language in our schools has a particularly adverse effect on foreign children. Their language difficulties are probably underestimated in many cases by the teachers, for in the normal classroom situation, statements by foreign pupils represent proportionately high linguistic achievements, whereas long periods of silence due to language inability remain unnoticed. In virtually all lessons it can be seen how great the language barrier is: for example, uncomprehending, apparently "stupid" pupils can immediately participate in class after a quick consultation with a teacher who speaks their language. In the long term, the cumulative effect of this language handicap will have severe consequences.

In this connection, too, there are as yet no answers to the problem. For example, should school beginners be taught to write a language, the sound of which they have not been familiar with from an early age and in which they can make the necessary differentiation between sounds only with extreme difficulty? If not, what consequences would a later changeover to the German language, with its different alphabet, entail? What effects are there from cumulative learning deficits in "linear" subjects such as mathematics, where the understanding of a certain content is essential for that which follows? What would be the consequences if, for the sake of full comprehension, subjects were taught in the mother tongue and then followed by repetition of the content in German, as an additional language exercise? Which subjects and aspects could be left out of the curriculum for foreign children so that the workload does not exceed their capabilities?

With regard to the organization of teaching, as pointed out above, foreign children are dependent on individualized learning. A further problem in this connection is that the majority of classes contain only a small group of foreign pupils, often of different nationalities. Approximately half of the school children do not live in urban agglomerations but in towns with less than 100,000 inhabitants. The presence of foreign children in classes represents a trial for the expectations connected with the differentiation and individualization of teaching. As soon as a group becomes too heterogeneous (or is recognized as such), so that the same instruction is not meaningful for all pupils, the teacher is faced with the task of diagnosing the various individual learning prerequisites and then developing appropriate measures. If this is taken seriously, these demands rapidly exceed that which is practicable, so that the teacher is confronted with the alternatives of either relinquishing the basic idea of adequately meeting individual needs, or grouping the pupils in relatively homogeneous units, or giving up direct guidance of the learning process. The results would be: either to continue teaching in the usual way, which presents great difficulties for some of the pupils; or to use forms of extreme differentiation; or to use programmed or computer-assisted instruction or to develop broad, many-faceted forms of instruction that are not preprogrammed, i.e., to set up learning situations where those directly involved make the complex discrimination between that which is offered and their own prerequisites to a greater extent, and where they have the freedom to participate in the reorganization of the learning situation, either alone, with other schoolfellows, or with the help of the teacher.

In the area of educational assessment, again only selected, preliminary questions can be outlined here. The traditional method of assessing performance (school reports, marks from 1–6, etc.) has long been considered questionable, even under normal conditions at German schools. With reference to the school problems of foreign children it is totally inadequate. The main problem is that not enough allowance is made for the conditions under which the children have to learn in school. It is thus not only problematical as a judgement on the momentary situation but also cannot be used for prognostic purposes. Here, the arguments employed in the debate in the FRG on reserves of talent, which took place years ago, are particularly applicable: average or even underaverage performance of a child who comes from a background which is not conducive to learning or even hinders it can be an indication of high potential, as this performance takes place under circumstances where such results are extremely improbable.

This fact must also be taken into account not only in giving marks and making school reports but also when using these, for example, in the selection of applicants for a place or job. A juvenile of foreign nationality who manages to attain a final school certificate — even with the worst possible marks — has probably, in the majority of cases, given proof of the mastering of extremely difficult subjects, and it is very likely that favorable prognoses can be deduced as to the further development of such an individual. Justice cannot be done to the foreign school leavers if a simple horizontal comparison of their performance is made with that of their German peers, but this would only be possible if biographical details are also taken into account. Similarly, it is obvious that there must be great flexibility in other areas of educational assessment, e.g., the minimum requirements for starting professional training. These must be redefined when applied to foreign juveniles according to their individual school record and circumstances (33).

Further problems in this field are: a) lack of diagnostic aids for the estimation of functional language competence; b) the separation of diagnosis and treatment: information on performance and performance deficits rarely contains recommendations for the next meaningful learning step; c) lack of diagnostic procedures for identifying disturbances in learning ability and behavior, particularly those which are specific to the various groups of foreign children and juveniles. Here also, national specificity must be regarded (34).

As a whole, the influence of school and class instruction must not be overestimated, of course. It is possible that these are of secondary importance compared with familial influences and contact with the peer group — perhaps in some cases of only marginal importance. As an example which may stand for other factors, I should like to draw attention to the role of the parents in families of foreign nationals here. Many of them effectively intervene in the school career of their children — however problematic this may be — in many ways. Many juveniles are taken away from school immediately after the period of compulsory education ends, regardless of exams still to be taken or the child's own wishes. This may be due to economic necessity on the part of the family or fear of excessive alienation. Furthermore, many parents resist the attempts of teachers to include foreign pupils fully in normal classes. Many foreign parents do not seem to be aware that there is a necessity for schoolchildren to be exempted from chores at home in order to do their homework and exercises — in their case, an absolutely necessary precondition for continuing participation at school.

While those factors may be at least partially influenced by the school, this is not true concerning the uncertainty as to the future with which many foreign children have to live. The constructive influence of school must be viewed with particular skepsis if the foreign children and juveniles are unable to develop longer-term life and job perspectives, for only then does the effort involved in school work appear meaningful. The possibility of a limited stay as well as little prospect of getting either a job or an apprenticeship are factors which certainly lead to demoralization and lack of motivation, particularly at secondary school, not to mention the possible emergence of delinquent subcultures as a response to the situation (35).

The additional difficulties facing teachers with regard to foreign pupils have been mentioned above. However, as yet only a very small proportion of teachers have prepared themselves for their difficult task, and these mainly on private initiative. Furthermore, very few specific aids are available. Thus, schools and teachers tend to deal with the additional difficulties related to the foreign children in the form of daily "crisis management." It is the exception rather than the rule that the presence of children and juveniles from other cultures is regarded as an enrichment. Their presence could be the occasion for changing school and teaching in such a way as to solve the — admittedly very difficult — problems, many of which have been known and discussed for a long time, but which have appeared for the first time in their full acuteness and severity since the advent of the foreign children and juveniles.

Acknowledgements. I wish to express my gratitude to S. Avineri, C. Fried, M. Smith, and A. Zolberg for their very helpful comments; to S. Bernhard for her editorial advice; and to G. Custance for her very competent translation and technical help.

NOTES (See BIBLIOGRAPHY)

(1) For a comprehensive statistical overview see Trommer and Köhler (1981).

(2) This paper is a revised version of Hopf (1981).

(3) Statistisches Landesamt Berlin II A 82/100.

(4) Calculated from Grund- und Strukturdaten (1982/83), p. 221.

(5) cf. Boos (1978), p. 73f; Hohmann and Stahr (1976), p. 166; and Spies (1979). An example of consequences that cannot be anticipated is the changes in the situation in schools that may result from making the payment of children's allowances dependent on proof of school attendance.

(6) For an account of the complicated and continuously changing legal situation of guest workers in Germany see Siewert (1980).

(7) Calculated from Grund- und Strukturdaten (1982/83), and Sekretariat der ständigen Konferenz der Kultusminister (1982).

(8) cf. Boos (1978), p. 3.

(9) Calculated from Grund- und Strukturdaten (1982/83), pp. 27, 55, 221. In the composition of these quotas there are considerable differences between nationalities; cf. Sekretariat d. Kultusministerkonferenz (1982).

(10) cf. Seidel (1979), p. 70.

(11) cf., e.g., Heinemann (1975).

(12) Op. cit. (1970), p. 38.

(13) Lee (1966) refers to a bimodal distribution of migrants, which comes about through the positive selection of those who react to the positive factors of the target country, and a negative selection of those who react to the negative factors of the country they come from. Sander (1977, p. 176) speaks of a "brain drain" in the countries of origin. For these and similar statements there is as yet no grounded data base, but they do possess a certain plausibility (cf. Geck (1979), p. 136; also see Künne (1979), p. 161f.).

(14) cf. Hoffmann-Nowotny (1973).

(15) cf. Boos et al. (1979), e.g., on the generation conflict in alien families. Of interest in this connection are the recent findings with respect to Berlin, according to which 41% of the Turks would agree to their sons remaining in the FRG were they themselves to return home (with 13% refusing); in contrast, the Yugoslavians and Greeks who gave a positive reply were only 3% and 2%, respectively (with 64% and 44%, respectively, refusing); cf. Der Regierende Bürgermeister von Berlin, Dec. 1980, Vol. 1, p. 140f.

(16) cf. Boos (1979), p. 62f.

(17) cf. Bayer et al. (1977).

(18) cf. Eideneier (1976); Cimelli and Liebe-Harkort (1976); Figge and De Matteis (1976); and Meese et al. (1980).

(19) cf. Hopf, Krappmann, and Scheerer (1980).

(20) cf. Schrader et al. (1976).

(21) Members of the second generation thus defined can be of differing ages, according to the age and time of arrival of their parents. The greatest number of people joining their families arrived before recruitment stopped in 1973. Since then, the foreign population has remained roughly constant.

(22) cf. Rist (1978), p. 62.

(23) On the many imprecise concepts used in this connection, see the detailed compilation by Katsarakis (1974), p. 51f; Esser et al. (1979), p. 5f; and Taft (1953).

(24) cf. Seidel (1979), p. 73.

(25) cf. the report by G. Franger and J. Vink in ISS Informationsdienst 2/1979, p. 68f, on problems of growing up in the family, which also draws attention to the discontinuity of even monocultural socialization.

(26) cf. Cummins (1979).

(27) Apart from the specialist journals and periodicals and the usual bibliographic aids, the following are of particular use;
— the information service with respect to foreign nationals of the Institut für Sozialarbeit und Sozialpädagogik, Frankfurt, 1979f;
— the special bibliography compiled by the research group ALPHA, Neuss (cf. Lernen in Deutschland 11/1980, p. 23f); and
— the bibliographies of the Institut für Sozialarbeit und Sozialpädagogik.

(28) cf. Hopf, Krappmann, and Scheerer (1980).

(29) cf., e.g., Bayer et al. (1977).

(30) Mixed classes exhibit fewer problems than classes of only foreign pupils. cf. ibid., p. 509.

(31) Such findings have resulted from research on secondary modern schools carried out by the Max-Planck-Institut für Bildungsforschung (Roeder et al.).

(32) An indirect verification of this can be found in Künne (1979), with regard to the Yugoslavian guest workers (cf. p. 161f and passim).

(33) In connection with this and other aspects, affirmative action, as discussed and practiced in the United States, might be very worthwhile considering (e.g., see Cohen, this volume).

(34) cf. Bayer et al. (1977).

(35) See Albrecht and Pfeiffer (1979). In this connection, ideas derived from programs like MANOF in Israel might be particularly helpful.

BIBLIOGRAPHY

Albrecht, P.A., and Pfeiffer, C. 1979. Die Kriminalisierung junger Ausländer. Befunde und Reaktionen sozialer Kontrollinstanzen. München: Juventa.

Bayer, W.; Gärtner-Harnach, V.; et al. 1977. Psychologische Untersuchung der Schulsituation der Kinder ausländischer Arbeitnehmer. Abschlußbericht der Forschungsgruppe "Kinder ausländischer Arbeitnehmer" an der Fachhochschule für Sozialwesen Mannheim. Mannheim o.J.

Bingemer, K.; Meistermann-Seeger, E; and Neubert, E. 1970. Leben als Gastarbeiter. Geglückte und mißglückte Integration. Köln: Westdeutscher Verlag.

Boos-Nünning, U. 1978. Berufsfindung und Berufsausbildung ausländischer Jugendlicher. Die Darstellung der Schwierigkeiten und Empfehlung zu ihrer Überwindung. Berichte und Materialien der Forschungsgruppe ALFA. Nr. 11. Neuss: Pädagogische Hochschule Rheinland.

Boos-Nünning, U., and Hohmann, M. 1975. Zur Situation deutscher Lehrer von Kindern ausländischer Arbeitnehmer. Ergebnisse einer qualitativen Untersuchung. Bildung und Erziehung **28**: 43–52.

Boos-Nünning, U.; Hohmann, M.; Reich, H.; and Kuhs, K. 1979. Materialien zur Fortbildung griechischer Lehrer. Neuss: Publikation ALFA.

Der Regierende Bürgermeister von Berlin (Hrsg.). 1980. Befragung deutscher und ausländischer Haushalte zur Ausländerintegration in Berlin. Vorgelegt von Socialdata, Institut für empirische Sozialforschung. Berlin, Dec. 1980 (2 vols).

Cimilli, N., and Liebe-Harkort, K. 1976. Sprachvergleich Türkisch-Deutsch. Düsseldorf: Schwann.

Cummins, J. 1979. Linguistic interdependence and the educational development of bilingual children. Rev. Educ. Re. **49**: 222–251.

Deutscher Bildungsrat (Empfehlungen der Bildungskommission). 1970. Strukturplan für das Bildungswesen. Bonn: Bundesdruckerei.

Eideneier, H. 1976. Sprachvergleich Griechisch-Deutsch (Publikation ALFA). Düsseldorf: Schwann.

Esser, H.; Gaugler, E.; Neumann, K.-H.; et al. 1979. Arbeitsmigration und Integration. Sozialwissenschaftliche Grundlagen. Materialien zur Arbeitsmigration und Ausländerbeschäftigung, vol. 4. Königstein: Hanstein.

Figge, E., and de Matteis, M. 1976. Sprachvergleich Italienisch-Deutsch. Düsseldorf: Schwann.

Geck, H.-M. 1979. Die griechische Arbeitsmigration. Eine Analyse ihrer Ursache und Wirkung. Königstein: Hanstein.

Griese, H.M. 1981. Jugendliche Gastarbeiterkinder: Situation und Problematik. Eine Literatur- und Forschungsdiskussion. Zeitschrift für Pädagogik **26**: 441–456.

Heinemann, M. 1975. Die Assimilation fremdsprachlicher Schulkinder durch die Volksschule in Preußen seit 1880. Bildung und Erziehung **28**: 53–69.

Hoffmann-Nowotny, H.-J. 1973. Soziologie des Fremdarbeiterproblems. Eine theoretische und empirische Analyse am Beispiel der Schweiz. Stuttgart: Ferdinand Enke.

Hohmann, M., and Stahr, I. 1976. Ausländische Schüler an allgemeinbildenden Schulen in Nordrhein-Westfalen. Daten und Analysen. Pädagogische Rundschau **30**: 165–182.

Hopf, D. 1981. Schulprobleme der Ausländerkinder. Zeitschrift für Pädagogik **27 (6)**: 839–861.

Hopf, D.; Krappmann, L.; and Scheerer, H. 1980. Aktuelle Probleme der Grundschule. **In** Max-Planck-Institut für Bildungsforschung, Arbeitsgruppe Bildungsbericht (Hrsg.). Bildung in der Bundesrepublik Deutschland. Daten und Analysen, vol. 2, p. 1113–1176. Stuttgart, Klett/Reinbek: Rowohlt.

Katsarakis, N. 1974. Probleme kultureller und gesellschaftlicher Integration griechischer Arbeitnehmer in der BRD. Exemplare Untersuchung im Bereich des Freizeitverhaltens. Dissertation, University of Aachen.

Kühn, H. 1979. Stand und Weiterentwicklung der Integration der ausländischen Arbeitnehmer und ihrer Familien in der Bundesrepublik Deutschland. Bonn, September 1979: Memorandum des Beauftragten der Bundesregierung.

Künne, W. 1979. Die Außenwanderung jugoslawischer Arbeitskräfte. Ein Beitrag zur Analyse internationaler Arbeitskräftewanderungen. Königstein: Hanstein.

Lee, E.S. 1966. A theory of migration. Demography **3**: 47–57.

Meisel, J.M. 1978. Emigratión sí — Einwanderung, nein! **In** Wuppertaler Arbeitspapiere zur Sprachwissenschaft **1**: 1–37.

Ministerium für Arbeit, Gesundheit und Sozialordnung Baden-Württemberg. 1979. Viele Ausländer wollen in Baden-Württemberg bleiben. Stuttgart: Materialdienst Ausländische Mitbürger in Baden-Württemberg.

Rist, R.C. 1978. Guestworkers in Germany. The Prospects for Pluralism. New York: Praeger.

Roeder, P.M. 1979. Kompensatorischer Sprachunterricht. **In** Lexikon zum Deutschunterricht, E. Nündel, ed., p. 178–196. München: Urban und Schwarzenberg.

Sander, A. 1977. Entgegnung zu Schmidtke, H.-P. Ausländerkinder und Sonderschule. Zeitschrift für Heilpädagogik **28**: p. 176–180.

Schmidtke, H.-P. 1977. Ausländerkinder und Sonderschule. Zeitschrift für Heilpädagogik **28**: 170–175.

Schrader, A.; Nikles, B.W.; and Griese, H.M. 1976. Die zweite Generation. Sozialisation und Akkulturation ausländischer Kinder in der Bundesrepublik. Kronberg: Athenäum.

Seidel, H. 1979. Ausländische Arbeitnehmer in der Bundesrepublik Deutschland: Ein statistischer Überblick. Deutsch lernen **1**: 52–76.

Sekretariat der ständigen Konferenz der Kultusminister der Länder in der Bundesrepublik Deutschland. Ausländische Schüler in der Bundesrepublik Deutschland 1970 bis 1981. Bonn, August 1982.

Senat von Berlin, June 1978. Bericht zur Lage der Ausländer in Berlin.

Siewert, P. 1980. Zur Entwicklung der Gastarbeiterpolitik und der schulpolitischen Abstimmung der Kultusministerkonferenz. **In** Bildung in der Bundesrepublik Deutschland. Daten und Analysen. Max-Planck-Institut für Bildungsforschung, Arbeitsgruppe Bildungsbericht (Hrsg.). Stuttgart: Klett/Reinbek: Rowohlt.

Spies, U. 1979. Ausländer in Berlin (West) 1972 bis 1977. Analyse der altersstrukturellen Entwicklung der ausländischen Bevölkerung aus Anwerbestaaten. Berliner Statistik **33 (2)**: 33–42.

Statistisches Landesamt Berlin. 1976. Zur Entwicklung der im Grund- und Hauptschulalter stehenden ausländischen Kinder in den Bezirken von Berlin (West). Berliner Statistik **30 (7)**: 174–184.

Taft, R. 1953. The frame of reference concept applied to the assimilation of immigrants. Hum. Relat. **6**: 45–55.

Trommer, L., and Köhler, H. 1981. Ausländer in der Bundesrepublik Deutschland. Dokumentation und Analyse amtlicher Statistiken. München: Deutsches Jugendinstitut.

Minorities: Community and Identity, ed. C. Fried, pp. 159–175
Dahlem Konferenzen 1983. Berlin, Heidelberg, New York, Tokyo: Springer-Verlag.

USES AND FUNCTIONS OF THE MEDIA

L. Huth
Communications Consultants
6301 Linden-Leihgestern, F.R. Germany

Abstract. This paper presents a typology of functions of the media for the Turkish minority in Germany. The present situation and the scattered results of research in this field are discussed with reference to this typology. Finally, the paper tries to draw some conclusions concerning both research tasks and a future media policy.

PRELIMINARY REMARKS

The uses and functions of media in connection with minorities cannot be discussed without particular reference to the minority group or groups in question. I shall therefore limit my remarks to the problems related to the migrant workers and their families in the Federal Republic of Germany — especially as far as the Turkish part of the population is concerned.

Any attempt to outline the present situation and to answer the questions involved in the title of this paper suffers from the limited knowledge we have in this field:

- Little genuine research has been done on the question of minorities and media in Germany. In particular, studies in the situation of minorities in Germany almost entirely disregard the existence and possible roles of the media.

- As a rule, media appear — if they do at all — under the heading of "leisure activities." This approach has two disadvantages: (a) It limits the functions of mass media to that of a "pastime;" and (b) it leads to construct rather rough categories for empirical research such as "reading" vs. "going for a drink," leaving open more subtle options and distinctions such as "books, magazines, or newspapers?" or "in the mother tongue or in German?"
- Since these studies differ in their empirical bases, in their categories, and in the years in which they were carried out, they are hardly comparable.
- Studies dealing with the minorities and media complex naturally will come to mention media functions. Theoretically, we may find a difference between *intended functions* (by the producers, politicians, etc.); *desired functions* (by the audience or public), which may or may not be fulfilled, and which we mean by the word "uses"; *assumed functions* (assumed again by producers, politicians, various institutions, and critics), and *empirically traced functions*. The less we know about the second and the last kinds of functions, the more we are tempted to talk about the other two — and this is largely the state of the art. In addition, the various functions are often enough not kept strictly apart in discussions.

TYPOLOGY OF MEDIA FUNCTIONS

Talking about media functions presupposes some kind of systematic relationship between a particular medium or media output and an observable audience reaction. This relationship, however, is no one-to-one correspondence*. The mere presence of "Hürriyet" in a German railway bookstall may reassure the Turkish person that even here he does not live on a remote island, while the German person may be filled with awe. Needless to say, this is a very rough differentiation, and in fact, we shall have to take into account as many different functions of one medium or program as we have social groups. For our present purposes we shall (idealizing) mainly concentrate on media functions with regard to the (Turkish) minority.

Furthermore, we shall deal only with those functions which are of interest with regard to the minority problem. No doubt, a great number of cultural products — such as Western films, crossword puzzles, or the weather forecast — offer gratifications without immediate relevance to the questions at hand.

* The theoretical background for this consideration is provided by the uses and gratifications appoach as formulated by (21).

Bearing these limitations in mind, we may arrive at the following groups of functions media can (or should) have for the (Turkish) minority in Germany* (Fig. 1):

"Function" is a normative term insofar as it refers to approved or desired functions; accordingly, we may formulate corresponding dysfunctions to those functions mentioned above: (a) assimilation and alienation, and (b) isolation and formation of a subculture. While, on an abstract level, preservation of the national and cultural identity *and* integration and participation of the minority in the institutions of the host country are the undisputed parallel aims of all considerations in Germany, single measures frequently will meet with the objection that they rather fulfil one of these dysfunctions.

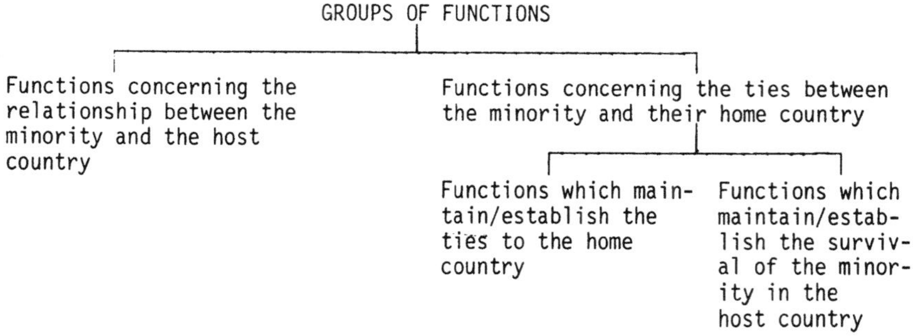

FIG. 1 — Functions of the media for the Turkish minority in Germany.

The following functions (either intended or assumed) are collected from the literature consulted (Tables 1 and 2):

We have little to no information whether and to which extent the intended functions are in fact fulfilled. Only with phono and video cassettes may we judge by their economic successes that they fill a need gap left open by the "traditional" media. Apart from home films and home music, there exists a genre of music cassettes which hitherto has entirely slipped the attention of critics and researchers: nostalgic songs which thematize the situation of Turkish migrant workers abroad. They indicate an emerging identity of the minority different both from being Turkish and from being German.

Media uses can be discussed under three different headings:

1. Media use as part of *leisure activities* (within the time budget);

* These functions are developed in (12).

TABLE 1 — Host country-directed media functions for the (Turkish) minority.

Function	Assumed/Intended to be Provided by:	Suggested by:
General information about the host country		
Practical ("how to") knowledge about the host country	Radio and TV minorities programs*; German trade union journals; information by other German institutions and local communities; information by Turkish banks (economic information)	
"Access" knowledge	Radio and TV minorities programs; local community information, e.g., "Berliner Bär"	Bubenik-Grüneisl/Bubenik (5) (media centers); Rissom et al. (23) (regional radio and TV)
Increase minorities' understanding for host country	Radio and TV minorities programs	Göktürk (9); Aktan (2); Rissom et al. (23)
Destroy Germans' prejudices	ZDF minority program	Göktürk (9); Aktan (2); Rissom et al. (23) (radio and TV in general); Eulgem et al. (8) (regional media)
Articulation of minorities' interests	Newspapers; radio and TV in general	Göktürk (9)
Address minorities in campaigns (e.g., crime prevention, health)		Stelzle (25)
Motivate minorities to use German language media	ZDF minority program (by bilingual presentation and formal resemblance to German programs)	Röll (24)
Encourage personal contact with Germans	Radio program "Rendezvous in Deutschland"	Rissom et al. (23); Huth (11)
Counter-information to third-countries' media		Politicians (cf. Stelzle (25))
Supply of arguments for home country representation abroad	"Türkiye'nin Sesi" Radio Ankara shortwave program	

*In the Federal Republic of Germany the broadcasting competence lies with the Länder (counties). In principle, each Land has its own radio station and programs. As concerns TV, each of these stations has a share in one joint TV program (first channel, ARD); the second channel is an independent foundation established by convention of the Länder (ZDF = Zweites Deutsches Fernsehen); the third consists of the different programs of the single Länder stations — a kind of regional program.

TABLE 2 — Home-directed media functions for the (Turkish) minority.

Function	Assumed/Intended to be Provided by:	Suggested by:
Maintain/establish contacts within the national minority	European editions of Turkish newspapers	Rissom et al. (23) (access programs; radio and TV in general)
Increase understanding between national minorities	Special radio and TV programs (cf. Stelzle (25))	
Home-culture entertainment	Turkish video cassettes (cf. Klitzke (17,18)); Turkish films in public cinemas	Klitzke (17,18) (undubbed Turkish films on TV, using one stereo channel)
Maintain cultural, religious, and ethnic identity	Radio and TV minority programs; TRT (Turkish radio) programs; Turkish newspapers; Turkish phono and video cassettes	
Prepare for/encourage return to home country	TRT programs; ARD radio programs	

2. Media use for particular purposes, especially for *information gathering;* and
3. Media use and media avoidance in connection with particular needs *(uses and gratifications approach).*

Aspect 3 will be discussed in a separate chapter in connection with the various media at the minorities' disposal.

Media Use As Leisure-time Activity

(Studies: (3, 14, 15, 20, 26)). Owing to differences in age groups, in categories, and in dates of inquiry, the results of these studies do not allow far-reaching conclusions. A further methodical problem lies in the fact that in some cases media use is tested within the larger categories of "indoor" and "outdoor" activities, and the shares vary considerably between nationalities: Turkish youths 71% indoor, Spanish youths 56% indoor; and between sexes: males 42% indoor, females 90% indoor (cf. (3), p. 48).

The ARD/ZDF study ((6), p. 466) gives the following most recent inter-media comparison (Table 3):

TABLE 3 — Comparison of media use.

MEDIA USE BY FOREIGN MINORITIES

	Total+	GR	I	YU	E	TR	D
Television	68	71	66	67	71	68	77
Radio	54	53	55	52	54	55	69
Papers/Magazines in German	31	31	39	42	36	20	76
Papers/Magazines in mother tongue	32	28	23	26	19	40	./.

"Daily reach" of different media (n% of national minority use this medium on an average day).

AVERAGE USE OF ELECTRONIC MEDIA DURING WORKING DAYS IN MINUTES

	Total+	GR	I	YU	E	TR	D
Television	112	102	112	114	109	113	119
Radio	91	97	85	110	93	83	136

+Total of foreign minority groups.

Cinema is used by 1% of the minority population on an average day, as is the case with the German population; and 14% read a book on an average day, which is slightly less than Germans do.

TABLE 4 — Preferences given to different content types in television.+

	Total	GR	I	YU	E	TR	D
News	55	63	53	64	65	49	90
Current affairs	21	30	17	31	25	16	43
Sports	45	45	45	47	52	42	44
Films, TV plays, Police films	61	60	62	60	64	61	59
Entertainment, Quiz, Music	54	58	53	61	62	52	59
Programs in mother tongue	68	74	51	65	65	80	./.

+The figures represent the percentage of those who at least occasionally view TV who view the program type regularly.

With regard to different program types, Darkow and Eckhardt ((6), p. 467) report the following preferences (Table 4):

Darkow and Eckhardt arrived at the conclusion that, apart from the role of the programs in the mother tongues, the minorities' viewing habits are not much different from those of the German population. I doubt this: first, the categories in some cases comprise rather heterogeneous individual programs; and second, the figures do not tell anything about the viewing situation. Of Turkish people, for instance, it is known that they tend to have the TV on with the sound turned off, etc.

Information Gathering

Non-systematic observation suggests that minorities and the German population differ in their uses of media for particular purposes. In particular it has repeatedly been claimed that minorities rely more heavily on interpersonal communication than Germans do. Evidence, however, is scarce. We know of three studies touching the problem of news flow. Two of them were carried out in 1968 among Italians and Spaniards in connection with questions related to the acceptance of the TV programs for these groups. We give the figures below (14, 15) and confront them with figures from a study on the media use of Germans (16), which up to a point are comparable. The question put to Italians and Spaniards was: "How do you come to know about events in Italy/Spain?"; the question put to Germans — ten years later — was: "Please name information sources on 'Political events in our country and in the world'" (See Table 5).

TABLE 5 — Information sources.

	Italians	Spaniards	Germans
Friends/Colleagues	56%	55%	./.
Letters from home	82%	63%	./.
Personal communication	./.	./.	19%
Radio	74%	89%	39%
Television	57%	86%	73%
Daily newspaper	./.	./.	70%
Magazine	./.	./.	11%
Specialized journal	./.	./.	4%
Daily newspapers/magazines	48%	80%	./.

There exist special education services for foreign children in Germany (MBSE). Amongst other things, a study of the Berlin Senate ((4), p. 70) tried to find out how far these possibilities are known. The figures show that more than half of the families whose children did not take part in these courses did not even know of their existence. Among those who did take part, more than half were informed by relatives, and more than one third by the labor exchange. Among those who were informed but did not take part in the courses, there are considerable differences between the nationalities: While about half of the Turkish group again relied on relatives and the labor exchange, there still remains a relevant factor — more than half of the Yugoslavs received their information from radio, TV, and newspapers. A comparison to Germans' information habits in the field of "job" shows that here, too, interpersonal communication is of major importance (41%) ((16), p. 136). It seems reasonable to conclude that information habits depend on various factors — two of them being national group and field of topic.

MEDIA FOR THE TURKISH MINORITY

For reasons of space, it is not possible to give a survey of the scope of media available to the Turkish minority in Germany. Instead, it might be useful to present a classification by which the various media can be discussed. In this, we shall draw upon the suggestion of Aktan (2), who distinguishes between media from the home countries and media from the host countries. Within each group, he differentiates between general media designed for the respective home population and special media designed for the minority abroad. We wish to add media from third countries, with the same subdivision. So, we arrive at six separate media groups (Table 6):

TABLE 6 — Types of media groups.

	Home Country Media	Host Country Media	Third Countries Media
General Media	1	3	5
Special Media	2	4	6

General Home Country Media
According to our information, neither Turkish radio nor TV programs designed for the population in Turkey can be received in Germany. This situation may alter with the introduction of satellite radio and TV. Turkish magazines and newspapers are imported into Germany at rather high costs. In about 200 cinemas in Germany, Turkish films can be seen; a change will presumably occur, since the video cassette market for Turkish films is rapidly increasing.

Special Home Country Media
Three Turkish daily newspapers (Hürriyet, Tercüman, Milliyet) issue European editions printed in Frankfurt (cf. (1), p.67; and (2), p. 80). The daily circulation of these three papers amounts to ca. 150,000 to 200,000. About one half of the content is identical with the home edition; within the European part, about one half of the articles deal with Turkish problems; reports on the minority lie between 17% and 33%. The shortwave program of TRT, "Tükiye'nin Sesi" (Voice of Turkey), broadcasts about 17 hours daily to Europe and the Near East ((2), p. 77). Its main purpose is to maintain the ties of the minorities to their home country; the news programs are identical with those broadcast in Turkey.

General Host Country Media
The latest figures concerning the use of these media are given above. Contrary to earlier studies and criticisms, there seems to be a tendency on the side of the minorities to make greater use of the general host country media. Increasing knowledge of German, longer residence in Germany, and uncertainty about the return date or intent to stay in Germany for an indefinite period may be counted amongst the determining factors. A particular problem is the *image of the minority* in the general host country media (cf.(7,8,10)). The two major complaints are that the minorities' image in these media is rather negative, and that the minorities are usually seen from a "German" perspective. Furthermore, there seems to be a tendency to distinguish between "good" and "bad" foreign minorities: in the late 60s and early 70s it was the opposition between legal and illegal migrant workers; in the late 70s it seems to be the

opposition between workers and people seeking asylum. Heine (10) points out that there is unequal treatment of positively depicted Southeast Asian refugees and others with a less favorable image.

Special Host Country Media

There is quite a number of periodical or occasional publications of organizations, ministries, and local authorities with practical information for minority members. A few German daily newspapers regularly provide special information for minorities. Those media with the widest distribution are the special programs on wireless and TV. On radio, there are 40 minutes daily in Turkish, Serbo-Croatian, Italian, Spanish, and Greek; besides, there are some regional programs. These programs are produced in Germany and are meant to serve three purposes: useful information for life in Germany, news service, and items to maintain the ties to the home countries (cf. (23), p. 25). On the third TV channel, the weekly program "Ihre Heimat — unsere Heimat" (Your home country — Our home country) is broadcast each Sunday between 16.30-17.30 h. Two blocks, one with a Portuguese, a Spanish, and a Turkish part, and the other in Greek, Italian, and Serbo-Croatian, are transmitted alternatively. The second TV channel (ZDF) broadcasts 45 minutes for each national group every fortnight. In both cases, the German channels do not produce reports on the respective home countries by themselves but take over material from their stations. This has led to some dissatisfaction in the past; the main objection being that the home country stations prefer to give tourists' impressions rather than information for their groups abroad. Although the figures of the studies are not strictly comparable, there are strong hints that the acceptance of these programs has been stagnating if not decreasing. During the last years, there have been some efforts to produce phono cassettes for the members of the minority; the titles comprise Turkish literature, practical information ("Let's meet in the town library," "Law for foreigners," "Information for baby care"), and minority literature; depending upon the subject, they are produced in Turkish, German, or in both languages.

Third Countries' Media

Both the general programs of the countries surrounding Germany and world services such as BBC External Broadcasting are receivable by the minorities in Germany. Additionally, there are said to be particular minority programs from Eastern European stations. How far these programs technically reach and to which degree they are used we could not find out with certainty.

CONCLUSIONS

Considerations concerning the functions of the host country media in the future ought to focus — to my mind — on the subsequent problems:

1. We have to take it as given that there are *two distinct publics* in Germany: the minority (foreign) population, and the majority (German) population; within these two groups we may distinguish further sub-publics (by nationality, degree of accommodation, outlook on life, etc.). The identity (or identities) of the several nationalities in Germany seems to be no longer simply one of, e.g., Turkish people temporarily abroad, but one of a national minority in Germany. We have, however, little information on how far this new kind of identity has developed.

Accordingly, the first question which comes to mind is whether the host country media ought to be *integrative* or *group media*. Integrative media are either indifferent to the separation of the public, or aim explicitly at integration. Group media may either address themselves to a particular minority as a minority in Germany, or to the minority as a part of the home nation. I should add that radio and TV in Germany are conceived as integrative media, while the press is indeterminate. The situation is similar to the one in Britain or Austria; it is different from the one in the Netherlands. It may, with the introduction of cable television, undergo considerable change. At first sight, it would seem sensible to say that those functions which maintain the minorities' relations to their home countries would best lie with the home country media. Looking at the above table of media functions, this need appears to be amply catered to by Turkish newspapers and the TRT external broadcasting service. Yet, the host country's radio and TV programs are predominantly sought for "contact with home" (cf. (6), pp. 469, 472), whereas the host country-directed functions of these programs rank lower. There may be various reasons for this: attractiveness of the electronic media, difficulties with the reception of TRT programs, cost of Turkish newspapers, etc., and — on the other side — little host country-directed content, higher credibility of other media, lack of interest, etc.

2. Altogether, we know too little about *which channels (media) serve which communication needs*. The studies relating to the problem of information gathering, or the conclusion that watching TV is the most popular leisure-time activity, or the figures giving the daily reach of different media are hardly more than rough hints. The rapidly increasing popularity of Turkish video cassettes seems to point to a need which is not satisfied by any other medium. But we do

not know whether cassettes are used instead of other media or whether it is a case of "the more, the more." And again, we do not know whether a particular section of the minority is attracted by video cassettes.

Apart from the fact that too little research has been done, there lies an additional difficulty in the methods available. Current social research methods may suggest, e.g., that the German local newspaper is of little importance to the minority at present; there might, again, be a number of reasons for this: lack of language knowledge, costs, too little relevance to the members of the minority, too little interest/acquaintance with the medium. Research can hardly predict whether more information for the minority in a local newspaper will result in an increased use of this medium. What is required, therefore, are experiments and projects which take into account that a change in media use habits on the part of the minority will require some time.

One reason for the stagnation in the popularity of special radio and TV programs may lie in the fact that the "second generation" with better knowledge of the German language (and decreasing knowledge of the mother tongue?) will turn to the general host country media; again, this is hardly more than a guess, since other reasons might be as important. On the whole, however, we ought to depart from the notion that any particular national minority is a homogeneous group — both in research and in future planning.

3. Members of minorities repeatedly object that the German general media project an *unfavorable image of minorities*. In some respects, this is not true: As far as radio and TV are concerned, you will rarely or never find a police film in which a foreigner turns out to be the criminal; current affairs programs meticulously try to avoid giving the impression that redundancy or housing problems are due to the presence of foreigners. On the contrary, there are quite a number of programs designed to fight prejudice and misleading information. It may even be argued that this kind of treatment of the minority could lead to what communication researchers call the "boomerang effect."

The area of factual reporting contains different problems: In one discussion, I recently encountered the opinion that the mere mentioning of the number of foreigners in Germany is apt to create hostile feelings against the minorities — regardless of intention. This is true, insofar as any new information will as a rule be individually linked with the existing opinions and attitudes. The request, however, not to print facts which might encourage adverse reactions remains problematic. A particular point of complaint in this respect is crime reporting. There is the empirically corroborated impression that criminal activities of members of the minorities are given undue attention; and there is the reproach that newspaper headlines tend to name the nationality of the criminal if he/she is other than German. No doubt, criminal activities of foreigners are more likely to gain attention by certain papers than do other problems related to foreigners. But this, to a certain degree, is the feature of a particular kind of press, which in any case will jump at spectacular events. Although the makeup of headlines has been criticized continuously and although many journalists acknowledge the complaint as justified, there seems to be little indication of change. Presumably, journalistic practices and habits are hard to overcome.

While the complaint of a comprehensive overt, anti-minority attitude in the German general media can hardly be maintained, there are some less conspicuous traits which deserve attention. A program with the intention of opening Germans' eyes to the disadvantages of minority life in Germany focuses on the more gruesome aspects of their living conditions: children playing between dustbins, overcrowded flats, etc.; it furthermore tends to show deplorable living conditions in the minorities' home countries to create an understanding for the foreigners' motivation for leaving home. I do not know whether these purposes of the producers are realized, or whether this kind of presentation provokes comments such as "They live here like they have been used to living at home; they deserve no better." But there is some evidence that this kind of depiction is resented by minority viewers, as it severely damages their self-image and their identity. Program makers seem to assume that the actual viewers of a particular program are exactly those at whom they aim their message; yet, this is unrealistic: programs *on* minorities — be they foreigners, disabled, or workers on the dole — will usually be programs *for* these groups who scrutinize the program as to whether it represents them properly or not. A good program must satisfy the divergent needs of information (for the German population) and of identity (for the minority).

4. There is another aspect to the treatment of minorities in the German general media: *Minorities* may appear as a *theme* of a program, but hardly ever as a *background feature* (cf. (23), p. 54). You will rarely find Turkish or Italian persons as members of the public in a TV entertainment show; in vox-populi-interviews, foreigners will hardly ever be questioned. While it has become the custom to have foreigners on the panel when minority problems are discussed, in discussion groups which deal with general problems such as "Comprehensive or grammar school?", foreigners usually are not represented. The same

seems to be true of fiction programs unless they are dedicated to a minority problem.

5. The last point leads to the problem of *access*. There is the general criticism that when minorities become the theme, it is the German perspective that dominates. The theoretical justification for opening up media to members of the minorities can therefore be derived from the consideration that a sincere and authentic representation of the views, aspirations, and experiences of the minorities can best be put forward by someone who personally shares these experiences; and this, as a rule, will be a member of the group. The question of access may be discussed on different levels. One of them, participation in an existing program under the control of professional media personnel, I have already mentioned. In a farther-reaching sense, access may mean that a group may take responsibility for producing a program item. This kind of access, of course, will not be better for national minorities than it is for any other group — and this is poor enough (cf. (19)); at present it is restricted to one ZDF program, but the introduction of cable TV with a larger number of channels available may bring about a change.

Access may also be discussed with respect to the representation of the minorities' members within the permanent staff of radio and TV stations and of newspapers. This is still very rare, but with the growing up of the "second generation" one may be optimistic to a degree. There is, however, one problem with minorities' members as journalistic staff: again, there seems to be a tendency to grant certain staff members exclusive competence for "minority questions" with the unwarranted result that this may give rise to "minority corners" in newspapers and radio and TV programs. This tendency has already become visible where German staff members have for some time dedicated themselves to minority problems. A better way, certainly, would be to have minority members in a variety of departments.

Finally, access means "power and ability to feed information into existing media systems." While German organizations have PR departments which flood the media with their press releases, the national minorities have nothing comparable. The foundation of a press agency of the national minorities has therefore been strongly suggested at a workshop held by the Otto Benecke Foundation last year. From my own experience I know that on a regional level this is as difficult as it is necessary.

6. Mass media can no longer confine themselves to the tasks of providing information to their clients and of entertaining them. Especially with the new

possibilities of regional and local electronic media, program makers will get involved in organizing activities around their programs. While local newspapers hold festivals and competitions (cf. (22)), representatives of the electronic media are rather reticent in this respect (cf. (13)). Certainly, to help minority members and Germans find personal contact is one of the tasks of mass media. Well-meaning words, however, will not be sufficient.

To conclude, I should like to stress one point: The contribution of mass media to the needs of national minorities in Germany can be more substantial than it is at present. It should, however, not be forgotten that the media and the people working in their institutions can hardly do more than operate within the given political framework; mass media cannot function as a remedy against contrary political decisions.

REFERENCES

(1) Acikalin, M., and Klitzke, D. 1980. Zeitungen und Zeitschriften. **In** Das Medienangebot für die Bevölkerung aus der Türkei in Berlin (West). Eine Dokumentation, ed. D. Klitzke, pp. 67–79. Berlin: Technische Universität.

(2) Aktan, M. 1981. Gastarbeiter und Massenmedien — Ghettosituation im massenkommunikativen Bereich. Migration **1.1:** 73–93.

(3) Bayern. 1980. Ausländische Jugendliche in bayerischen Großstädten. Ergebnisse einer Ümfrage in Augsburg, München und Nürnberg. München: Bayer. Staatsministerium für Arbeit und Sozialordnung.

(4) Berlin. 1980. Untersuchung der Determinanten der beruflichen Ausbildungsbeteiligung von ausländischen Jugendlichen in Berlin (West). Berlin: Senatskanzlei.

(5) Bubenik-Grüneisl, M., and Bubenik, A. 1975. Nicht stimmberechtigt. Medium **5.12.:** 13–15.

(6) Darkow, M., and Eckhardt, J. 1982. Massenmedien und Ausländer in der Bundesrepublik. Erste Ergebnisse eines ARD/ZDF-Projekts. Media Perspektiven **7/1982:** 462–473.

(7) Delgado, J.M. 1972. Die ‹Gastarbeiter› in der Presse. Eine inhaltsanalytische Studie. Opladen: Leske.

(8) Eulgem, B.; Kaczmarzik, R.; and Weisser, E. 1980. Meldungen und Filmberichte der 'Berliner Abendschau' über den türkischen Bevölke-

rungsteil in Berlin (West). **In** Das Medienangebot für die Bevölkerung aus der Türkei in Berlin (West). Eine Dokumentation, ed. D. Klitzke, pp. 107–135. Berlin: Technische Universität.

(9) Göktürk, E. 1981. Der mögliche Beitrag der Massenmedien zur Integration der ausländischen Arbeiter in der Gastgebergesellschaft. Migration **1.1**: 65–71.

(10) Heine, E. 1980. Ausländer in der veröffentlichten Meinung — Perspektiven einer Integration. **In** Autorengruppe Ausländerforschung an der FHSVR Berlin: Untersuchung von Straftaten, Problemen des Strafvollzuges und der Resozialisierung bei Ausländern in Berlin. Teil 3: Reaktionen auf Straftaten von Ausländern in Berlin, pp. 580–609. Berlin: FHSVR.

(11) Huth, L. 1981. Fremde unter uns. Die Ludwigshafener TV-Klausur. epd/Kirche und Rundfunk 20 (14. März 1981), pp. 1.

(12) Huth, L., and Gökce, O. 1981. Medien und ausländische Mitbürger aus Italien und den ehemaligen Anwerbeländern. Forschungs- und Diskussionstand. Linden-Leihgestern: ComCon.

(13) Huth, L., and Krzeminski, M. 1981. Zuschauerpost - ein Folgeproblem massenmedialer Kommunikation. Tübingen: Niemeyer.

(14) Infratest. 1968a. Die Bedeutung der Fernsehsendung des Zweiten Deutschen Fernsehens 'Cordialmente dall'Italia' für die italienischen Arbeitnehmer in Nordrhein-Westfalen. München: Infratest.

(15) Infratest. 1968b. Die Bedeutung der Fernsehsendung des Zweiten Deutschen Fernsehens 'Aquí España' für die spanischen Arbeitnehmer in Nordrhein-Westfalen. München: Infratest.

(16) Infratest. 1978. Kommunikationsverhalten und Buch. Endbericht. München: Infratest.

(17) Klitzke, D. 1981. Türkçe Video Kasetler — oder Das Geschäft mit dem mangelhaften Programmangebot. Migration **1.1**: 94–102.

(18) Klitzke, D. 1982. Türkisches Video hierzulande: Tendenziell ein Integrationshemmnis. Funk-Korrespondenz **30.37**: 50–56.

(19) Krzeminski, M. 1979. Vermittlung sozialer Erfahrung im Fernsehen. Eine Falluntersuchung zum ZDF-Jugendmagazin DIREKT. Berlin: Spiess.

(20) Mannheim. 1980. Bericht zur Situation ausländischer Jugendlicher im Alter von 14 bis unter 20 Jahren in Mannheim. Mannheim: Stadt Mannheim.

(21) McQuail, D.; Blumler, J.G.; and Brown, J.R. 1972. The television audience: a revised perspective. **In** Sociology of Mass Communications, ed. D. McQuail, pp. 135–165. Harmondsworth: Penguin.

(22) Otto Benecke Stiftung, ed. 1982. Ausländische Mitbürger in der Fremde daheim? Chancen der Massenmedien. Baden-Baden: Nomos.

(23) Rissom, H.-W.; Brüning, J.; Nax, W.; Renckstorf, K.; and Thissen, R. 1977. Ausländische Arbeitnehmer und Massenmedien in der Bundesrepublik Deutschland. Köln: Unesco.

(24) Röll, H.H. 1981. 'Nachbarn in Europa' — Eine Sendereihe des Zweiten Deutschen Fernsehens für ausländische Mitbürger und Deutsche. **In** Visodata 80. AV-Medien- und Datensysteme für Bildung und Kommunikation. Dokumentation, pp. 547–561. München.

(25) Stelzle, W. 1981. Das Ausländerprogramm der ARD. Auf dem Weg zu einem Hörfunk-Familienprogramm für nationale Minderheiten. ARD Jahrbuch 80: 55–67.

(26) Walz, H.D. 1980. Sozialisationsbedingungen und Freizeitverhalten italienischer Jugendlicher. Eine empirische Untersuchung in einer süddeutschen Kleinstadt. München: Deutsches Jugendinstitut.

Minorities: Community and Identity, ed. C. Fried, pp. 177–190
Dahlem Konferenzen 1983. Berlin, Heidelberg, New York, Tokyo: Springer-Verlag.

MINORITIES' INFLUENCE ON THE MAJORITY: REACTIONS OF THE MAJORITY IN POLITICAL, INSTITUTIONAL, AND SOCIAL SCIENTIFIC SPHERES

C. Wilpert
Institut für Soziologie, Technische Universität Berlin
1000 Berlin 10, F.R. Germany

Abstract. This paper introduces the theoretical question of minority influence on the majority. In the analysis which follows, emphasis is placed on the reactions of the majority to the minority even before formal demands or ethnic mobilization occur. Attention is given to the political, institutional, and social scientific spheres which, advertently or not, interpret the nature of the minority in question and provide an ideology which justifies the limitations set and the problems the minority faces.

INTRODUCTION

The question of a minority's influence on the majority addresses the complex problem of dialectics in social change. The terms "minority" and "majority" in this context signify social relationships of subordination and dominance. The majority determines the norm. Influence, of necessity, however, implies changes in this relationship and thus a change in the social-political reality, a step toward greater equality and less subordination. It is unlikely that a universally applicable theory exists which may either predict conditions for influence or define spheres of influence achieved.

At different levels these concerns are the field of the sociologist, political scientist, and historian, as well as the social anthropologist and social psychologist. I cannot claim to be able to integrate or incorporate the divergent and overlapping views of these disciplines, nor even to adequately approach some of the rich and revealing content of historical processes; therefore, I will limit my contribution to pointing out some of the theories and attempt to apply some examples as evidence of the nature of influence and reactions within the context of the Federal Republic of Germany.

SOME EXPLANATIONS FROM SOCIAL SCIENCE ABOUT MINORITIES' INFLUENCE ON THE MAJORITY

The various social science disciplines offer some theoretical frames of reference to aid us in understanding the influence of minorities and the reactions of the dominant group.

As Walzer (26) emphasizes in his historical analysis, national and ethnic pluralism has been the rule, not the exception. It was only in 18th and 19th century Europe that certain assumptions about democracy and cultural unity arose and became the basis for a new social order, preparing the way for the monocultural ideology of the nation-state.

Traditional sociological approaches to minority questions have also been grounded in the concept of nationalism and the assumed necessity of cultural homogeneity for the maintenance of national stability. Most of the theory and research in this area addresses the situation of minorities within a superordinate indigenous framework (14). The immigration countries of Australia, Canada, and the United States have been the prime references for these models. In the classical tradition, the minorities (i.e., immigrant groups) were assumed to assimilate progressively following certain stages of adaptation to the new society. Social problems expressed in terms of social isolation or deviancy were explained by the cultural differences of the minority groups. Contemporary sociological analyes of minority-majority interactions in both immigration countries and countries with a migrant labor population are more likely to assume that the structural opportunities in the receiving society determine the cultural processes (12). At the same time, this theoretical approach views labor migration within a framework of international stratification which results in the "Unterschichtung," an underclass of migrants, unable to break down the barriers to social mobility which create new anomic conditions.

Political economy also offers some partial theories related to the position of the minority in the labor market. In the internal colonialism model, the minority is a low-skilled, flexible reserve cadre which is kept in place by an alliance between the dominant bourgeois capitalists and the indigenous working class.

According to the interpretation of the split labor market theorists, a historical accident produced a correlation between ethnicity and the price of labor. The indigenous working class must oppose and fight its exploitation by the capitalists as a cheap source of labor, a by-product of which is the exclusion of minority workers from the labor market (4). The significance of these approaches for potential minority influence on the majority is the assumption that minority group formation is a result of the type of social division of labor which prevents the access of certain groups to social and legal rights, better jobs, job security and advancement, adequate housing, and education. This position of inferiority may be justified by the dominant group's reference to a deterministic ideology which explains the minority's inferior position as being caused by genetic factors such as descent or racial origin (18).

The social anthropological approach explains minority influence in terms of the process of setting and maintaining boundaries. Social boundaries between groups are formed as a reaction of one system to another. While Barth (3) emphasizes the significance of the boundaries of ethnic groups as categories of ascription and identification, Wallman (25) points out that the significant characteristics which differentiate the group have a subjective as well as an objective basis and may change according to the context and the need to maintain boundaries.

The persistence of an ethnic group in cases of contact with other groups (e.g., immigration, labor migration) implies the structuring of interactions governed by a systematic set of rules which allow interethnic social encounters in some areas and prevent them in others. In this way ethnic boundaries are constraints on the kinds of roles an individual is allowed to play. According to Wallman (25), each side may manipulate the assumed ethnic differences to achieve particular ends and to preserve itself as a group in opposition to the other.

A major social psychological approach roots minority group relations in the relevance the group membership has for the individual's social identity. Tajfel (23) notably defines this as that part of an individual's self-concept and its evaluation derived from membership in social groups which are salient to him. The concept of social identity provides insight into the reaction of the majority and the minority to each other. The significant difference is, however, that a minority member may not easily divest himself of his group membership which may determine and limit his social advancement.

According to this theory, the significance of the membership group is maintained through social comparison which focuses on valued group characteristics. Changes or reactions of the dominant group to the minority accordingly result from unfavorable social comparisons which occur when a "superior" group perceives "its superior status to be threatened by another group, or, when there is a conflict of values within the superior group" and the inferior status of the minority is no longer considered inevitable or legitimate (22). Once the legitimacy of the status quo is questioned, the minority may be open to mobilization and the majority may be split. Potential conflict may increase around the crystallization and rigidity of the social identity of certain segments of the dominant society.

Nevertheless, none of the above theoretical approaches are adequate in themselves to account for the historical moves of minorities toward influence, boundary shifts, and less subordination. For example, certain elements in the historical process which contributed to objective conditions, and in turn made the mobilization of minority resources possible, are overlooked. Some immigrant groups at certain periods were able to transform their manpower through unionization and coalitions in order to protect members of their own group. At other times, objective conditions allowed for the use of other resources — for example, through the ownership of property or territorial concentration a basis existed for "people power," political machinery, and ethnic politics; a world language or ties to supranational cultural or religious bodies facilitated relations with world powers of sympathetic neighboring countries. Economic resources were gained through ethnic enterprise. However, these examples, taken from the history of immigration countries characterized by minority diversity, do not necessarily provide a set of conditions applicable to contemporary minorities in countries with a predominant monocultural national ideology.

In the case of a migrant minority population confronting a strong majority, analysis of the role of the minority must take the pre-migration situation into account. That is, the basis for influence is rooted in the need for migrant labor or immigrants, which causes the dominant society to seek certain manpower outside its national borders. From the first encounter, boundaries are set in the form of selection and entry conditions and of limitations to social and legal status.

The boundaries set by the dominant group find their justification in attempts to construct a consistent ideology. In the process, presumedly unique cultural or national origins are used, just as race in the past, to define a social category which differentiates attributes associated with descent and cultural distance.

These attributes are then used to compare, evaluate, define, and explain the position and the lack of integration of the minority. In the following section some processes are illustrated with respect to the categorization of migrant worker minorities in the areas of policy formulation, indigenous education, and social services, as well as in the social sciences with primary reference to the particular German case.

SOCIAL CATEGORIZATIONS AND SOCIAL COMPARISONS IN THE AREAS OF POLICY, SOCIAL SERVICES, AND SOCIAL SCIENCE AS CONTRIBUTIONS TO THE IDEOLOGY OF BOUNDARY MAINTENANCE

Policy Issues

Immigration policy and legislation governing alien status are the decisive areas which give or restrict the access of migrants to the social and legal rights of the receiving society. In addition to the boundary-setting role of policies, other goals must be met. In the case of migrant labor, the needs of the employer for a dependable pool of labor has received priority (8). In attempting to justify differential treatment of certain subgroups, for example, toward citizens of the European Community countries, policies are contradictory while attempting to be consistent. In the American case, this same attempt is most clearly seen in the immigration quota system which upheld a long period of oriental exclusion, specifically of the Chinese and Japanese, together with the preferential treatment of North Europeans. World events and reactions to communism created contradictions in ideologies which led to the lowering of barriers for refugees from China. The criteria for differentiation shifted from race and color to political ideology.

Within the Federal Republic of Germany, such contradictions and reactions have arisen on a variety of levels since the process of "guestworker" recruitment was initiated, and it continues today. The original "guestworker" concept of rotation was a good example of strict boundary maintenance and the restriction of foreign entry purely to manpower needs. This attractive limitation of the foreign influx to short periods and limited rights became too costly both for the employer and for the state. The former lamented the retraining costs of on-the-job learning, and the latter was required to maintain an expensive apparatus of selection procedures in a number of labor-exporting countries.

Further contradictions arose during the ensuing so-called family reunification period. Although this period is often referred to as though it grew out of a conscious official policy decision, it resulted instead from the changing conditions of international labor migration. Among other reasons, it was due to the widespread demand for female migrant labor, the impossibility of lengthy family separations, and the stagnant economic conditions in the home countries. German economic concern about large remittances sent abroad rather than the goal of uniting families brought about a change in the allocation of family allowances for those children in the home country. Previously, family allowances were the same for all children of workers, whether in Germany or in the home country. In fact, family allowances for all economically active residents were raised in 1974, but non-European Community foreigners did not receive this bonus for children in the home country. In addition to this, when the family allowance regulation was changed, other factors were also operating to influence the steady inflow especially of Turkish children into Germany. Thus, this change in family allowances is often referred to as a political "mistake," since it is assumed to contribute to the significant growth in the Turkish population in this period.* True or not, it is evidence of the contradictions arising in the process of seeking an economically sound policy while striving to maintain boundaries and ideological consistency.

Another striking illustration may be found in the controversy surrounding the definition of family membership in the case of foreign workers from non-European Community countries (16). According to the West German constitution, the right of the family to be together must be respected. Yet in 1981 a decision was made no longer to allow the entrance of foreign youths aged 16 and over to join their parents in Germany. Previously, youths were recognized as dependent family members up to the age of 18, as was the case for natives and members of the European Community. These changes were particulary directed at the Turks; they, in the meantime, represented almost one third of the total foreign worker population.†

* In the same period, the fear of rising infrastructural costs led to stricter controls on the admittance of family members to "überlastete Siedlungsgebiete" (settlements with over 12% foreign worker population) (cf. (8), p. 307).

† Berlin was the first state to enact this new directive in November 1981, followed two weeks later by a cabinet decision of the SPD-FDP government. This is considered to have been directed at the Turks because, of the two largest foreign worker groups and non-European Community members, the Turks (32.4%) have considerably more

Recent social policy legislation in Berlin provides payments to mothers on the birth of each child. The purpose is to enable them to stay at home (and out of the labor market) after childbirth. Residents from non-European Community countries are purposely excluded. Behind this legislation is an attempt to counteract the German zero birthrate, while at the same time, other programs are especially concerned with the high birthrate of the Turkish mother.

These boundary maintenance features of official policy rest upon the rationale that certain groups, regardless of their formal fulfillment of previously agreed upon criteria permitting equal treatment, may no longer be treated equally because a higher good, the welfare of the natives, is endangered.

Conflicts of this nature may be found in educational policy as well. In Berlin, the formal policy of integration has produced a program of preparatory classes for foreigners taught by foreigners and including only a few hours of German instruction weekly. At the end of two years, pupils transfer to a regular class which again, due to the housing market and resulting residential segregation, is likely to be Turkish. Apparently liberal policies include their contradictions as well. The recognition of Turkish as a second language in the curricula of some secondary schools would appear to raise the status of the mother tongue, yet it may, in fact, limit the social mobility of those who may not enter into higher education without the necessary requirements of English and French.

The latest concern confronting school administrators is the existence of Koran classes for Turkish children outside the framework of the school. Since Islam is not a religion organized in the same hierarchic manner as is the Christian church in Germany, there is no official Islamic representative with whom politicians may deal. Currently consideration is being given to a change in the Berlin school law to accommodate the state's need to control the content of what is taught about the Islamic religion. If it is achieved, this would be a noteworthy example of the influence of minorities on the dominant society (7).

Interpretations of the Minority and the Emancipatory Goals of Social Workers and Teachers: Contributions to Boundary Maintenance

Social workers and teachers, agents of the welfare and educational systems in Western bureaucracies, have a double role which is significant for their

children than the Yugoslavs (14%). In 1981 they still had over one third of their children in the home country, and the majority of the Turkish children left behind were under 15 years of age (cf. (15)).

interaction with the new minorities and migrant families in these countries. While the social worker's job is to assist needy families and the teacher's to prepare pupils for the future, they also represent selection and control mechanisms. These particular roles and their direct contact with the minority place them at the interface between the majority and the minority. In the maintenance of boundaries they contribute their authoritative share to the selection of significant differences.

The experience of these social workers and educators is particularly salient for the reinforcement of the dominant ideology that cultural differentiation is a hindrance to integration and a justification of social boundaries. In Berlin both professions have been deeply concerned about the emancipation of young Turkish women and girls, and in public debate they have testified to the cultural conflict caused by the patriarchal structure of the Turkish family. Their witness was given wide publicity during the Fall 1981 debate about the limitation of rights of resident Turks in Germany to be joined by their newlywed spouses. As one social worker expressed it:

> "As soon as young Turkish girl reaches puberty her father locks her up in the house. He has a great fear of the loss of her virginity: this could either lower the bride price or mean that no marriage may be arranged for her. These children then sit home — without schooling, without training, without a youth group, or peers — until the marriage arrangement has been perfected" (10).

A similar associative feedback system arises with the question of the high birthrate among Turkish women:

> "As a married woman it's not any easier. They sit at home and suffer from homesickness from morning until evening. Her main occupation is the time demanding preparation of Turkish meals and giving birth to children. For a Turkish man a pregnant wife is proof of his masculinity. They've never heard of birth control. What isn't born is aborted. But a Turkish woman who knows the right address is an exception. Most of them do it themselves with a knitting needle and soap."*

Such a concern for the subjugation of these women and their ignorance about birth control methods provides legitimacy for their exclusion from motherhood allowances while justifying the development of birth control education for this specific group.

* Here Grunenberg (10) quotes a German medical doctor who was trying to organize a Turkish women's group. Similar observations may be found in other articles based on reports by social workers.

School teachers identify their role with the emancipatory goals of German education. Confronted with interpretations of the reactionary nature of Islam and its subjugation of Turkish women, the covered heads of elementary school girls symbolize an affront to emancipation and democrary. In a case where a teacher felt called upon to request a pupil to remove her scarf, the girl refused, explaining that it was required by her faith, whereupon the teacher replied, "You are in a German school" (5). Clearly, in the educator's view, elementary school is charged with a mission to promulgate accepted German values. Apart from the pedagogical problems posed by this incident, there is a more general issue: that is, the role teachers have in defining the particular values of German and Turkish society. In the public view their witness reinforces and justifies the necessity and "logic" of boundaries in terms of the significant cultural differences they assign to the group in question.

I have focussed on these interpretations of the values and lives of the minority primarily because of their use in justifying limitations of rights. These problem-oriented interpretations ignore the social conditions of the dominant society as well as the widespread interest among the minorities for good education and training. Recently social scientists have brought attention to similar phenomena with respect to immigrant minorities in other countries. These phenomena offer further insights into the categorization process and the justification it provides for boundary maintenance (9, 21, 28).

In a study of French social workers in Lyon, France, Grillo refers to the similar perception and categorization of North African immigrants as a kind of "folk" anthropology (9). Social workers define their mission in terms of the evolution of North African women. As he puts it, the social worker's outlook is based on a two-culture model. The transition from one to another is characterized as openness, evolution, becoming like the French. Finally, Grillo raises a question which brings us to the next section: that is, whether the views he has minutely examined "have been shaped by formal academic anthropology, or whether anthropology itself is shaped by the popular tradition, or rather by the assumptions on which the tradition rests and which anthropology shares" ((9), p. 86).

Social Science As the Verification of "Folk" Anthropology

Before these problems were raised with respect to contemporary European minorities, anthropologists and sociologists were delving behind the meanings given to race, culture, and values in science and in everyday life. The first generation of Mexican American social scientists in the United States critically reviewed the previous generations of anthropological and sociological studies and challenged the "cultural mystique" of these studies which they saw as preoccupied with the passive, fatalistic, and familial values of Mexican culture (5). The major flaw in these studies was the nature of the generalizations drawn from limited observations by American cultural anthropologists and used to explain the lack of intergenerational mobility among the Mexican Americans (13, 19, 24). In fact, one sociologist went so far as to attempt to explain the high delinquency rate of Mexican American youth in Los Angeles with references to overconformity to the cultural patterns of the Mexican American (11).

Cultural deterministic assumptions may also be found in some of the German sociological studies about the second generation and its future in this country. One study is particularly significant in the light of its wide distribution to educators, social workers, and social scientists (20). Its treatment of the culture-specific educational styles of foreign parents has been applied in the interpretation of other studies. In a statistical review of reported crimes in German cities over a five-year period, the increase in criminal acts committed by foreign and German youth according to age groups is compared. Since the increase in crime rate among foreign youth appears higher in certain age groups, the authors discuss some assumptions – however, one assumption made but not expressly stated is that foreign refers to Turkish youth, the largest foreign group. The authors also attempt to associate the above-mentioned study (which finds the cultural, authoritarian, and feudal origins of Turkish migrants especially in conflict with the primary democratic values of Western industrial society) with the frequency of juvenile delinquency among foreign youth.

According to the authors, of all groups Greeks and Turks differ most from Germans in respect to certain key values. These differences are said to relate to work attitudes, individual achievement, and mobility associated with value differences in family socialization, norms, and sex role attitudes. In this way a far-fetched connection between increasing rates of criminality among young Turks and cultural conflict is implied. In reality, the foreign group could not be distinguished according to nationality, nor was any attempt made to ascertain significant cultural values with respect to criminal behavior.* (1, 27)

* My own study found high mobility orientations and positive attitudes toward school and individual achievement among Turkish and Yugoslav youth (cf. (27)).

Such studies may be considered out-of-date.* However, the resemblance between "folk" anthropology and social science is noteworthy. Perhaps it is necessary to conclude that the boundaries between this commonsense interpretation, journalism, and social science are less rigid than those between the minorities and the dominant society. Of course, the difficulty of producing strictly objective social science is not new. The problem lies in its pretense to objectivity and the authority it assumes and feeds back to the dominant system (17).

In the German context, the term "foreign worker" usually means Turks, and a Turk, according to the above logic, implies a culturally distant, nonassimilable Anatolian Moslem peasant responsible for the cultural conflicts and resulting deficient socialization of his children. These boundary-legitimizing, social categorizations and evaluations associated with being Turkish, plus the size of the Turkish population, raise the question about the material basis for future boundary shifting in Germany with respect to this group. In order to approach an answer to this question three significant factors deserve consideration.

One concerns the long-range effects of transformations taking place in the labor market: whether the new Turkish minority in Germany will continue to be indispensable due to its substantial adaptability to the labor market and stabilize as an ethnic working class, or whether the next generation, due to social marginalization and changing labor needs risks becoming the prophesied "social-time-bomb." In the case of Turkish migrants, their stake both in the home country and in Germany, aided by the functioning of their families as cohesive units which are more capable of affording unemployment, may contribute to their endurance and flexibility (29).

The second factor has to do with the nature of the settlement process of Turks in certain urban areas and regions. Their dense concentration and use of social networks as sources of information, support, and solidarity may, while fostering greater reliance on original cultural resources, also symbolize resistance to Germanization and provoke further definitions of differences and reactions from the dominant society. A third factor resulting from the settlement process and the high urban concentration of Turks is the territorial basis provided by property ownership, ethnic enterprise, and ethnic media, since this is a potential starting point for greater minority influence.

* For example, in a study in Berlin on foreign youth, no significant differences in delinquency according to nationality were found (2).

The final outcome of these considerations are unknown, but in the meantime, kebab and köfti stands already mushroom outside ethnic areas challenging currywurst, pizza, and hamburgers.

REFERENCES

(1) Albrecht, P.A., and Pfeiffer, C. 1979. Die Kriminalisierung junger Ausländer. Befunde und Reaktionen sozialer Kontrollinstanzen. München: Juventa Verlag.

(2) Autorengruppe Ausländerforschung. 1981. Zwischen Getto und Knast. Reinbeck bei Hamburg: rororo aktuell.

(3) Barth, F. 1969. Introduction. **In** Ethnic Groups and Boundaries. Boston: Little, Brown.

(4) Bonacich, E. 1980. Class approaches to ethnicity and race. **In** The Insurgent Sociologist: Race and Class in Twentieth Century Capitalist Development, Special Issue, vol. X, no. 2, pp. 9–23.

(5) Brischetto, R., and Arciniega, T. 1973. Examining the examiners: a look at educators' perspectives on the chicano student. **In** Chicanos and Native Americans, eds. R.O. de la Garza et al., pp. 23–42. Englewood Cliffs: Prentice-Hall.

(6) *Der Tagesspiegel,* May 8, 1980. Schulverwaltung mißbilligt Tragen von Kopftüchern im Unterricht.

(7) *Der Tagesspiegel,* July 11, 1982, p. 14. Islamische Religionslehre jetzt auch Verfassungsproblem.

(8) Dohse, K. 1981. Ausländische Arbeiter und Bürgerlicher Staat — Genese und Funktion von staatlichem Ausländerrecht. Königstein/Ts.: Verlag Anton Hain.

(9) Grillo, R.D. 1982. Social workers and immigrants in Lyon, France. **In** Nation and State in Europe: Anthropological Perspectives, ed. R.D. Grillo, pp. 73–87. London: Academic Press.

(10) Grunenberg, N. 1982. Was tun mit den Türken? *Die Zeit,* January 29, 1982, pp. 3–4.

(11) Heller, C.S. 1966. Mexican-American Youth, Forgotten Youth at the Crossroads. New York: Random House.

(12) Hoffman-Nowotny, H.-J. 1973. Soziologie des Fremdarbeiterproblems. Stuttgart: Enke-Verlag.

(13) Kluckholm, F.R., and Strodbeck, F.L. 1961. Variations in Value Orientations. New York: Row, Peterson & Co.

(14) Lieberson, S. 1972. A societal theory of race and ethnic relations. **In** Sociological Perspective, ed. P. van den Berghe, pp. 38–51. London: Basic Books.

(15) Mehrländer, U. 1981. Situation der ausländischen Arbeitnehmer und ihrer Familienangehörigen in der Bundesrepublik Deutschland, pp. 341–355. Bonn: Der Bundesminister für Arbeit und Sozialordnung.

(16) Münscher, A. 1979. Ausländische Familien in der Bundesrepublik Deutschland. Familiennachzug und generatives Verhalten. München: Verlag Deutsches Jugendinstitut.

(17) Pitt-Rivers, J. 1973. Race in Latin America: the concept of "Raza". Eur. J. Sociol. **14**: 3–31.

(18) Rex, J. 1982. The political and economic context of race relations. Keynote Address to the Meeting of the Research Committee on Racial and Ethnic Minorities of the International Sociological Association, Mexico City, August 1982.

(19) Romano, O.I., and Octavio, I. 1968. The anthropology and sociology of the Mexican-Americans. **In** El Grito. A Journal of Contemporary Mexican-American Thought, vol. 2, no. 1, pp. 13–26.

(20) Schrader, A.; Nikles, B.W.; and Griese, H. 1976. Die zweite Generation. Sozialisation und Akkulturation ausländischer Kinder in der Bundesrepublik Deutschland. Darmstadt: Athenäum Verlag.

(21) Swetland, C. 1982. Pornography and the Moslem Woman. Paper delivered at the 10th World Congress of Sociology, Mexico City, August 1982.

(22) Tajfel, H. 1974. Social identity and intergroup behavior. Soc. Sci. Inf. **13(2)**: 65–93.

(23) Tajfel, H. 1978. The structure of our views about society. **In** Introducing Social Psychology, eds. H. Tajfel and C. Fraser. Harmondsworth: Penguin Books.

(24) Vaca, N.C. 1970. The Mexican-American in the social sciences 1912-1970. Part II: 1936-1970. **In** El Grito. A Journal of Contemporary Mexican-American Thought, vol. 4, no. 1, pp. 17–51.

(25) Wallman, S. 1978. The boundaries of "race": processes of ethnicity in England. MAN **13(2)**: 200–217.

(26) Walzer, M. 1980. Pluralism — a political perspective. **In** Harvard Encyclopedia of American Ethnic Groups, ed. S. Thernstrom, pp. 781–787. London: Harvard University Press.

(27) Wilpert, C. 1980. Die Zukunft der Zweiten Generation. Königstein/ Ts.: Verlag Anton Hain.

(28) Wilpert, C. 1981. Bedeutung von sozialer und kultureller Identität für ausländische Frauen und Mädchen. **In** Situation der Ausländerin. Beiträge des Instituts für Zukunftsforschung Berlin, ed. U. Welzel. München: Minerva-Publikation.

(29) Wilpert, C. 1983. Returning and remaining — contexts and dynamics of decisions to return among Turkish migrants in Germany. Research Commission on Migration of the International Sociological Association, in press.

Standing, left to right:
John Rex, Wolf Hütteroth, Diether Hopf, Lutz Huth, John Gumperz, Declan Quigley.

Seated, left to right:
Czarina Wilpert, Michael Hechter, Jack Brand, Alan Murray, Marianne Heiberg.

Minorities: Community and Identity, ed. C. Fried, pp. 193–217
Dahlem Konferenzen 1983. Berlin, Heidelberg, New York, Tokyo: Springer-Verlag.

MINORITY DEMANDS, MAJORITY REACTIONS?
Group Report

A.D. Murray, Rapporteur
J. Brand, J.J. Gumperz, M. Hechter, M. Heiberg, D. Hopf, W. Hütteroth, L. Huth, D. Quigley, J. Rex, C. Wilpert

ON APPROACHING A PART OF THE AGENDA ON HUMANE OPTIONS AND MINORITY GROUPS ENTITLED DEMANDS AND REACTIONS (REX)

There were serious difficulties faced by this group in addressing a part of the total problem posed in an overall workshop agenda which was preconceived and which presupposed a particular theoretical model of majority-minority or state-minority relations. Any attempt to deal with the overall agenda therefore must discuss these difficulties openly.

The model implicit in the general workshop agenda posited a situation in which a minority made demands upon the society in which it was situated and in which representatives of that society responded positively or negatively to those demands. In fact, the situations in which minorities, whether regionally located minorities or those resulting from immigration, find themselves are not usually of this kind. They face structural situations of a kind which other groups of this workshop will describe as non-humane and the actual minority-majority processes which occur are often of a kind which is too complex to be described under the heading of demands and reactions.

Certain central items on the agenda drawn up for itself by our group tended to merge into a general discussion of minority group structure and strategies.

These included the "capacity of the minorities to formulate demands," "the actual demands made," and the "strategies pursued by minorities and the possibilities of their success." This discussion was concerned with nonideal minority situations where "non-humane" majority policies were being pursued and various kinds of conflict between minorities and other societies resulted.

Further difficulties for this group resulted from the fact that although it commanded a variety of forms of relevant expertise, there was not any single shared paradigm within which problems could be posed even though there were several significant attempts made to offer and seek acceptance for such a paradigm.

These observations should not, however, be taken to mean that the group did not attempt to address itself to the "demands and reactions" questions as posed. It was able to agree on a provisional agenda, a number of discrete issues were discussed, possible approaches and paradigms were offered, and at least one attempt was made at a late stage to offer a general conceptual framework. There was, however, a considerable reinterpretation of the agenda, the paradigms and approaches as offered met considerable resistance, the specific issue discussions left many disagreements unresolved, and there was insufficient time to revise and gain agreement on the conceptual framework. Such developments are inevitable and should properly occur within any serious scientific work. What the detailed report which follows does offer is a basis on which progress might be made in the documentation and analysis of the problem in future scientific work.

INTRODUCTORY NOTES AND DEFINITIONAL PROBLEMS

The group recognized, before drawing up an agenda, that some crude attempts at definition should be made, and this was done on the basis of the range of interests of group members.

a) One group of participants was primarily interested in regional, national, or ethnic minorities within long-established states. In the European context, this group had members with primary interests in the Basques and Catalans in Spain and in the Scots and Welsh in Britain. Our working group also had members with interests in ethnic and regional minorities outside Europe in the Middle East (especially the Kurds and Armenians) and in Nepal. Discussion was limited, however, to minorities in Western industrialized countries, and these minorities are referred to in this report as *regional groups.*

b) A second group had interests in the industrialized European countries where immigration has been a relatively recent phenomenon, due essentially to labor shortages in these countries. These immigrant populations, because of the drying-up of labor sources within Europe, were drawn from further afield. The immigrant groups are differentiated from the majority group in various obvious ways, such as skin color, religion, dress, and by social relations based on a rural rather than urban background. Two categories can be distinguished in the overall category of *migrant groups*: the *immigrant groups* with a legal right of abode, such as most of the groups from the New Commonwealth in Britain, and the *guestworker groups* who, at least initially, were recruited on a temporary basis without a legal right of permanent abode.

c) Other participants were interested in minority communities in the United States, where the country itself is an "immigrant nation," where minorities are numerous and diverse, and the original community was weak and overwhelmed, and where the dominant Anglo-Saxon culture was itself imported. The other key factor in the United States situation is the presence of the black minority, with an original forced migration, the crucial experience of slavery, and a second migration leading to dispersal across the national territory. The special circumstances of this group are so significant that, although they share many features with regional and migrant groups, and both common features and differences were raised in the discussion, the *focus* of discussion was on the other groups. It was therefore decided to exclude this group from the report rather than to present a very superficial analysis.

There was also an emphasis on the importance of the wider sociopolitical context in which demands are made on the minority. An important factor in the decision to hold a conference on this sort of topic in this period of time is the questioning in Western industrial societies of the role of the state. There is no longer agreement about the right of the state to coerce its inhabitants into following certain patterns of behavior, and there is therefore considerable debate about *when* the state can use violence against its inhabitants, and *how much* violence it can use. The decline of a clearly-defined status in the community for individuals and the loosening of strong family ties are features of Western societies which weaken the hold of the dominant culture on individuals. There is a much wider range of cultural/sub-cultural choices available to all members of Western societies. These include religious affiliations, alternatives to the legally sanctified marital relationship, and the nuclear family (Heiberg).

In certain cases the state may attempt to interfere more than in the past in the lives of its citizens (e.g., demands made on Sikhs for the wearing of crash-helmets on motor-bikes, or demands made on Moslem girls to participate in mixed bathing in schools). In particular, the universal introduction of compulsory education, with its demands on all children to learn the national language and identify with the national culture, may be perceived by minorities as a threat to their identity (Hütteroth). If the state and its institutions are perceived by minorities as monolithic and even predatory, there is (a) a need for identity defined against the state, which the minority group can satisfy, and (b) a basis for demands which are grounded in the desire to continue customs and practices which are familiar (Brand).

Two riders to this view were registered, one noting that national and ethnic pluralism has been the rule, not the exception, until the 18th and 19th centuries; the other emphasizing that the Western nation-state has only recently become intrusive — even 19th century states tended to leave their ethnic and regional minorities to their own devices.

It was further recognized, both during the discussion and at the completion of the group's deliberations, that certain key issues were not discussed. The concept of group identity was not discussed, since this was primarily the concern of Saifullah Khan et al. (this volume). There was no debate about a theory of collective action, of which ethnic collective action might be a sub-category (Hechter and Gumperz). Most importantly, participants did not have the opportunity to present a "state of the art" perspective based on their research interests or theoretical concerns. The fact that the overall "goal" of the workshop was a normative goal, with implications for the policies pursued by governments, meant that the group did not feel it proper to spend time on an introductory debate on theoretical perspectives.

The group gained by these omissions considerable time for discussion of the topic, but it also lost in terms of clarity and mutual understanding. An agenda was arrived at, in the form of a series of questions, with very little disagreement, but replies to the questions, as is evident from the report, were given at many different levels. As a consequence, several participants were asked for written comments on parts of the agenda, which have been incorporated in the report at appropriate places. Where group members brought to the agenda an approach based on a theoretical perspective which clearly differed from those of other participants, or was formulated at a different level of abstraction from other approaches, the members concerned were asked to provide a brief statement of their position in the form of an *afterword* to the group discussions. These *afterwords* have been included as appendices to the main report, and it is suggested that the reader turn to the appendices before moving on to the next section (Summmary of Discussions). These summaries of positions should also be read in conjunction with the papers written by group members, especially those by Gumperz, Rex, and Wilpert (all in this volume).

The questions as they appeared on the agenda were: (a) What are the preconditions for the emergence of demands by minorities? (b) What are the conditions which determine a minority group's ability to make demands (collective action and organization)? (c) What are the types of demands that the minority makes? (d) What determines variations in these types of demands? (e) How do minorities strategically pursue demands, and what determines their success (majority and state reactions as constraints on success)? (f) What are the state's reactions to these demands, and what determines these reactions? (g) What are the criteria by which we can describe state reactions as humane?

SUMMARY OF DISCUSSIONS

Preconditions for the Emergence of Demands

This part of the discussion centered more on immigrant minorities and migrant workers than on regional minorities. The discussion focused on (a) the conditions and contexts of the immigration process which determine differences in the distribution of resources and rights, (b) the validity and reality of multiple identities, as well as (c) the situations which ascribe identities and provide a new basis for the organization of identities, and (d) the combinations of original networks and new experiences which influence identity-choices. Finally, the question was raised as to why ethnic instead of other organizational choices were selected by minorities in order to achieve goals.

Attention was drawn to certain universals of the immigration process as it has occurred in the industrial West since World War II (Rex). Immigrant workers were recruited or made their way as a response to labor-shortages, and in the majority of cases arrived as single men. They therefore have certain basic human needs — for jobs, the right to stay in the country, the right to house their wives and children, medical care, police protection, education — which they share with the working population of the home country. The satisfaction of some of these needs is periodically or permanently blocked, and it is at these times that demands emerge which lead to collective action. Further types of needs were added by the group: (a) the need for status, and where this is imposed by the majority, or under attack by the majority, different demands may occur; and (b) certain cultural needs by an ethnic group, and when these are threatened, there may be mobilization around demands. Certain needs (e.g., halal meat for Muslims) are constituted by group membership. Furthermore, it was pointed out in this connection that many minority group members tend to create new networks of relationships in the host country, which, although they build on commonality of background, nevertheless reflect goals and values that are quite different from those of the home country. Such networks, which are very important among Chinese and Koreans in the United States, for example, tend to center around religious or voluntary organizations. They serve to provide essential social and economic support in the host country and to compensate for existing economic and social inequities, in addition to serving cultural needs (Gumperz).

In discussion of the relationship between needs and demands, the concept of multiple identity was introduced ((Gumperz), cf. also Wallman, this volume). A Yorkshire Pakistani might also think of himself as a Muslim or as a Mirpuri — depending on the context. It was argued that it is the context which is the significant determinant of the emergence of demands and the types of demand which emerge. The context itself is determined by social relations — whether occupational, religious, of kinship, or other — and it is what is significant in those social relations in respect to the majority group that determines the demands that are made (either by the majority on the minority, or vice versa).

It was added that the ascription of identity by the majority group may lead to a change in the significance individuals apply to their own identity, e.g., Kurds in Germany are ascribed Turkish status, though they see themselves an oppressed minority in Turkey. It was pointed out (Sondhi) that some groups may *use* the ascription at particular moments to mobilize around an issue which affects a number of minorities (e.g., the 1982 change in the British nationality laws).

Much of this initial discussion focussed on the formation of minority groups rather than on the conditions which lead to the emergence of demands. Some preconditions for emergence were advanced: (a) the existence of a differential distribution of resources between the minority and the majority group; (b) a shared, historically-based cultural experience among the group; and (c) a shared experience, normally of deprivation, in the host country. The question was also raised as to why the immigrant worker joined an ethnic organization rather than another organization. One reason emphasized was that the members of the organization in question may not share the worker's particular form of deprivation (for example, white trade unionists in Britain did not prosecute the demands of Asian workers effectively because of divisions in the labor market, and the *I.W.A.* emerged as an alternative pressure group) (Rex). Furthermore, organization to achieve goals is facilitated by the basis language, shared background, and shared symbolic values which provide for communication. Shared language in itself is a rich resource for mobilization, but the shared values brought from the homeland and the commonly owned symbolism (in the broader semiotic rather than narrowly linguistic sense) provide an additional means towards achieving solidarity in order to push demands.

The Conditions Which Determine a Minority Group's Ability to Make Demands
Discussion was initially on migrant groups, and some theoretical perspectives were proposed. It was suggested at the outset (Rex) that there was some similarity between the problem of the formation of minority groups and the formulation of demands by these groups and the Marxian problematic of class-in-itself and class-for-itself. This problematic, however, involved the difficulty that the creation of social bonds and ideologies and goals is conceived of as a rationalistic and artificial process. When these bonds, etc., already exist, they preempt the processes which are rooted in the notion of the formation of a class-for-itself. Amongst these preemptive bonds none is more important than the bond of ethnicity. In the case of immigrant groups, the basis for the formation of social bonds is provided by kinship and other structures in the homeland. These bonds do not disappear in the immigrant situation, but they are seriously fractured. In these circumstances such factors as patron-client relations, the associations of immigrants, religious and political ideologies, and structures serve to reunite and reorganize the immigrant community. Groups bonded in this way may be thought of as "making demands" and, insofar as they are not met, as organizing for conflict and/or negotiation. In some circumstances the group may simply be seen as illegitimate and may itself regard the representations of authority in the society as illegitimate, in which case organizations for a power conflict may occur independently of making demands.

An alternative formulation proposed (Huth) was that choices take place which are based on loyalties within the group, and especially in relation to the situation faced by the group. If people's self-image is denied or attacked by the majority, then the community draws together; it mobilizes to protect itself from attack, e.g., in England groups have remobilized after anti-immigrant speeches by leading politicians, after demonstrations by an extremist organization such as the National Front, or following police mobilization against a local community, as in the Southall riots of 1981. It was added that it was at this point, once the immediate threat is diminished, that it becomes more difficult to mobilize the group, and splits in the leadership often occur. The urgency for solidarity diminishes, so the opportunities for division over tactics are greater (Heiberg). A further example of attack on minorities (Wilpert) was in the 1970s in both Britain and Germany when changes in the law occurred which removed previous rights and resulted in ethnic mobilization.

The focus of debate shifted to an attempt to grapple with the question: *what are the variables which determine a minority's ability to mobilize around demands?* Why does one group organize around a demand, and another not in a similar situation? Why are minority groups inert at one moment and active at another? In particular, were there alternative solutions to the free-rider problem? (Hechter)

An offer was made of a preliminary list of major variables in answer to Hechter's questions. These were (a) the real nature of the situation faced by the group (the extent of deprivation, hostility from the majority, etc.), (b) the existence of a power-base, and (c) the likelihood of success (Wilpert).

Some examples were added of the histories of enthusiasm in minority group movements, their build-up, their peaks, their down-turns in terms first of Hechter's theory. People (in, e.g., political movements) may initially perceive their rewards as highly worthwhile, support may then peak, leading to the achievement of some goals (e.g., the leaders gain seats in government), but support may then fall away, as the individuals in the group perceive that their own rewards are minimal. There may also be situations where the enthusiasm can be generated around a set of demands in which the nature of the demands is less important than the psychic rewards of participation in the struggle. It was added (Brand) that it was an observable phenomenon in the context of Scotland that people joined the Scots nationalist bandwagon because it was an interesting, exciting thing to do. In opposition to this view, it was argued (Hütteroth) that in most parts of the world individual obligations to the group are fixed and immutable; the free-rider problem is limited to the most prosperous Western societies.

Types of Demands

Three sets of variables which influence the types of demands a minority might make were advanced:

1) types of minority — migrant (immigrant or guestworker), regional/national, the special case of blacks in the United States.
2) types of political arrangements — centralized or decentralized — nation-state, federation or co-federation.

3) types of needs — legal (especially laws affecting immigration, nationality, and the family), political, economic (whether specific to the minority or shared with sections of the majority), and demands relating to language/culture maintenance.

Discussion covered issues in each of the above sets of variables but was organized primarily according to the types of minority.

Under regional/national minorities, it was accepted that political demands, which could be classified according to three general categories, predominated. There were also demands organized around equality — equal treatment and equal opportunity for members of the minority, together with the right to continue a separate cultural tradition (language, religion); for separation, amounting to secession to form a separate state; and between the two for some form of regional autonomy, ranging from a regional assembly with limited powers to a separate statehood within a federation.

The oscillation which occurred in some groups, as in the nationalist movement in Europe between political and economic demands was pointed out. In regional, as opposed to immigrant, minority groups, it was possible that a group might make demands in order to maintain a position of privilege (the case of Catalan agitation against the distribution of taxes on regional industries to other, less industrialized regions, was raised). The tremendous flexibility of a well established group was pointed out; such groups commonly have demands in all the four categories indicated above and will switch from one type of demand to another according to political conditions. The same process may occur within a particular category, so that a group may move from a political demand for home rule to a demand for independence and back again over a period of time.

The question of the significance of class differences in relation to regional groups was raised (Rex). The bourgeoisie may procure the leadership of nationalist groups in pursuit of opportunites of advancement, while the working class provides the ranks of the movement but may have different demands. Examples from Spanish and British nationalist groups were advanced to support the hypothesis. Alternatively, the members of a bourgeoisie in a minority may become aware of exclusion from possibilities of political power and may join the minority group bandwagon in order to gain power for purposes of economic gain (cf. the French entrepreneurs who attacked the English control of the business environment of Quebec through gaining political power which gave them control of the bureaucracy and thus the opportunity to use state capitalism to the benefit of their class) (Zolberg). It was pointed out that this demand can be described *either* as a demand for institutional change *or* for mobility and economic opportunity.

The tendency of regional groups to attempt to maximize their constituency by raising demands which would attract wider support both from within the minority and from sections of the majority was noted (Brand). Several such European minorities have adopted the rhetoric of class conflict (it was pointed out that the reverse has occurred, as when the Communist Party in Scotland adopted the nationalist banner). Other European regional groups have made ecological issues central issues among their demands. Against these examples, the case of the Welsh nationalists was cited. They took a deliberate decision which had the effect of limiting their constituency by focussing on a language demand which affected only the 25% of the population who spoke Welsh.

In turning to demands made by migrant groups, the primacy of demands concerning legal status was advanced (Wilpert, Rex). Although the correspondence between needs and demands was questioned, particularly in relation to needs as perceived by the majority (Huth), the right of residence and right to legal status was a major demand among all such groups. Differences also exist at the level of equality of such rights, as in the case of Germany, where the definition of the family unit as including all the children up to the age of 18, with the rights to family unity which this implies, is no longer applicable to the same extent to the dependents of all migrant worker groups. This right was recently withdrawn for those children from non-European Community countries who are 16 years of age or older. This most notably affected the two largest populations, the Turks and Yugoslavs. Currently, consideration is being given to lowering this entry limit to 6 years (Wilpert).

The importance of language/culture demands was emphasized, particularly where the group originated from a society with a strong religious and cultural heritage. In cases such as Algerians in France and Turks in Germany, relatively frequent movement between host countries and home was possible. In other cases, such as Punjabis in Britain, there was long experience of diaspora in many parts of the world. The persistence of the "myth of return" was stressed as a feature common to all groups, extending in some cases (especially where maintained by marriages arranged through the home country) to the second or third generation.

A plea for the recognition of the complexity of such demands was entered (Gumperz). Demands for language maintenance were symbolic demands subject to many levels of interpretation, in relation both to the group itself *and* to the majority's response. A demand for bilingual education may emerge, but the real problem may be built-in discrimination, by the institutions of the majority culture, in school curricula, methods of examination, selection, interviewing, etc. Majority group interpretations of the demands and formulations of a response may miss the real point, and solutions may be put forward which relate to quite different problems.

It was recognized by the group that there was a need for further research as a precondition to fruitful discussion of demands. In particular there was need for clarification in these areas: a typology of demands; a typology of circumstances leading to demands; a typology of structures of movements. At this point, a framework for organizing the discussion of demands and reactions was advanced (Hopf), which is summarized in the concluding section.

Strategies of Minorities to Pursue Demands

Two broad categories of strategy were proposed: (a) extralegal, which could be further divided into violent and nonviolent strategies; and (b) legal strategies, within which there could be variations between groups which used the existing channels of communication within the state, groups which built up powerful organizations in order to achieve demands, and groups which found ways of fulfilling demands from within their community networks or postponed the fulfillment of their goals until return to the home country.

Regional groups were discussed; the interplay of social and political conditions within the groups and in the majority society were emphasized as a determining factor on the choice of strategies (Heiberg). In the case of the Basque movement, a correlation was indicated between the use of violence and the social and political conditions. The option of violent extralegal action was pursued (a) when the state issued a blanket refusal to negotiate on demands and used force to reinforce its refusal (Franco period), and (b) when the demands made (independence, revolution) were of such a character that they could never be met through any accepted democratic procedure although they received firm support from a minority of the population (Post-Franco period). It was also pointed out that even those who rejected violence as a means of pursuing demands could, nevertheless, provide violence with an implicit mandate if the existence of armed struggle was used to press for greater concessions from central authorities than would otherwise have been possible.

The factors conditioning the choice between legal and extralegal action were examined; if there is a symbiotic relationship between the moderates and the terrorists, why does anyone choose to be a terrorist, since there is extreme risk involved and little likelihood of reward? Three factors emerged from the ensuing discussion: (a) the existence of a historical tradition of violence in the pursuit of group demands in the society; (b) the influence of Third World national liberation movements, where success had been achieved through violent group actions; and (c) the availability of Great Power interest in and support for violent movements in societies with ethnic/regional divisions.

Further arguments were advanced that individuals joined violent movements for reasons of excitement; when typical routes to establishing a sense of identity in society, through a job or career, are blocked because of high unemployment or discrimination in employment, alternative contexts for establishing a very strong sense of identity in violent group activity become attractive. This may be especially attractive in certain periods of the life cycle of groups, e.g., the greater willingness of young men to take risks.

Governmental Reactions to Minority Demands and Some Criteria for the Evaluation of Reactions As Humane

As became clear in the previous sections, governmental reactions as well as demands and strategies are dependent on a number of factors. The factors already outlined as influencing the types of demand also influence the types of reaction. These points will not be repeated. In addition to these factors the following may also play a role: the nature of the demand, the ideology of the state, the amount of pressure from interested groups within and outside the country (e.g., pro-minority groups in the majority, the home country, international organizations), strength of host culture and its institutional manifestations, (e.g., bureaucracy), and socioeconomic conditions for the majority group in the country concerned.

In attempting to classify governmental reactions to demands, the following broad possibilities appeared: fulfillment, denial, attempting to understand (e.g., (a) opening channels of communication, (b) setting up research projects, (c) setting up commissions (or model projects), reinterpreting demands, (sometimes on the basis of a, b, and c above, but also, e.g., with respect to riots as an indicator of racial-social tension, commissions may be set up to investigate riots which result in more effective police control rather than attempting to deal with the source of the problem), penetrating the community in order to co-opt the leaders or divide the group, initiating affirmative action or positive discrimination.

In addition to these reactions many governmental strategies may not be in reaction to demands of the minority expressed as such, but arise from the desire of the majority society to control the minority, e.g., the creation of an institutional legal framework to deal with a minority religious group.

In spite of the difficulty of a detailed treatment of specific demands and governmental strategies, some attention was given to the criteria for the evaluation of these strategies. The strategies of governments may range from oppression of the minority group to the full inclusion of the minority in the participation process. In this sense, humane options would be those which allow a minority maximum opportunity for participation in political decisions, especially in those which affect its own fate. Some examples were given with respect to immigrant minorities from the areas of housing and multicultural education. Housing programs which force dispersal are as negative as segregated housing if the minority is not free to participate in the decision-making process.

SUGGESTED FRAMEWORK FOR EXAMINATION OF OUR TOPIC

An attempt was made to organize some of the more important topics discussed in the group and to introduce some hypotheses on the interrelationship between the aspects (variables) mentioned (Hopf).

There was discussion of *needs* (N) of minorities or members of minorities, especially in the section on "Preconditions for the Emergence of Demands." Examples were given from a number of countries of *barriers* (B) which impeded the fulfillment of the needs — barriers usually set by the majority or by the conditions under which the minorities are living. Barriers were seen as having at least two functions: (a) they lead to *demands* (D); and (b) where there are specific barriers to members of a minority, they can work in the direction of bringing the individual members of the minority closer together. Thus three aspects emerge:

N B D

If there are no needs or if there are needs but no barriers, demands may be absent. But this holds true only in the case of demands being made by the minority or individual members of it. There are also demands made by the majority. These may be made in the belief that they will be to the advantage of the minority, or may be reinterpretations of minority demands (e.g., if there is a belief that the members of the minority are not conscious of their situation). So, demands are subdivided into

D < by minority / by majority

and consideration must also be given to the agents which put forward demands, whether individuals or organized/unorganized groups.

Barriers can also be subdivided, e.g., (a) barriers which have the effect of denying the fulfillment of minority needs (e.g., non-resources, non-autonomy, non-self-fulfillment), and (b) barriers which are pressures to do certain things, (e.g., speaking the language of the host society, joining the religious congregation of the majority, etc.). Furthermore, the intensity of barriers or pressures has to be taken into account (e.g., very strong barriers may lead either to the non-emergence of demands or to the espousal of different (e.g., violent) means of trying to fulfill the demands).

The three aspects could therefore be further subdivided thus:

N	B	D	
-basic	-denials	-by majority	-by individuals
-cultural	-pressure for change	-by minority	-by spontaneous groups
-new	-intensity high / low		-by organized groups

There are also distinctions between types of *minorities* (M) (immigrants, migrant workers, regional minorities, etc.). For each "type" the different aspects mentioned above must be taken into consideration. In addition, *frame factors* (F) have to be taken into account (e.g., the legal situation, the economic situation, etc.), and these ought also to be divided into groups in order to make it possible to interpret a given demand and to discuss reactions to those demands. The suggested framework would then become:

| M/F || N B D || |

Finally, means or *strategies* (S) of reaching the intended *outcome* (O) were discussed, as well as the *reactions* (R) of the majority to the demands and/or the strategies (e.g., there might also be rections of the majority which did not arise in response to demands).

The final form of such a framework based on our discussion might look like the following:

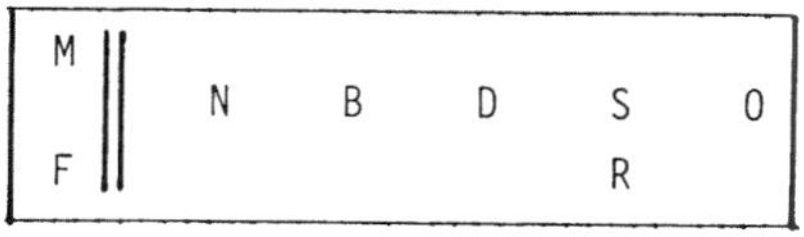

But this can by no means be considered as a one-way, unidirectional process: backwash effects of many kinds have to be taken into account. In addition, the framework looks much more rational than are many of the processes described, for example, there may be contradictory demands within a minority.

CONCLUSION AND AFTERWORDS

The framework advanced by Hopf and described above was welcomed by the group as an attempt both to summarize the group's discussion in a coherent form and to suggest a model by which further discussion of the topic "Demands and Reactions" might be organized. But although the framework met with no serious opposition, there was insufficient time for adequate discussion of the framework itself or of how it might work in practice with any of the examples of minority group demands raised during the group's deliberations. It is offered not as a neat conclusion to the work of the group but rather as an indicator of the complexity of the topic and of the number of variables which must be considered in the continued debate among scholars and practitioners in the field.

Indeed, there can be no clear conclusion drawn from the work of the group other than a recognition that minority group studies are still at an early and controversial stage. Much work is required both at the level of detailed studies of minority groups and the emergence of demands and at the level of the more theoretical debate about acceptable models of minority-majority relations. The problems which occurred in the early part of the group's discussions because basic terms — group, culture, identity, ethnic, network, demands — were used by different group members (from different disciplines) in different senses are perhaps to be expected in any interdisciplinary discussion. But they were highlighted because work in minority studies is still pioneering work, in a subject area in which the political and moral value-load is very high, whatever the discipline of the scholar concerned. In such a context, the exchange of ideas across boundaries — whether between nations or disciplines — was recognized by all participants to be essential to future work in the field.

As a first step towards clearing some of the ground it was agreed by the group that the conclusion to our report should be a series of "afterwords" in which some of the main theoretical perspectives which members brought to the discussion would be briefly summarized. Some of those perspectives are expounded at greater length in papers to be found elsewhere in this volume (see especially Rex, Wilpert, Gumperz, Hopf, and Huth), but the afterwords were written with the benefit of hindsight and relate very directly to the discussions summarized above.

Afterword 1 (Hechter and Gumperz)

There has been a tendency in this workshop to view the issue of minority demands and reactions in political terms as arising from the presence within a host society of preexisting collectivities that impose common attitudes and interests on their members. Although it is agreed that individuals within such collectivities differ in terms of such factors as economic resources, occupational status, educational attainment, and degree of assimilation to majority culture, it nevertheless tends to be assumed that minority issues and problems of intergroup relations are best approached by ignoring these individual differences and concentrating instead on the histories and shared characteristics of particular groups and on the differences among them. Such an approach has difficulty in accounting for aggregate behavior whenever individuals have a choice about the extent to which they identify with their group or a choice about the possible groups to which they may belong. The individual's freedom of choice is maximized under the conditions of rapid socioeconomic change and breakdown of traditional group sanctions characteristic of the urban industrial setting, where individuals have multiple options for group identification. The consequence of the possibility of multiple group affiliations is to alter the significance of the group to its members. In this kind of social setting, therefore, the tie of the individual with the group must always be perceived as problematic.

Analysis of intergroup phenomena such as those we were concerned with at this workshop can therefore not be cast solely in terms of group structure or values. The individual has to be the starting point. This does not mean that sociocultural background or particular state policies have no significance for the analysis. On the contrary, these are essential parameters that determine an individual's options, but they must be treated as constraints on choice and studied in relation to particular types of goals.

Our perspective has the following implications for the first two items on the agenda:

1) Group demands are predicated on the empirically demonstrable existence of commonalities in individual life experience. In the case of minority groups this is determined by the constraints that society imposes upon the individual members, e.g., the extent to which group membership implies limitations on the individual's choice of occupation, residence, and access to public facilities (most importantly education). These structural preconditions generate individual demands.

2) Despite the existence of common demands, participation in collective action cannot be taken for granted (because of such phenomena as the free-rider problem) but must be elicited by rewards and sanctions. Some analyses focus on the notion that elicitation depends on the existence of a shared rhetoric (or ideology) upon which appeals to collective action may be based. Others tend to place more emphasis on material rewards and sanctions as mediated through various monitoring mechanisms and on organizational structures. These two approaches are not at all in conflict.

3) Listing the demands of various groups is unprofitable, for there is no inherent limit on them and the meaning of these demands can only be interpreted in terms of the structural conditions that give rise to them. Similarly, variations in types of demands are partly a function of changes in structural conditions, partly a function of group solidarity, and finally, partly a function of the expected costs and benefits arising in the particular historical situation. At the most general level, these are determined by the nature of the system of social control (is there an independent judiciary; is the state otherwise restrained in its treatment of dissenting individuals; etc.). The same kinds of factors weigh heavily in the strategic formulation of a group's demands.

4) It follows from this that state reactions are humane not only to the extent to which they act to maximize group rights as such but also to the extent to which they maximize the individual's rights to choose among available options irrespective of sociocultural background.

Afterword 2 (Heiberg and Quigley)

Ethnic groups — be they territorially based or of immigrant origins — have become socially and politically important in the modern world mainly because of certain fundamental shifts inherent in the process of modernization itself. In the economic sphere this shift is linked to industrialization and urbanization, in the political one to state centralization and functional expansion, and in the social (and to some extent ideological) sphere to the transition from ascribed to achieved status. These shifts have thoroughly disrupted previous social formations and dislocated and dispersed the individuals within them. One of their consequences has been a radical rearrangement of the raw materials of the societies affected which has resulted in new economic, political, and, very importantly, ethnic alignments and cleavages emerging.

Another consequence is that a taken-for-granted sense of shared culture between co-residents of a society is no longer the case. Now, culture must be viewed as existing on two interrelated levels. The first comprises the visible surface elements of language, dress, folklore, ritual practices, and so forth. These ar the elements which ethnic groups tend to elevate into the emotive symbols of their cause and solidarity and to instrumentalize in a competitive struggle for material and political gains. However, the second level is culture in a deeper hidden sense. Here culture can be defined as a set of shared meanings, as a series of cognitive categories. On this deeper level culture must b be seen as a code, albeit at times ambiguous and contradictory, embedded in the minds of the members of a given culture through which experience is ordered and acted upon. It is the existence of this set of shared meanings that makes an ordered and *understandable* social life possible.

This implies that ethnic or cultural groups are critically different from other types of collectives — e.g., trade unions or political parties (and this helps to explain why ethnicity has historically been a far more effective mobilizer than, for instance, class). In an intricate, frequently tenuous, but nevertheless fundamental way, individuals tend to be linked into these groups through the basic cognitive structures of their minds. These structures determine perception and therefore are critical for the determination of an individual's goals, wants, and values.

This does not mean that individual behavior is rigidly fixed by culture. But it does mean that (a) individuals do not invent their culture entirely, and (b) this has very important implications for the understanding of ethnic action and demands. This approach differs from that of Gumperz and Hechter in a number of ways.

The contention that individuals act to maximize on the basis of rational choices appears to be doubly redundant: first, there is no agreed yardstick of rationality on which levels of maximization could be measured, neither for means nor ends. Second, the theory seems to hinge on the notion that "structural preconditions" generate individual demands. If maximization is a universal constant (which begs all the questions), the level of explanation then shifts to that of social structure. This is because the question changes from "why do individuals make certain choices given certain structural preconditions?" to "what are those conditions and how do they constrain individual behavior?"

In terms of an ethnic group's "demands and reactions," rational choice theory is unable to confront the following problems: (a) the specificity of ethnic groups in contrast to other groups, (b) the nature of the linkages between individuals in these groups especially with regard to the ritual and symbolic sphere, (c) the factors which underlie individual perception of wants, aspirations, and values, and (d) the cultural, economic, and political processes through which these groups are shaped and in which they operate.

The first demand of an ethnic group is to be recognized in terms of its self-proclaimed ethnicity. This is not as trivial as it appears. In most historical societies identity has *not* been expressed through ethnicity or shared culture, it has been expressed through kinship or ties which are essentially economic or political. In such societies ethnicity is not an issue. Common culture is either assumed — as in relatively self-contained communities — or ignored — as in predatory political expansionism. Ethnicity as a sociological problem, as a political reality, and particularly as a moral dilemma (what to do with minorities?) is essentially a recent phenomenon, and it was no accident that our group restricted ourselves to modern, Western, democratic societies.

Afterword 3 (Rex)

The work of this group has been bedevilled by a conflict of theoretical paradigms. This conflict, however, rests only partly on the incompatibility of the paradigms. It also results from the fact that theories have been formulated at differing degrees of abstractness. The theoretical position suggested in this note is in fact compatible with others formed on a more abstract level but seeks instead to relate itself to the level on which minority groups are formed and develop demands.

The key terms in this approach are "institution" and "group" (including the notion of quasi-groups, such as class). These concepts may be subject to a more ultimate analysis in methodologically individualist terms, but it is also the case that they can be held constant as the context within which further meaningful individual and group actions occur. The model for such analysis is set out in Max Weber's *Economy and Society*, especially in the first chapter.

Any society develops standardized ways of responding to the needs of its members and to the further functional exigencies which arise in the course of these responses. The societies in which immigrant minorities settle have sets of responses of this kind which we refer to as institutions. Not all individuals and groups, however, have their needs met by these institutions and necessarily develop their own counter-responses or their own sets of institutions and strategies for countering and modifying the institutional demands of the main society.

In any case individuals do not seek to attain their goals merely by individual action. They pursue them collectively through group action, and an analysis of groups is a necessary part of sociology, cross-cutting that which is made in terms of institutions. This group analysis has to take account of shared cultural meanings, norms, and forms of organization which enable individuals to act as responsible or representative individuals.

One tradition of political sociology deriving from Marxism considers not merely the formation of organized groups of this kind, but the larger question of the formation of classes. A so-called "class-in-itself" consists of a mass of individuals who share the same experience of having to fight for the attainment of their goals within an institutional system. A "class-for-itself" comes into being when these individuals become conscious of their situation or "interests" and develop shared meanings, norms, and organizational forms for the attainment of their ends.

In Marxian sociology of a certain kind a problem emerges as to why those who have common interests do not become a "class-for-itself." This is sometimes attributed to false consciousness, and one particular form of false consciousness is that which results from the meanings, norms, and organizational forms deriving from ethnicity preempting those of class.

In the approach adopted here, minority group formation is seen as analogous to class formation. It is not seen as a form of false consciousness. Individuals develop common interests as a result of their position in their sending societies (including their relations to economic and political institutions) and in their relations to the institutions of societies of settlement. Just as the formation of classes for themselves is always historically imperfect, however, so also is the formation of "minority-groups for themselves."

Thus we must assume that individual minority members have interests deriving both from their position in sending societies and in relation to the institutional set of the society of settlement. Within this complex situation any number of individuals may have similar sets of materials, and it is to be expected that those with the same set of interests will develop forms of collective consciousness to pursue these interests. Central to these in an immigrant situation are the goals of legal and political security within the system, adequate economic prospects, adequate housing, and adequate education for children.

It is not denied that the security which comes from sharing a common ethnicity and culture, independently of interests, is not itself an independent bonding factor. But such ethnic bonds gain significance when they are related to a sense of shared interests.

The interest of this group is in demands and reactions. Amongst the most important demands are those which are collectively formulated by "minority-groups-for-themselves." This is not to say that the complexes of interest on the basis of which groups formed are not many and varied, nor that all individuals understand their relationship to the institutions of the society of settlement solely in terms of shared interests. But the formation of groups and the formulation of collective demands is inevitably of central importance both in itself and as a marker for the formulation of individual notions of interest.

Afterword 4 (Wilpert)

Social scientific questions arise to a great extent out of the pressures of particular historical circumstances, and this, consequently, leads to the emphasis on one particular theoretical perspective over another; perspectives which in more abstract global terms may not necessarily be mutually exclusive. For this reason it is necessary to make some comments about the relationship between individual choice and ethnic group membership from the background of the European social context.

In the framework of the "guestworker" migration, the historical period is short, at the most some twenty years; the potential reality of the settlement process of the foreign worker nationalities into the industrial North is at its beginning. The unclear legal status of non-European community members does not permit the perception of alternative choices for group or national membership: the range of choice is limited. In this context allegiance to a rational choice model may confuse the analysis and divert attention from the real structural constraints in membership choice imposed by the labor migration history and the conditions set by the labor importing country, and it may, as well, ignore the relative role of ideologies with respect to national group loyalties and their legitimacy in contemporary Europe.

The perspective Hechter and Gumperz summarize (this report) with respect to intergroup phenomena and group membership is based on the assumption that individual choice is maximized under conditions of rapid socioeconomic change characteristic of the urban industrial setting. They believe that in this context individuals may have multiple choices for group affiliations which then alter the significance of the group for its members. This is clearly an optimistic view of contemporary labor market conditions and opportunities for social mobility amongst newly arriving migrant workers within the urban industrial setting. Their position formulated here may best be understood within the history of ethnicity studies in the United States (American) context. Their perspective reflects a critique of the mystification of ethnicity as both a factor of political mobilization and in the determination of group choices and shared values. Their argument rests on the assumption that the urban setting stimulates socioeconomic mobility, forging new common experiences and new group formations. It is the universality of this assumption which needs to be questioned with respect to contemporary European socioeconomic development.

In a number of ethnic minority situations, an individual may not be able to divest himself of his group membership. He may be restricted and evaluated primarily on his membership to a particular group. Due to these constraints the minority must rely on members of his group both for aid in solving problems and for understanding the world around him.

It is this particular situation within which the majority of the immigrant worker groups find themselves. It is the combination of social conditions and the social and cultural resources of the particular ethnic minority which will influence both the minority's capacity to mobilize and the new social and cultural forms which may develop. Thus, in addition to common structural conditions and a shared historical immigration experience, the social and cultural resources found in existing kin and local networks, a common language, shared signs and symbols facilitate communication and provide a natural basis for social organization and problem-solving. Although certain individuals may be exposed to situations where they are free to disregard their ethnic group membership in their social interactions, this will be the exception rather than the rule in the European context. National origins and structural position are highly correlated in the contemporary urban setting.

Furthermore, ethnic group social organization need not imply ethnic collective action. Collective action is an unlikely alternative for groups with questionable legal status or groups whose orientation is primarily towards the home country. Thus, it is not yet on the agenda for the two largest minorities in the Federal Republic of Germany, the Turks and Yugoslavs. Instead of consistent collective action, some form of intermittent collective movements may stir, which may loosen a chain of reactions, question legitimacy, raise consciousness, and at times stimulate some forms of limited social change.

Intergroup phenomena certainly may not be understood solely through an analysis of group structures and values, neither may they be restricted to the question of individual choice. Thorough analysis is necessary of the context of individual choice: the structural conditions which create barriers or opportunities. This includes socioeconomic processes influencing the distribution of goods or position on the labor market, as well as processes of ideology-building and -questioning among the dominant group. One level of response of the latter may be caused by insecure social comparisons and contradictions in the dominant ideology. Barriers to individual choice in group affiliation are likely to be more rigid than assumed in the rational choice perspective.

Special Note: The paper, "A Theory of Ethnic Collective Action," by M. Hechter et al., was discussed at the workshop but has not been republished in this report as it has already been published elsewhere. The rational choice theory of collective action (of which collective action by ethnic minority groups is a particular case) propounded in the paper strongly influenced the form in which the agenda for our group's discussion was drawn up. Although the theory was never adopted by the group and was strongly attacked by individual group members, some parts of this report are not readily comprehensible without reference to the theory. The reader is therefore initially referred to:

Hechter, M.; Friedman, D.; and Appelbaum, M. 1982. A theory of ethnic collective action. Int. Migr. Re. **16(2)**: 412–434.

SUGGESTED READING

(1) Berreman, G., ed. 1981. Social Inequality. New York: Academic Press.

(2) Cohen, A.P., ed. 1982. Belonging. Manchester: Manchester University Press.

(3) Cohen, A.P. 1974. Two-Dimensional Men. London: Routledge & Kegan Paul.

(4) Cummins, J. 1979. Linguistic interdependence and the educational development of bilingual children. Rev. Educ. Re. **49**: 222–251.

(5) Erickson, F., and Schultz, J. 1982. The Counselor as Gatekeeper. New York: Academic Press.

(6) Gellner, E. 1973. Scale and nation. Phil. So. Sc. **3**: 1–17.

(7) Gellner, E. 1964. Thought and Change. London: Weidenfield and Nicholson.

(8) Gumperz, J.J., ed. 1982. Language and Social Identity. New York: Cambridge University Press.

(9) Hechter, M. 1979. Internal Colonialism. London: Routledge.

(10) Hoffmann-Nowotny, H.-J. 1973. Soziologie des Fremdarbeiterproblems. Eine theoretische und empirische Analyse am Beispiel der Schweiz. Stuttgart: Ferdinand-Enke-Verlag.

(11) Holzner, B. 1978. The construction of social actors: An essay on social identities. **In** Phenomenology and Sociology, ed. T. Luckmann. London:Penguin.

(12) Karabel, J., and Halsey, C. 1979. Power and Ideology in Education. New York: Oxford University Press.

(13) Pitt Rivers, J. 1973. Race in Latin America; the Concept of 'Raza'. Eur. J. Soc. **14**: 3–31.

(14) Rex, J. 1983. Race Relations in Sociological Theory, new rev. ed. London: Routledge.and Kegan Paul.

(15) Rex, J., and Moore, R. 1970. Race Community and Conflict. London: Oxford University Press.

(16) Rex, J., and Tomlinson, S. 1979. Colonial Immigrants in a British City. London: Routledge and Kegan Paul.

(17) Saifullah Khan, V. 1981. Some comments on the question of the second generation. **In** Report on Immigrant Children and Youth in Norway. Oslo: Ministry of Local Government and Labor.

(18) Schrader, A; Nilkes, B.W.; and Griese, H.M. 1976. Die zweite Generation. Sozialisation und Akkulturation ausländischer Kinder in der Bundesrepublik. Kronberg: Athenäum.

(19) Smith, A.D. 1981. The Ethnic Revival. Cambridge: Cambridge University Press.

(20) Tajfel, H. 1981. Human Groups and Social Categories. Cambridge: Cambridge University Press.

(21) Rokkan, S., and Urwin, D., eds. 1982. The Politics of Territorial Identity: Studies in European Regionalism. London, Beverly Hills: Sage Publications.

(22) Wallman, S. 1978. The boundaries of 'race': Processes of ethnicity in England. MAN **13(2)**: 200–217.

Minorities: Community and Identity, ed. C. Fried, pp. 219–227
Dahlem Konferenzen 1983. Berlin, Heidelberg, New York, Tokyo: Springer-Verlag.

STATES AND MINORITIES

M. Walzer
Institute for Advanced Study
Princeton, NJ 08540, USA

Abstract. The heterogeneity of ethnic and religious groups is a constant feature of political life, and it has been encompassed within four different sorts of political structures: empires, federations, multinational states, and nation-states. Small and weak groups have sometimes found safety, and they have sometimes found danger, within each of these structures. For the members of such groups, there is no single best structure, but one can specify helpful principles appropriate to each of the four.

INTRODUCTION

Imagine a purely political community, a state where citizenship or subjection was the only basis of collective identity, the only general bond, and all other social relationships were private, personal, or familial. In such a community, there would be no minorities. Every man and woman would have the same standing vis-a-vis the political authorities, whoever the authorities were. Everyone would feel equally at home. There might be social and economic differences among the citizens, but there would be no alien, outcast, or pariah groups. No such community, however, has ever existed in the world of states.

Citizens and subjects are never only citizens or subjects — how much easier politics would be if they were! They are always something else. They share a culture or a religious commitment, or an ethnic or national history with some but not all of their fellows. These additional identities are shaped by politics, but they are never wholly determined by it. Since the boundaries of the political community change over time, and since individuals and sometimes groups of individuals move across whatever boundaries exist, men and women whose culture or religion developed within one community commonly find themselves living in another. Every state presides over a heterogeneous body of citizens or subjects and over a multiplicity of larger and smaller groups.

It is with the political life of these groups that I am concerned here — and more particularly with the sheer survival and minimal well-being of the smaller groups. How does the state cope with the heterogeneity of its subjects? How does it relate to its minorities? What are the common patterns and the theoretical possibilities of coexistence? In the brief compass of this paper, I shall offer little more than a list of patterns and possibilities. But though the list is short, it is meant to be exhaustive of the arrangements consistent with the reality of the *state*, that is, of a single political center and a more or less effective sovereign power. Hence there is no discussion here of feudal systems or of anarchist theories. I should stress that my subject is the *coexistence* of groups under state authority, and while repression is conceivably a form of coexistence, expulsion and genocide are obviously not. These latter two represent "final solutions" to the problem of heterogeneity, but they have no place in moral or political discourse.

EMPIRES

The dominant political structure through most of human history has been the multinational empire. The classical city-state is an aberration in the ancient world, and the nation-state is a modern novelty. Most men and women most of the time have lived under the rule of a distant and alien center. The empires are multinational in a triple sense: first and most obviously, because they incorporate the territories and populations of many nations; second, because they move nations about for the sake of internal security (as in the Babylonian exile of the Jews) and so create new juxtapositions and interminglings of diverse peoples; and third, because they facilitate the voluntary movement of individuals, particularly into the great capital cities (such as Alexandria and Rome in classical times or, more recently, Baghdad, Constantinople, and Vienna), where all sorts of people mix, more or less freely, with one another.

Empires are cosmopolitan in character; hence they are the enemies of nationalism — or at least of every nationalism save that of the imperial nation itself. Imperial rule is commonly an unstable mixture of repression and tolerance: repression of every movement for independence; tolerance, sometimes, for the multiplicity of cultures, religions, languages, and customary ways of life that make the empire what it is. Faced with the cruelties of modern nationalists, we tend to idealize the old empires, to emphasize the tolerance and forget the repression. This is surely a mistake, for the conquests by which empires are established bring with them a human loss that is infinitely cruel: think of Rome and Carthage, or Spain and the Aztecs, or modern Russia and the Baltic peoples. And to maintain those conquests requires at least an intermittent cruelty, a reminder from time to time of the reality of imperial power. But it is also true that the empires often bring with them a kind of local peace and that, under the rule of their bureaucrats, people of different religious and ethnic groups live side by side for many generations without killing one another. Indeed, these bureaucrats are especially likely to extend their protection to the smaller and more vulnerable groups, partly on the principle of divide and rule, partly because such groups stand in greatest need of protection and are most likely to be grateful for it. The nostalgia for empire is probably strongest among their members, who rarely prosper once national liberation has been accomplished. The Chinese of Indonesia, the Asians of Uganda, and the Jews of Iraq, for example, were all better off in the old empires than in the successor states.

Distance is the key concept for an understanding of imperial rule. Ideally the bureaucrats come from far away, like Roman proconsuls in Egypt or British regents in India; they are indifferent to the quarrels of the nations they rule. They are without zeal in their repressiveness, amenable to corruption, and generally reluctant, so long as things are quiet, to interfere much in the everyday affairs of their subjects. No doubt, distance can also breed a brutal incomprehension of local ways, but this is likely to be an undiscriminating incomprehension, making for a generalized local resentment rather than for specific forms of ethnic and religious conflict. It is self-government that makes for conflict: when distance is dispensed with and the crucial question of political life becomes, Who *among us* shall rule over the others? Who shall rule right here, on this ground, in these villages and towns, whose inhabitants suddenly realize how different they are from one another?

FEDERATION

The most obvious successor to the multinational empire is the multinational (or binational) federation, for this sort of regime reproduces the pluralist or cosmopolitan character of the empire under the aegis of voluntarism. Czechoslovakia and Yugoslavia are useful examples of states effectively created by imperial rule but redesigned by their founders as free associations. It cannot be said that the redesign was entirely honest or entirely successful, for in the years immediately after independence, Czechoslovakia was effectively ruled by the Czechs, Yugoslavia by the Serbs. In practice, these were centralized states, but the *idea* of the federation was that some decisions, at least, would be made at local levels, while power at the center would be democratically shared. With regard to some aspects of government, each group would rule itself; with regard to other aspects, each group would yield its claims to a majority of individuals drawn, it was hoped, from all the groups. Arrangements of this sort have worked fairly well in Switzerland, and not well at all in Lebanon and Cyprus.

There are not many genuine federations in the world today, but it is worth remembering that the idea dominated socialist thought in the late 19th and early 20th centuries. Indeed, the impact of the old empires — Hapsburg and Romanov above all — on socialist thought would be hard to overestimate. Here were practical examples of economic interdependence and political internationalism, and the idea of replacing them with a multitude of nation-states, each with its own "national bourgeoisie," each cultivating its own narrow parochialism, was abhorrent to socialist leaders who had grown up in Petersburg, e.g., or in Vienna (or even, as with Rosa Luxemburg, in Warsaw). It was only necessary, they thought, to destroy the autocracy, democratize the central government, and allow some degree of local autonomy. The solidarity of the workers would sustain the unity of the empire against secessionist movements. The Soviet Union today is a fraudulent version of this old socialist model: an empire disguised as a federation. The most attractive versions remain theoretical — as, for example, in the work of the Austrian socialists Otto Bauer and Karl Renner, who proposed to reconstitute the Hapsburg empire as a single territorial state and a series of nonterritorial national "corporations." Each constituent nation, whether its members were numerous or few, concentrated in a single place or widely dispersed, would have the right to organize a corporation, tax itself, and run its own educational and cultural institutions. Minorities would be protected and the unity of the whole guaranteed, again, by worker solidarity.

This is the oppositionist form of imperial nostalgia, and its appeal is easy to understand. The various socialist programs represent an effort to maintain, indeed to build upon, the heterogeneity-in-unity of the old empires. For if capitalism had socialized production, bringing large numbers of workers into the factory and submitting them to a common discipline, the empires had socialized habitation, fostering an extraordinarily complex intermingling of peoples. What possible reason could there be to undo the intermingling, drawing boundaries and moving populations? But the socialists underestimated the force of national feeling outside the imperial elites and the thin educated stratum from which their own elites were recruited. Had they come to power in, say, the Hapsburg empire, they would have had to reestablish the pattern of cruelty and bureaucratic distance, as the Russians did, or they would have had to offer something more than local or corporate autonomy. For the quiet living together that marked the old empires had been, it turned out, a product of popular passivity. When the people were mobilized in the struggle against autocratic rule, they were also mobilized for nationalist politics. They argued in their own languages, recalled the heroic traditions of their ancestors, and demanded self-government for themselves, among themselves. The last thing they wanted was to replace the imperial bureaucrat with the Pole, Lithuanian, Russian, Turk, Greek, Jew, or whatever, who lived down the street or in the next village, whose language they did not know, whose character and commitments they distrusted.

Hence the great strains under which contemporary federations function and the difficulties of finding an appropriate constitutional structure – visible most recently in the debates that have gone on in Spain and Great Britain. We think of these two as the home countries of old empires, but in fact they are themselves mini-empires, incorporating diverse nations, and moving, under democratic pressure, toward something like a federal structure. The purpose of the move, in the eyes of its advocates, is to enhance the power of minority groups such as the Basques, the Catalans, the Welsh, and so on, without breaking up the old imperial union. Federalism is a stopping point between empire and national independence, and for all its difficulties, it is an attractive stopping point for small and weak nations, who have lost some of the conventional attributes of nationality (language, for example) and have come to feel more or less at home in the empire. Over much of the world, however, it has not proven an effective substitute for full independence.

MULTINATIONAL STATES

There is one more alternative to the independent nation-state that must be considered at some length: the community of immigrants (the United States, Australia, Argentina, and so on) whose citizens come from many nations. The immigrants have left their old territorial base behind them and, with rare exceptions, have established no new base in their new country. Each nation lives dispersed among all the others. Together, they constitute not an empire, not a federation (for whatever federal arrangements exist have no ethnic significance), but a multinational state, unitary in form, pluralist in content. In principle, all citizens are equal, and the state deals with them as individuals, not as members of groups; but they are free to organize groups and to sustain, if they can, their national culture.

But this principle stands in an uneasy relation to the two dominant doctrines of the multinational state. The first of these describes the state as a "melting pot," that is, a political structure within which a new nation is being formed: e pluribus unum; from many, one. The multinational immigrant community is a temporary creation, to be replaced eventually by a unitary nation-state incorporating and transcending the diverse nationalities of the immigrants. But then there is bound to be competition among the different groups that strive to dominate this process — and hostility toward any groups that resist the incorporation and transcendence or whose members, because of their race or religion, do not fit one or another desired outcome. The second doctrine of the multinational state describes it as a permanent regime, a system of "ethnic pluralism," a union of groups, rather than of individuals, with equal rights. But then it looks as if the state ought to act to sustain the groups and enhance their internal life. Ethnic pluralism strains toward a kind of corporate federalism, even though this latter arrangement is probably not compatible with the individuality of the immigrants (who came to their new country, most of them, one by one) and with the degree of "melting" or of cultural homogenization with which they have already made their peace.

These two doctrines suggest some of the current and future difficulties of the multinational state. But its condition right now — the practical culture, legal rights, and ideological commitments of American citizens, for example (leaving aside the blacks and focusing for the moment on European and Asian immigrants) — seems to many observers to constitute an ideal arrangement: heterogeneity without autocracy on the one hand and without nationalist zeal on the other. This is a pattern founded in its own way on distance — not the distance of the bureaucrat from the people he governs, but the distance of the people themselves from their original homeland and their ancient traditions. The current balance of melting pot and pluralist ideas is a function of the shared deracination of their adherents. So I must consider now the political life of men and women who are not immigrants, who live where their ancestors lived before them, and who govern themselves, as their ancestors probably did not, by themselves.

NATION-STATES

Ideally, the nation-state brings together all the members of a single national group, and no one else, in a unified political structure. For reasons I have already given, however, no actual nation-state is quite like that. Migration and conquest have so mixed populations that it is impossible to draw a set of boundaries that encloses exactly the right people. Hence what distinguishes the nation-state is that the largest number of its citizens are members of a single nation, from which the state as a whole takes its character. All other citizens are members of minority groups, and they are unlikely to recognize themselves in the official history of the state or to feel their culture confirmed in its symbols, rituals, and celebrations. In the multinational federation and the multinational state, there are many larger and smaller groups; politics is likely to be a matter of shifting coalitions. But the politics of the nation-state is dominated by a single group (at least, this is true of linguistic, cultural, and religious politics: I do not mean to deny the reality of socioeconomic divisions).

Minority citizens are inevitably vulnerable in such a setting, but it would be wrong to assume that they are inevitably badly treated. Persecution and ostracism (and expulsion and massacre, too) are common enough in the history of nationalist politics, and there are many splendidly cosmopolitan imperial cities that are now dreary and provincial state capitals. But the nation-state is a setting for self-government, and self-government at least sometimes takes liberal and democratic forms. When it does, a number of options open up for the members of minority groups. First of all, they are commonly invited to assimilate into the majority, to adopt its culture, identify with its history, or even, if there is no racial bar, to raise children who will be indistinguishable from its own children. Acceptance of this invitation imposes a real loss upon the minority — a loss of collective identity and way of life and even, as assimilation proceeds, of collective memory. At the same time, it holds open to individuals the possibility of full membership in both halves of the nation-state and active and equal participation in national culture and politics. But perhaps, given the loss, we should call this first option liberal and democratic only if a second option is also available: to maintain a separate group life on the margins of the political community while enjoying the same civil and political rights that the majority enjoys. This is to assimilate to the state but not to the nation, to make of oneself (or to try to make of oneself) a *permanent* minority. Nation-states with long traditions of cultural homogeneity are uncomfortable with such permanence, but the existence of "national minorities" is by no means uncommon in the world today, and it does not look to be a transient existence. At various points in the twentieth century, efforts have been made to protect national minorities through international treaties. But such treaties have never yet provided effective help. In a liberal democracy, citizenship itself is the best protection, at least against the usual forms of persecution and harassment. It cannot protect against the experience of cultural dissidence, however: this is the inevitable experience of national minorities in a nation-state.

The greatest risk for national minorities is that they can be denied citizenship altogether — like the metics in ancient Athens, or like guestworkers in Europe today. Then they have no part at all in the government of the community in which they live, and so they are particularly vulnerable to majority tyranny. They are outside both the nation and the state. Alienage is easier within an imperial or multinational setting, where citizenship is less central, and ethnic and religious groups are likely to sustain a more lively internal life and to welcome newcomers even if the newcomers are not legitimized by state authority (thus, "illegal aliens" in the United States blend more easily into the general population than guest workers do in Europe). But the centrality of citizenship, when it exists, is a real advantage for the citizens: historically, self-government has been more direct, politics more engaging, rates of political participation far higher, in the relatively homogeneous nation-state than in any of the more heterogeneous structures. It is a hard question for contemporary nation-states whether the centrality of citizenship can be maintained while new immigrant groups are admitted to full democratic rights.

CONCLUSION

The empire, the multinational federation, the multinational state, the nation-state (I omit the miniature states of the contemporary world, city-states and island-states, such as Singapore today, which are probably best assimilated to the nation-state): these four exhaust the possible political arrangements for ethnic and religious coexistence. Generalized statelessness is not a real alternative, though one can imagine a global empire or a global federation or even — given a radical increase in forced and voluntary migrations — a global multinational state. In practical terms, the fate of groups large and small will be worked out within the four structures that I have described. No one of these is best, though committed democrats will prefer the last three to the first. Each of them "fits" a certain set of historical circumstances; each of them has its own advantages and disadvantages. Sometimes there is a choice between or among them; sometimes no effective choice is available. The members of minority groups have lived in peace and prosperity in all of them — and have been oppressed and persecuted in all of them. This last point suggests that there are other factors than political structure at work in shaping the fate of minorities. Indeed, there are many other factors. But the four structures fix the terms of political struggle, and they define the goals we must work for if we are to secure the existence of small and weak groups: in empires, bureaucratic tolerance; in multinational federations, genuine autonomy; in multinational states, a balance of pluralism and individuality; in nation-states, universal citizenship.

BIBLIOGRAPHY

(1) Kedourie, E. 1960. Nationalism. London: Hutchinson.

(2) Liebman, L., ed. 1982. Ethnic Relations in America. Englewood Cliffs, NJ: Prentice-Hall, Inc.

(3) Lijphart, A. 1977. Democracy in Plural Societies. New Haven: Yale University Press.

(4) Nairn, T. 1977. The Break-up of Britain: Crisis and Neo-Nationalism. London: New Left Books.

(5) Patterson, O. 1977. Ethnic Chauvinism: The Reactionary Impulse. New York: Stein and Day.

(6) Ronen, D. 1979. The Quest for Self-Determination. New Haven: Yale University Press.

(7) Van Dyke, V. 1977. The individual, the state, and ethnic communities in political theory. World Polit. **29**: 343–369.

(8) Walzer, M. 1981. The distribution of membership. **In** Boundaries: National Autonomy and Its Limits, eds. P.G. Brown and H. Shue, pp. 1–35. Totowa, NJ: Rowman and Littlefield.

Minorities: Community and Identity, ed. C. Fried, pp. 229–246
Dahlem Konferenzen 1983. Berlin, Heidelberg, New York, Tokyo: Springer-Verlag.

PATTERNS OF INTERNATIONAL MIGRATION POLICY: A DIACHRONIC COMPARISON

A.R. Zolberg
Dept. of Political Science, Graduate Faculty
The New School for Social Research
New York, NY 10011, USA

Abstract. Global conditions shape international migrations by determining the dispositions of individuals toward movement as well as of the states in which they live or seek to enter toward potential population flows. The paper focuses on the policy responses of states in three historical periods, the early modern, the XIXth century, and the contemporary, with special emphasis on the role of western states in patterning labor migrations and refugee flows. An attempt is made to account for the shift from a trend toward liberal exit and entry policies in the XIXth century to one involving highly restrictive entry policies on the part of potential receiving states, and a combination of restrictive exit with forced departure in other parts of the XXth century world.

INTRODUCTION

Today as in the past, international migrations form identifiable patterns shaped by changing world conditions which determine not only the dispositions of individuals toward movement, but the policies of states toward the movement of individuals across their borders. It is quite evident that contemporary flows would be very different if all the states of the world allowed individuals under their jurisdiction to choose freely whether to stay home or move abroad, and if they admitted freely all those originating within the jurisdiction of another state. The posture of states toward exit in fact ranges very widely on both sides of some hypothetical indifference point toward the one extreme of enforced immobilization and toward the other of expulsion. As for entry, the general drift in the XXth century among the states most international migrants aspire to enter has been toward severe restriction, coupled with exceptional measures for (a) the procurement of specific categories of labor, on a temporary or permanent basis; (b) the accommodation of constituency claims for family reunion; and (c) the admission of some refugees. In this paper, which summarizes a forthcoming book-length work, I shall provide a brief analytic overview of historical changes in patterns of state action in the sphere of international migration so as to foster a contextual understanding of contemporary trends.

The theoretical apparatus governing the inquiry has been discussed extensively elsewhere (18,19). In short, states may be considered as instruments through which a variety of groups — ranging from ruling elites to organized interests (business, labor, "natives") — seek to achieve certain goals by way of the adoption of given migration policies. These refer not only to ordinary legislation and administration regulations governing immediate movement, but to a vast range of formal and informal public action affecting exit or entry. It should be noted that (a) public action includes, as already suggested, the use of force by the state itself, or tolerance by the state of violence directed against target groups; (b) policies include public inaction (i.e., laissez-faire or benign neglect); and (c) on the receiving side, the relevant sphere of policy includes naturalization as well as immigration stricto sensu.

It is assumed further that the goals the groups under consideration seek to achieve by securing such policies can be understood in terms of some instrumental calculation, on the basis of existing beliefs and available information, of how certain interests are likely to be affected by the population movements the policy is designed to prevent or to bring about. The determinants of such policy are multifaceted because the human beings involved are simultaneously assessed in two very different ways: as market actors, and within this most prominently as workers, and as actors within political institutions and cultural arenas. Each of these perspectives, which may be referred to as the "economic" and the "moral," determines a distinct structure of interest alignments with respect to policy alternatives. Most familiarly, assessment of the economic impact of the flow on the country of origin or of destination links migration policy to more general economic strategies; and the interest alignments that come into play are closely related to the general configuration of class relations in the relevant country. On the receiving side, for example, a given group of immigrants can be viewed simultaneously by employers as a welcome addition to the labor pool, by indigenous workers as unwelcome competition, and by welfare agencies or local taxpayers as "free riders" who constitute a burden on welfare facilities; and similar considerations arise in the country of origin as well. Moreover, the economic perspective on international migrations also entails assessments of the value of a particular flow from the respective points of view of the sending and receiving country.

But migrants are never viewed exclusively from an economic perspective. They are also seen as members of racial, ethnic, religious, and linguistic groups, as well as persons imbued with distinctive political dispositions and preferences. Concomitantly, the actual or potential role of the migrants as members of society evokes their assessment on a scale of moral value, referring to their impact on regime-maintenance and national integration. There is no gainsaying that, if it is difficult to assess with precision the economic effects produced by given population movements, it is well-nigh impossible to do so in the sphere presently under consideration. Nevertheless, it is very much the case that all established migration policies and all attempts to change them are predicated on some such assessment. How do the scales on the basis of which groups are judged to be assets or liabilities arise? "Prejudice," "xenophobia," and other individual-level explanations only go halfway.

An alternative starting point is the observation that constituted societies commonly strive to achieve and maintain a certain degree of cultural homogeneity, and that the codes which delineate the preferred culture often take the form of an emphasis on distinctions between "us" and "others," whereby the latter function as a negative anti-cultural model to be avoided. The "others" are always mythical; but the myth is often thought to be incarnated by actual groups, whose known attributes may in fact have contributed to shaping the myth in the first place. If such groups happen to be physically located within the state, they may be confined, segregated, or even expelled; should they be outside, little action is required unless, by some play of circumstances, they start flowing across the border; the most obvious response is then to keep them out altogether. But, paradoxically, situations have repeatedly arisen whereby an immigrant population is simultaneously highly desirable from an economic point of view, and highly undesirable from the perspective presently under discussion. This is so because, by and large, "cheap labor" comes from the periphery. As will be demonstrated below, the ensuing dilemma has been resolved variously in different historical epochs.

THE AGE OF ABSOLUTISM AND MERCANTILISM

From the middle of the fifteenth century on, as a consequence of internal transformations within Europe and its outward expansion, "for the first time, the world began to be one migratory network dominated by a single group of technologically advanced and culturally similar states" ((3), p. 96). This is also when international migration, properly speaking, emerged as a distinctive phenomenon, since its distinctiveness is predicated on the organization of global space into territories controlled by sovereign states that have the right to control the movement of people across their borders. Such migration was almost entirely controlled by sending and receiving states. Migration from one European state to another (including to foreign colonies) was very limited. Since population was considered a scarce economic and military asset, rulers deployed considerable efforts to police their territorial boundaries and to confine subjects within them. This process was acknowledged in, and reinforced by, the formation of international law in the seventeenth century ((7), pp. 91–96; (11)).

As I have documented in previous publications, the three largest components of the Western network during this period were: (a) the importation of an estimated 7.5 million West African slaves to work plantations, initially in Europe or its outlying islands, and then in the New World; (b) the relocation of two to three million Europeans in the New World colonies, mostly under some form of bondage; and (c) the expulsion or flight of approximately one million persons from the newly-forming European states in the XVIth and XVIIth centuries (17).

Coercion was probably a necessary concomitant of labor importation under prevailing colonial conditions. On the one hand, transportation costs were very high in relation to the workers' potential output and were usually assumed by the employer in advance; similarly, the employer (or intermediaries, such as brokers) must pay in advance for a substantial share of the estimated labor output. Under such circumstances, it was imperative to secure control over the actual performance of the labor. This was particularly difficult to achieve under colonial conditions, since the abundance of open land normally afforded any person the possibility of engaging in subsistence agriculture as an alternative to working for others. Bondage, in one form or another, thus arose as a solution to prevent workers from taking advantage of the labor-scarcity conditions that prompted their importation in the first place. For Europeans, initially mostly Spaniards and later mostly British, the typical arrangement was indenture, under which the prospective employer or a colonial labor agent bore the costs of transfer and settlement in exchange for a seven-year contract. Imperial authorities sought to regulate such movements to avoid a counterproductive drain of valuable population from home and the autonomous development of colonies in competition with the metropole; but if they leaned too far in this direction, as was the case with France in the XVIIth century, then the very development of the colonies was jeopardized.

Cultural considerations came into play early on. For example, France prohibited the settlement of its own Protestant subjects in its North American possession, Acadia and New France, probably out of fear they might be disloyal in the event of a conflict with Britain and the Netherlands; similarly, Britain in practice prohibited the settlement of its own Catholic subjects in the New England colonies exposed to the possibility of conflict with France; Protestant dissenters, however, were not ruled out as their bitter anti-Catholicism insured they would function as reliable guardians of empire. With respect to Africans, debate still rages as to whether white racism was a precondition for the emergence of chattel slavery, or whether slavery fostered racism as a rationalization. What matters most, for the present purpose, is that the extreme segregation associated with the Northwest European form of slavery (British and Dutch) can be thought of as a rational (in the sense of instrumental) solution to the problem of importing economically valuable but culturally undesirable labor. What such segregation entails, institutionally, is the erection of an internal boundary which prevents the group under consideration from becoming incorporated into the receiving society.

The contradictory dictates of economic and political considerations also came into play within Europe itself, where the cultural dynamics associated with the process of state-formation sometimes led rulers to willfully divest themselves of otherwise valuable population groups or to persecute them to the point of driving them into flight. In a competitive international setting, state-building entailed strenuous efforts to render populations more homogeneous within, so as to emphasize national differences and thus promote identification among subjects with the interests of their rulers. The main cultural instrument to that effect was religion. It is noteworthy, in this respect, that official expulsion of Jews living in England and France — as against local attacks on them — coincided with the take-off of their respective monarchies in the late XIIIth century. From such a perspective, it stands to reason that by expelling the unconverted Jews in 1492, the newly united Spanish monarchy was merely trying to "europeanize" Iberian society. A similar fate was subsequently meted out to the Iberian population of Arab descent (the *moriscos*) who not only were considered "unassimilable," like the Jews, but thought to constitute a strategic liability because of their location on the Mediterranean coast vulnerable to attack by Spain's Moslem enemies — a situation akin to the one which arose with respect to U.S. citizens of Japanese descent on the West Coast in World War II. In the same manner, in the late XVIth century Spanish authorities knowingly jeopardized the economic viability of their valuable Netherlandish possession (approximately present-day Belgium) by driving Protestants, who figured prominently in that region's manufacturing and commerce, out of the country. Numerous forced population exchanges occurred during the Thirty Years' War and in its aftermath, when the principle of conformity to the prince's religion was generalized throughout German Europe. The last massive wave of refugees were the Protestants who fled France when Louis XIV revoked the Edict of Nantes (1685), another major instance of a compulsive striving for cultural homogeneity in the name of raison d'état, with little or no regard for the economic consequences — let alone the human costs — of such action.

The immigration policy of European states during this period was generally governed by acquisitiveness. Sovereigns aspiring to economic development welcomed foreign artisans and often sought out valuable population groups, sometimes to the point of conceding to such foreigners a significant degree of religious toleration. A similar orientation prevailed with respect to the importation of mercenaries who were in some sense the original guestworkers. In the XVIIth century, the generous asylum policy of the newly independent Netherlands toward foreign Protestants and toward Jews contributed significantly to bringing about its spectacular economic development; and this example, widely broadcast by contemporary thinkers, in turn fostered the emergence of greater religious tolerance throughout the region in the XVIIIth century.

THE LIBERAL EPOCH

In retrospect, the nineteenth century constituted a deviant episode in the history of international migrations, when a group of states in one part of the globe by and large relinquished control over both exit and entry, so that a large segment of the overall flow came to be governed mostly by "push" and "pull," in relation to which the migrants themselves made their choices, much as they might do with respect to relocation from one part of their country of birth to another. It should be noted, however, that if during this period freedom of exit came to be established as a sine qua non of liberal regimes, this was not the case with respect to freedom of entry; even where laissez-faire was practiced in fact, receiving states insisted on their theoretical right to regulate admission and, hence, to restrict it.

The transition to new migration policies and concomitant population movements reflected the interrelated global changes known as the demographic, industrial, and democratic revolutions; beginning in the latter part of the XVIIIth century, it was largely completed by about 1830. One major outcome was the emergence of an unprecedented pattern, the voluntary and permanent transfer of a large mass of free individuals from a variety of European countries to independent or self-governing countries under the control of people of European descent. Two other significant changes were the waning of European refugee movements, reflecting a generalized trend toward more liberal regimes in that region, and the abolition of the slave trade, together with the substitution for it of another form of labor importation from the non-Western world.

It is now generally recognized that the vast movement of northwest Europeans to North America got under way initially more as a response to the push of economic necessity at home than to the pull of opportunity abroad (14). The turning point occurred at the end of the Napoleonic Wars when, with the internal labor supply of England assured, emigration came to be seen as (a) a way of lowering welfare burdens, (b) a safety valve in the face of perennial social unrest, (c) protection for Britain against being swamped by proliferating Irish Catholics, and (d) a mechanism for populating the empire with culturally appropriate settlers, especially where it was necessary to counterbalance other European stocks (French Catholics in Canada, Dutch Calvinists in recently conquered South Africa). By the late 1820s, all legal and administrative obstacles to exit were swept away. Efforts were made initially to direct the flow exclusively toward British possessions, but in the face of continuing evidence that many of the emigrants preferred the United States and were willing to pay their own way to get there, these were abandoned, except for special programs to populate Australia and South Africa. As it was, at this time the United States was emerging as the primary locale for British investment overseas, so that the flow of population from the British Isles contributed to higher returns on capital by increasing the supply of U.S. labor and hence lowering its cost. Concurrently, the growth of bulk exports from North America to Britain, timber and cotton, provided shippers with an incentive to recruit emigrants as west-bound ballast and also resulted in a steady lowering of the costs of passage. By the middle decades of the century, processes similar to the ones that triggered off the exodus from the British Isles engulfed the western parts of the continent, where state after state began to view emigration as a solution rather than as a problem. In 1854, for the first time, arrivals in the United States from Germany surpassed immigrants originating in the British Isles. It is noteworthy that France, which experienced much less population growth in the XIXth century and did not undergo an abrupt industrial take-off, remained the singular exception to the European pattern. In the second half of the century, emigration from the Iberian peninsula to South America resumed and expanded as well.

Contrary to conventional thinking on the subject, the United States did not become a "nation of immigrants" until the early 1850s, three-quarters of a century after independence. Reliable estimates indicate that in the 1780–1830 period, immigration contributed only about 6% of U.S. decennial population growth ((9), p. 62). In 1831, when Alexis de Tocqueville conducted field work for *Democracy in America*, the bulk of the white American population, then numbering approximately 10.5 million, consisted of highly fertile English-speaking Protestant Americans, of British descent (mostly English and Scottish or Scots-Irish), of several generations' standing. From the founding onward, U.S. policy in this sphere can be characterized as "selective immigrationism," with a strong preference for self-capitalized farmers and artisans from the middle classes of European society. This was expressed (a) in state legislation which attempted to restrict the landing of paupers and criminals, (b) in federal legislation limiting the carrying capacity of ships and organizing the sale of public lands so as to foster commercial agriculture, and (c) in naturalization laws which prohibited the incorporation of nonwhites into the body politic. Although there were no longer any immigration restrictions based on religion, it is well established that Americans viewed themselves very much as a protestant nation and regarded Catholicism as incompatible with a republican political culture.

Given this orientation, it is not surprising that a sudden escalation in the number of entries from 1832 on, with an increasing proportion of Catholic arrivals, alien in language, as well as destitute, led many to question the impact of such immigration on the country's political development. There were repeated confrontations between immigrationists and restrictionists throughout the middle decades of the century, the most acute one in the 1850s when this issue contributed as much as the conflict over slavery to the destruction of the U.S. party system and the ensuing national crisis ((12), pp. 241–255). Yet the door remained open. Far from reflecting a consensus on the desirability of immigration, as many historians have maintained, this is attributable to vigorous intervention by a segment of the American business community — shippers, who thrived on expanding traffic; the railroad industry and operators of mines and mills, who depended on a vast influx of unskilled labor; real estate promoters — to keep it open in the face of increasingly antagonistic public opinion. The restrictionists were hampered by constitutional controversies over state rights, which rendered the enactment of effective federal regulation of entry nearly impossible at this time, and the immigrationist camp was reinforced by the fact that the newcomers quickly became citizens and were mobilized as voters, with an interest in bringing over more of their own kind. Overall, the country's immigration policy can be understood as one of the great triumphs of U.S. capitalism. It is noteworthy that in the face of labor shortages at the time of the Civil War, the U.S. government departed from mere laissez-faire and initiated steps to establish a system of labor importation under contract. This was rescinded in 1868, but the national government henceforth assumed a more active role in rationalizing the flow of economically suitable immigrants from Europe.

In the last decades of the century, the character of United States immigration underwent significant change; not only did the flow originate increasingly outside of northwestern Europe, but a larger proportion of it was temporary. In effect, with the advent of transatlantic steamships, many Europeans went to the United States — or to South America — as migrant workers rather than as settlers. On the eve of World War I, returns amounted to about one-third of arrivals; the major exception to this among the "new" immigrants were the Jews, whose relocation was attributable to political persecution as much as to economic distress ((10), pp. 141–163).

This pattern, consisting of the temporary migration of free workers from neighboring less to more developed countries in order to fill unskilled jobs in agriculture or industry, with little or no formal regulation, was crystallizing in many European regions as well. Britain had long depended on such a flow from southern Ireland, then part of the United Kingdom; in the middle of the XIXth century, something like one-fourth of the urban industrial population of England and Scotland was Irish-born. By the 1880s, large numbers of Belgians and Italians were migrating to France, and at this time a similar flow linked Poland and Germany. Such labor was cheap because it originated in less developed countries, but by the same token, the incoming workers were considered culturally inferior to natives. Wherever they occurred, labor migrations of this type left a residue of permanent settlement and triggered off negative reactions from native workers as well as from guardians of the established moral order. German assessments of the injurious cultural impact of Poles, as illustrated by Max Weber's analysis in 1896, fostered stricter regulation of the flow with emphasis on its temporary character; in these measures one can see the immediate ancestry of contemporary "guestworker" programs (4).

In the world's plantation belt, the importation of labor was not merely advantageous, but a necessity. In the first decade of the XIXth century, Britain, Denmark, and the United States prohibited the importation of slaves into their possessions. The critical decision was Britain's, who controlled about half of the total European traffic; it was brought about by the concurrence of a growth in the moral sensibility of elites to this issue with an unfavorable economic conjuncture in the West Indian sugar plantations (2). Britain subsequently imposed abolition on its competitors as well. Whereas in the United States the prohibition on importation had little effect on slavery and on the economic sector it served, because the rate of natural reproduction of the slave population was quite high; elsewhere the end of importation signified a rapid decline in the labor supply and, hence, spelled the end of the plantation system itself. The situation in the British empire was exacerbated by the decision of the reformist Parliament in the 1830s to end slavery itself. Plantation owners in the Indian Ocean and the Caribbean then began to bring workers under indenture from India; notwithstanding denunciation of this as a new system of slavery, official support was provided for these undertakings which were also looked upon by imperial authorities as a way of reducing social tensions in India. The traffic was extended later on to Southeast Asia as well as to southern and eastern Africa, and British India became the leading supplier of indentured plantation labor for other colonial powers (15).

The indenture system, which lasted well into this century, generated a large permanent residue of permanent Asian immigrants in the relevant territories and contributed, together with slavery, to the formation of tension-ridden plural societies in the world's plantation belt ((6), pp.303–312;(13)). The importation of Chinese as mine and railroad workers into California at the time of the gold rush was a U.S. variant of this general pattern; indeed, "coolies" were seriously considered in the post-Civil War South as a substitute for slaves or free blacks. Although the indenture system was organized by Chinese entrepreneurs and the contracts were not sanctioned by U.S. law, the distinctive status of the Chinese as "sojourners" rather than immigrants was highlighted from the very outset by prohibitions against their naturalization. Given American legal traditions, however, it was well-nigh impossible to deny citizenship to U.S. born offspring of Chinese sojourners. The fear of invasion by an alien race, together with the militant opposition of organized labor to "coolie" competition, rapidly led to the imposition of strict prohibitions on Chinese entry. This was replicated throughout the white dominions of the Pacific, including Canada, and subsequently applied to other Asians as well.

CONTEMPORARY PATTERNS

The contrast between the policy patterns of the liberal age and those of our own time is very sharp. Starting around the turn of the XXth century, one after the other the states of destination began to close their gates, restricting access to relatively small numbers of people of the receivers' choosing, and with some minor modifications and occasional but temporary exceptions, this still remains the basic stance of the affluent liberal states today. At the same time, the world began to produce refugees on a hitherto unprecedented scale, while a number of states, containing a substantial proportion of the world's population, erected more effective barriers than ever devised before against exit. Our interest in international migration should not blind us to the fact that it is an exceptional phenomenon, which occurs against a background of enforced immobility. Most human beings alive today are confined in the country of their birth as a consequence of restrictive exit or entry policies, or of some combination of both. It is also the case that among the migrations that have occurred in the XXth century, refugees outnumber voluntary migrants (1).

These trends emerged rapidly in the last decades in the XIXth century, when the market and the state became truly worldwide forms of social organization, and the globe was linked for the first time into a single network of rapid mass transportation (steamship and railroad) and telecommunications. The unevenness of world conditions was accentuated by a growing gap between a small number of capital-rich, technologically advanced, and strategically powerful countries, European or of European origin plus Japan, and the rest, whose internal conditions the leading countries affected more than ever by way of colonial control and of the transnational economic, social, cultural, and political processes they generated. In the demographic sphere, completing the revolution they had begun to experience in the XVIIIth century, the industrialized countries entered into a phase of much slower growth; whereas in the previous century and a half they had grown at a more rapid rate than the rest, the difference was now reversed; Europe reached its historical maximum proportion of world population around World War I, and the United States and Japan did so around 1950 ((16), p. 205). Whereas the achievement of peak rates of population growth by Western countries generally coincided with a rate of economic growth sufficient to absorb additional manpower and to foster a secular increase in per capita income — taking into account the redistribution of population among them by way of overseas migration in the XIXth century — this combination rarely occurred in other parts of the world. Hence, as world population mounted in the XXth century, an ever-larger proportion of them were poor.

Given the way in which the world is structured, the work any person is capable of performing will bring much higher returns in an affluent than in a poor country, especially if "collective goods" as well as individual ones are taken into account. It is therefore quite reasonable for many people living in poor countries — even if they are themselves not among the poorest — to aspire to relocate. The availability of such a vast reserve of cheap labor located abroad provides obvious opportunities for capitalists in the industrial countries, most prominently the possibility of cushioning the effects of the business cycle by procuring labor when it is needed and divesting themselves of it when it no longer is, without bearing the costs of maintaining it when unproductive. But the potential number of candidates to fill the ranks of cheap labor is much vaster than any conceivable demand in receiving countries, and for obvious reasons, many workers admitted on a temporary basis end up wanting to stay. It is evident that as ordinary residents, however, they become less valuable economically, and in the face of a tendency to settle, objections are also raised concerning their cultural impact on the receiving country.

The escalation of refugees is attributable in the first instance to the breakdown of empires and their concomitant replacement with a plethora of new states whose founders tend to adopt the national model fostered by earlier developers. As noted earlier, state building in Europe resulted in the formation of target groups whose elimination was thought to be necessary in order to achieve political integration; unleashed in highly populated and culturally more heterogenous regions in the XXth century this process has resulted in a proliferation of religious, racial, or ethnic minorities subjected to the most extreme forms of persecution, including not only expulsion but also immobilization and mass murder. Paradoxically, those who in fact become refugees — as a result of expulsion or flight — sometimes turn out to be the more fortunate members of the original victim groups. Because state-formation commonly encompasses an entire region, it also tends to generate international tensions and conflicts which interact with the process delineated earlier to produce further flows. The political transformations under consideration are sometimes accompanied by attempts to restructure society as a whole, and in the course of social revolutions or counterrevolutions entire classes are cast in the role of victim-groups.

In recent decades, the international community has come to acknowledge special obligations toward refugees, but the number of claimants is much greater than states are collectively willing to admit as permanent residents, so that some form of triage must come into play. Moreover, the circumstances under which people become refugees often involve a mix of political oppression and economic distress, which makes determination of the validity of individual claims difficult. And finally, once refugees are admitted, their impact is indistinguishable from that of other newcomers, so that the issues mentioned earlier move to the fore in this case as well.

The economic and political processes outlined may be considered as the global givens in relation to which the world's more affluent and liberal countries have determined their immigration policies in the XXth century. One very striking outcome has been a steady reduction of the long-standing distinctiveness of the "nations of immigrants," as these countries took steps to reduce the role of immigration in their societal development. This is well illustrated by the experience of the largest of them, the United States, where from the 1890s on, pressures mounted to drastically reduce the entry of "new immigrants" from southern and eastern Europe. Expressed with respect to the undesirability of specific nationalities and ethnic groups, the new restrictionism in fact represented growing opposition to mass immigration of any kind.

Political alignments were much as before; to the restrictionist camp was now added organized labor, and resistance to it came mostly from the business community and from the ranks of recent immigrants. Under the impact of World War I and ensuing upheavals, including the Soviet revolution, the restrictionists gained a succession of victories in 1917–24, as the result of which annual U.S. immigration from Europe was reduced, in relation to population size, to below the pre-1830 level. Few concessions were made to the plight of refugees in the interwar period. United States business adjusted to the exclusion of Asians and to the reduction in European immigration by turning to cheap labor from Mexico and the Caribbean; beginning in World War I, and again in World War II, government assistance was provided to insure a flow of temporary labor. In this manner, U.S. policy came to be divided into a segment pertaining to immigration by way of the main gate, and another governing labor procurement through the back door.

In 1952, the United States reaffirmed a very restrictive stance with respect to the main gate; however, many ad hoc enactments in effect implemented a somewhat more generous policy toward European refugees and family reunion in the post-World War II period. These reforms were formalized in 1965, when all traces of racial and ethnic discrimination were removed; however, quantitative limitations were maintained and extended to include the Western hemisphere, hitherto unrestricted in this respect. At about the same time, the long-standing temporary workers program, involving several hundred thousand annual entries from Mexico, was eliminated; instead, back door policy drifted toward benign neglect of the southern border, allowing for a massive flow of undocumented workers. Notwithstanding mounting concern with the latter in the 1970s exacerbated by high unemployment and fear of "hispanization," proposals to effectively restrict their entry by imposing sanctions on employers of illegal aliens have been resisted by the employers themselves as well as by U.S. citizens and residents of hispanic culture, so far (early 1983) successfully. In keeping with its foreign policy interests, since the 1950s the United States has been especially generous toward refugees from Communist regimes, most recently with respect to Indochinese and Cubans; concomitantly, there is considerable official reluctance to accord refugee status to persons originating in states considered friendly. Refugee admissions are likely to be curtailed because of mounting objections from local communities to the escalating welfare costs the settlement of large numbers of Cubans and Indochinese has entailed.

In the XXth century, the European states hitherto concerned mostly with emigration came to be countries of destination and were faced with the need to devise an immigration policy. For many of them, the issue first arose in earnest with the arrival of Jews from eastern Europe beginning in the 1880s. Germany reacted forcefully by resorting to administrative refoulement; Britain, hitherto more liberal than the others with respect to foreigners, devised a restrictive Aliens Act (1906). In contrast with these, France maintained an open door — not because of an absence of anti-Semitism, but because the state looked upon immigration in general as a device for increasing its demographic weight in the face of a declining rate of natural reproduction. For those same reasons, rendered more urgent by population losses in World War I, France continued to welcome immigrants in large numbers in the 1920s and absorbed much of the flow deflected from the United States by the new restrictive legislation; as of 1930, approximately 7% of the population was foreign-born, a much higher level than found elsewhere in Europe. Massive refugee flows were triggered off by the breakdown of liberal democracies in the interwar period, particularly in Germany; but when wartime conditions and the refusal of liberal countries to open their doors made it impossible for the Nazis to expel undesirables, they turned to a form of forced emigration that produces no refugees.

To a remarkable extent, the national orientations established in Europe before World War I guided policy-making after the region recovered from World War II (5,8). The situation differed somewhat in each of the principal cases. The United Kingdom remained inhospitable to general immigration while continuing to rely on the Irish; however, in the course of the negotiations leading to the establishment of the Commonwealth (1948), Britain extended to the citizens of its various member states a status that was tantamount to British nationality, which allowed them to settle freely in the British Isles. Within a decade, the ensuing flows of West Indian blacks and of Asians from the Indian subcontinent, albeit welcomed by some employers, had propelled immigration to the forefront of political controversy. In 1961, a Conservative government began to undo the permissive effects of the 1948 arrangements, so as to transform "new Commonwealth" populations into ordinary aliens, whose entry might be restricted; Labour subsequently joined in, and the task was finally completed in 1981. However, in the intervening period a sizeable nonwhite minority had become permanently established in Britain.

The German economy benefited in the immediate postwar period from the additional supply of labor provided by a steady flow of refugees and returnees from the East, but even before this ebbed, the Federal Republic revived the guestworker pattern by negotiating bilateral agreements with less developed European countries (Italy, Spain, Greece) and later with Turkey and Yugoslavia. By the early 1970s, foreign workers constituted about one-tenth of the country's labor force. Contrary to original intent, however, some of them were beginning to turn into permanent residents; Turks, who constitute only about one-third of the foreign total, have been singled out as a source of special concern. The energy crisis of 1973 brought an end to new labor immigration, but the question of whether established workers and their families will be allowed to stay — i.e., whether Germany will become an immigration country — is not yet settled. With relatively minor variations, German experience was replicated in Switzerland, Belgium, Luxemburg, and later the Netherlands, Sweden, and Norway. French exceptionalism prevailed until about 1972, after which policy shifted toward the continental norm; similar tensions have surfaced there, particularly with respect to Algerians. The socialist regime which came into power in 1981 has moved toward allowing established foreign workers to settle permanently, while erecting more rigorous barriers against new arrivals.

CONCLUSION

In the permanent face of a large mass of external population propelled into the migration stream by economic distress or brutal force, the countries of putative destination have steadily moved in recent times toward the adoption of extremely restrictive entry policies, for which Japan might serve as a model. The trend has been accelerated by the economic crisis of the past decade, but was already in the making before this occurred, and it is likely that the new restrictionism will outlast the current depression. As it appears impossible to operate guestworker programs without incurring permanent settlement, entrepreneurs are henceforth much more likely to relocate their capital in cheap labor regions than to import workers from the Third World. Immigration into the affluent societies is thus likely to remain limited to a small stream of family reunion and a cautious intake of refugees. It is obviously rational for affluent societies to erect walls in order to protect the desirable economic, cultural, and political conditions they have achieved, and there is no gainsaying that these states owe it to their own populations to provide them with such protection. But one should understand these policies for what they are: a collective device to prevent the redistribution of existing world resources to the benefit of the disadvantaged.

Notwithstanding their exploitative character, policies designed to procure cheap labor have been the source of one of the few openings in an otherwise impenetrable wall. Minimal justice requires that the foreign workers who contributed to the affluence of the receivers be allowed to stay and have an opportunity, if they wish, to bring in their immediate families so as to complete the process of relocation and begin that of integration into a new society. Such integration requires mutual adjustment by the newcomers and the host society, but nowhere are the numbers involved so overwhelming, nor the people involved so incapable of change as to make accommodation impossible.

Beyond this, in the worldwide process of triage, the highest priority must be granted to political refugees. Liberal states should, of course, wherever possible avoid engaging in actions that contribute to the generation of refugees, and they might be able to contribute, by way of diplomacy and assistance, to alleviate somewhat refugee-producing tensions in certain parts of the world. However, the flow of victims is certain to continue in the foreseeable future.

and the refugee camp is rapidly emerging as one of the major images of the age.

Refugee policy consists of two very different components, the one involving aid and assistance to refugees in "third countries," the other involving their admission to one's own country, together with assistance for doing so. Both are vital, but international relief should not be viewed as a substitute for asylum. To take in the victims of willful persecution, regardless of whether the persecutor is friend or foe, is one of the fundamental obligations of states founded on justice; and given the paucity of such states on the world scene, each of them must assume an equitable share of the common burden.

Acknowledgement. The research on which this paper is based was initiated in 1977 with the aid of a grant from The Rockefeller Foundation under its Population and Development Policy Research Program, sponsored jointly with the Ford Foundation.

REFERENCES

(1) Beijer, G.J. 1969. Modern patterns of international migratory movements. **In** Migration, ed. J.A. Jackson. Cambridge: Cambridge University Press.

(2) Davis, D.B. 1975. The Problem of Slavery in the Age of Revolution, 1770–1823. Ithaca: Cornell University Press.

(3) Davis, K. 1974. The migrations of human populations. Sci. Am. **231**: 92–105.

(4) Dibble, V. 1968. Social science and political commitments in the young Max Weber. Eur. J. Soc. **7**: 95–110.

(5) Freeman, G. 1979. Immigrant Labor and Racial Conflict in Industrial Societies: The French and British Experience, 1945–1975. Princeton: Princeton University Press.

(6) Furnivall, J.S. 1948. Colonial Policy and Practice. London: Cambridge University Press.

(7) Glass, D.V. 1967. Population: Policies and Movements in Europe. London: Frank Cass and Co.

(8) Kubat, D., ed. 1979. The Politics of Migration Policies. New York: Center for Migration Studies.

(9) Kuznets, S. 1966. Modern Economic Growth: Rate, Structure, and Spread. New Haven: Yale University Press.

(10) Piore, M. 1979. Birds of Passage: Migrant Labor and Industrial Societies. Cambridge: Cambridge University Press.

(11) Plender, R. 1972. International Migration Law. Leiden: A.W. Sijthoff.

(12) Potter, D. 1976. The Impending Crisis, 1848–1861. New York: Harper.

(13) Smith, M.G. 1965. The Plural Society in the British West Indies. Berkeley: University of California Press.

(14) Thomas, B. 1973. Migration and Economic Growth: A Study of Great Britain and the Atlantic Economy. Cambridge: Cambridge University Press.

(15) Tinker, H. 1974. A New System of Slavery: The Export of Indian Labor Overseas, 1830–1920. London: Oxford University Press.

(16) Wrigley, E.A. 1969. Population and History. New York: McGraw-Hill.

(17) Zolberg, A.R. 1978. International migration policies in a changing world system. **In** Human Migrations: Patterns and Policies, eds. W. McNeill and R. Adams, pp. 241–286. Bloomington: Indiana University Press.

(18) Zolberg, A.R. 1981. International migrations in political perspective. **In** Global Trends in Migration: Theory and Research on International Population Movements, eds. M. Kritz, C.B. Keely, and S.M. Tomasi, pp. 3–27. New York: Center for Migration Studies.

(19) Zolberg, A.R. 1983. Contemporary transnational migrations in historical perspective: patterns and dilemmas. **In** U.S. Immigration and Refugee Policy: Global and Domestic Issues, ed. M.M. Kritz, pp. 15–51. Lexington: D.C. Heath.

Minorities: Community and Identity, ed. C. Fried, pp. 247–254
Dahlem Konferenzen 1983. Berlin, Heidelberg, New York, Tokyo: Springer-Verlag.

ETHNIC PLURALISM: THE U.S. MODEL

S. Thernstrom
Dept. of History, Harvard University
Cambridge, MA 02138, USA

Abstract. The striking feature about American society is not its much-vaunted ethnic pluralism but its capacity to absorb newcomers. This has been much less true of nonwhite groups until fairly recently, but is happening increasingly of late.

INTRODUCTION

The ethnic composition of the population of the United States is, of course, extraordinarily diverse. Much of its diversity derives from the obvious fact that for at least the past two centuries America has been the focal point of global population movements, by far the largest receiver of migrants from other countries in the world. Over fifty million people have entered the United States since the first U.S. Census of 1790. Although the flow of newcomers was cut back sharply by restrictive legislation in the 1920s, it resumed on a very large scale after the passage of the liberal Immigration Act of 1965. Five million newcomers came legally in the 1970s, plus an unknown but undoubtedly large number of illegal aliens. In 1980 alone the figure was over 800,000, more than twice as many as for the rest of the countries in the world combined.

Voluntary immigration has been the most important but not the sole source of ethnic diversity within American society. We should also remember that 400,000 African blacks were brought to North America as slaves before the

United States barred further importations in 1808, a population that had grown to 26 million by the time of the 1980 Census. And there are sizable numbers of people whose ancestors did not come to the United States either voluntarily or involuntarily. Instead, the United States came to them in the course of its relentless expansion across the continent and into the Caribbean and Pacific. This category most obviously includes the native peoples Columbus erroneously called "Indians" in the delusion that he had reached the shores of Asia. It includes many others as well: French Acadians who were forcibly uprooted by the British before the American Revolution and dispersed to other colonies, Louisiana especially; Hispanics in the Southwest who were annexed when the United States stripped Mexico of her northern provinces from Texas to California after the Mexican War of 1846–1848; Puerto Ricans, Hawaiians, and various other Pacific islanders taken in America's final expansionist thrust at the close of the nineteenth century. Thus, the peopling of America, which Bismarck once called the most important fact of modern history, was accomplished not only through voluntary immigration but also through enslavement, invasion, and conquest (4).

The result of these processes has been a society that seems, at first blush, amazingly polyglot. The recently published *Harvard Encyclopedia of American Ethnic Groups* (4) identified more than one hundred such groups, plus 170 surviving Indian tribes, and the list could have been extended further if there had been opportunity for further research on some of the newest arrivals. Today the Elmhurst section of Queens, New York, is the home of more than 20,000 immigrants from 110 different countries. Is there a single "model" that can illuminate how all these groups interacted with each other and the larger society, and if so, can it be characterized as "pluralism"? I think we need at least two models, a separate one for most nonwhite groups, whose situation I will consider later in this paper.

As to the label "pluralism," it depends upon what we mean by that murky term. It is fashionable today in the United States to say that America is a highly pluralistic society and to deny that the melting pot ever melted. But what is "pluralism"? My dictionary defines it, not very helpfully, as a "condition of society in which numerous distinct ethnic, religious, or cultural groups coexist within one nation." The United States certainly qualifies by that unexacting standard, but so do a great many countries around the world. The real question is not whether the nation includes a variety of cultural groups of an ethnic character, but what role they play in structuring social life, how salient they are to their members and to others, and whether they are of diminishing, growing, or unchanging significance with the passage of time.

The concept of pluralism as it was first worked out by Kallen and others in the decade after the outbreak of World War I had the merit of addressing these questions. Although Kallen was sometimes clearly prescribing rather than describing, his theory can be taken as description or prediction and put to empirical test. He believed (perhaps only hoped) that the United States was or would become a federation of peoples, "a nation of nationalities." Political unity did not require cultural homogeneity, he insisted. Ethnicity would be the fundamental line of social division in American society. Each group would perpetuate and transmit much of its Old World culture over the generations, and would retain its distinctive character permanently. Its members would not melt away in the American soup, because no one can "change his grandfather."

EUROPEAN WHITES

If this is what we mean by a pluralistic society, Canada, Belgium, or Yugoslavia may qualify, but the United States does not — not with reference to the various European white immigrant-based groups. Most immigrants did, indeed, show some tendency to cluster together with their fellow countrymen, taking advantage of the freedom of association guaranteed them by the Constitution to form organizations for mutual aid, worship, and cultural maintenance to help them survive in a bewilderingly different environment. Periodic spasms of nativist hostility, as in the 1850s, 1890s, and the years from World War I to the Great Depression, reinforced the impulse to withdraw into a closed and comfortable circle of kinsmen. Since "no Irish need apply," the Irish found it imperative to stick together, and so did most other newcomers for a time. American society has long been "pluralistic" in that sense.

Strong though the tug of ethnic allegiances was, however, ethnic groups were largely transitory phenomena. The recent upsurge of ethnic awareness in the United States has, I think, made many of us forget the speed and thoroughness with which most groups were assimilated. The melting pot may not have eroded group differences as rapidly as some zealous 100 percent Americanizers may have wished, but their expectations were closer to the mark than Kallen's. Kallen failed to recognize that while you cannot change your grandfather, you can forget him. Millions of ethnic Americans did just that. Even in the first generation ethnic boundaries were not rigid but permeable, and the tendency towards fusion was strong. Ethnic consciousness has waxed, waned, and sometimes waxed again. Assimilation is not the linear, irreversible process some theorists took it to be. But at the risk of some oversimplification one can say this. With each passing generation, the core of activists most strongly committed to group maintenance shrank, and growing numbers lost touch with their roots. No major immigrant group was able to transmit a non-English mother tongue and the cultural baggage associated with it to a majority of the third generation. Successive generations became more and more like the old stock native white population not only in language and culture but in their educational level, occupational distribution, residential dispersion, and propensity to marry outside the group (5). The latter index of assimilation — the rate of ethnically exogamous marriages — is a particularly critical one, because the family is the key institution for the transmission of a sense of ethnic identity and culture. Of course, it is possible for the offspring of mixed marriages to identify ethnically with one of his parents, or even with both in different contexts. But the more complex the family tree, the weaker the identification with any one branch of it is likely to be. Hence the significance of the fact that 80–90% of the immigrant generation in most groups married someone from the same homeland, but a distinctly lower proportion of their children did so, and an even lower proportion of their grandchildren. A 1963 survey of American Catholics revealed that, of those under 30, 82% of the Italians and 65% of the Poles had married outside the group (1).

Some more recent evidence, from national polls conducted in the years 1973–1978 (see Table 1), points in the same direction (1). Respondents were asked to identify the countries their ancestors had come from, and those of mixed ancestry were asked which country they felt "closer" to. When we compare people born before World War I with those born since World War II, we see enormous changes. The older generation in the four Catholic groups lagged far behind white Protestants in rates of college attendance. In the younger generation, the Italians, Poles, and Irish were actually somewhat ahead of white Protestants, probably because they were more heavily concentrated in cities, which offer greater opportunities for higher education. Only the German Catholics lagged a bit behind, presumably because they were less urbanized than the others. The proportion reporting mixed ancestry — whose parents or grandparents had married outside the group — was tiny for the most recent arrivals in the pre-World I generation, the Italians and the Poles, and modest for the older immigrants, the Irish and the Germans. It was dramatically higher for every group in the younger generation. The changes suggested by these figures are all the more impressive when we realize that 11% of Catholics

TABLE 1 — Generational differences among selected ethnic groups, 1973-1978.

	birth cohort	percent who attended college	percent of mixed ancestry	percent married across religious lines
Catholics				
Italian	pre-WWI	6	2	21
	post-WWII	49	40	40
Polish	pre-WWI	11	7	20
	post-WWII	51	41	35
German	pre-WWI	12	21	41
	post-WWII	32	38	51
Irish	pre-WWI	26	27	18
	post-WWII	59	52	40
White Protestants				
	pre-WWI	21	25	11
	post-WWII	39	42	24

and 25% of white Protestants were of such mixed origin that they were unable to name any foreign country they felt "closer" to. The boundaries dividing Americans into groups according to national origin have weakened enormously over the course of the past century.

It is also significant that the salience of religious boundaries, which theorists of the "triple melting pot" thought would replace nationality as the principal source of ethnic cleavage, has declined precipitously as well. No less than 40% of younger Catholics in the 1970s were married to non-Catholics; in 1972, 48% of Jews married non-Jews.

The "ethnic revival" of the past decade or so, the period of the alleged "rise of the unmeltable ethnics," may seem evidence to the contrary (3). I suspect it was mainly media hype that did not much affect masses of people. To the extent that it was real, it might best be seen as the last gasp of groups nearing extinction. Assimilation had proceeded so far that spokesmen for the ethnic revival felt secure enough in their Americanism to assert that they wanted their differences noted and tolerated as well. It was hard to find any real differences, since the original Old World cultures have been so thoroughly eroded away, and few third or fourth generation ethnics are prepared to put in the hard work to learn Estonian or Serbo-Croatian and to study the history of their ancestors. Instead they settle for what Herbert Gans calls "symbolic" or "expressive" ethnicity, which makes few behavioral demands upon them (2). Ethnic consciousness of this highly attenuated kind may survive long into the future, but it is hard to consider it a very important phenomenon.

NONWHITES

The generalizations above, it need hardly be said, do not apply to most nonwhite groups — Afro-Americans, Indians, Mexicans, Puerto Ricans, the Chinese, the Japanese. Race has made, and with some continues to make, a difference. How much of a difference and what kind of a difference, though, has varied a good deal from group to group and over time. Some students of ethnic matters sharply distinguish racial from ethnic groups, on the grounds that racial visibility is ineradicable and inescapable (6). White immigrants, the argument goes, were free to change their mode of dress, speech patterns, and table manners and "pass" for "plain Americans." Dark-skinned people cannot; physiognomy is destiny. This is too simplistic. The distinctive physical traits of a racial group may not change over time — though they can as a result of interbreeding — but their social significance certainly can.

The clearest demonstration of this from American history would be the case of the Japanese and Chinese. Their arrival on the West Coast in the latter half of the nineteenth century provoked hysterical fears of the "Yellow Peril" that led to a complete ban on further immigration from an "Asiatic Barred Zone." They were denied the right to become naturalized citizens, barred from owning land and entering many occupations, forbidden to marry whites, attacked by lynch mobs, and otherwise treated about as badly as blacks in the Jim Crow South. During World War II the Japanese had the unique experience of being forcibly confined in desert camps for four years, losing property for which they later received only partial compensation. And yet today both the Chinese and Japanese groups rank well above American whites on every measure of socioeconomic status. They are as physically distinctive as their ancestors, but no one worries about the Yellow Peril any more. Orientals are no longer a stigmatized racial minority but a rapidly assimilating ethnic group. Their marital assimilation is now so rapid as to call into question how many generations they would survive as groups without renewal through further immigration. By 1972 a third of the young Chinese and half of the Japanese were marrying whites (5).

None of the other principal nonwhite groups has made such spectacular recent progress as the Orientals or gained such general social acceptance by whites. But it can be said that in the past few decades the significance of racial differences has been declining. Impressive educational and economic advances have been made by Mexicans and Puerto Ricans, for example, advances obscured by aggregate figures for the entire group that include many raw newcomers that show up when the data are broken down by generation. And more assimilated Hispanics are entering the marital melting pot. In 1970 a third of all second-generation Puerto Rican women, and 44% of those who were high school graduates, were married to non-Hispanic whites; the figures for Puerto Rican males were even higher, and only a bit lower for Mexicans. Despite the separationist rhetoric of some Hispanic activists and federal support for bilingual education programs that may impede assimilation, it is likely that these groups will continue to be absorbed much like their white European predecessors.

The situation of Afro-Americans is the most difficult to assess briefly. As late as World War II a majority of them lived in a South committed to white supremacy and strict separation of the races, and the minority in the North faced intense prejudice that kept them out of many desirable jobs and neighborhoods. Within the past four decades anti-black prejudice has declined very sharply, and formal segregation has been eliminated. Black people have won full equality before the law — in some respects a favored position before the law. In their educational attainment, occupations, and incomes they have moved considerably closer to parity with whites than they were a generation ago. But they are still well behind and are certainly not in the marital melting pot. Although the rate of interracial marriage doubled between 1963 and 1970, the rise was only from 0.6% to 1.2%. More current figures would undoubtedly show some further increase, but not a very dramatic one. In this respect, the black-white divide remains the major fault line in American society.

The major problem confronting black America today, though, is not race per se but class (7). The black middle class has grown very rapidly, and its members have made astonishing gains. But a large black "underclass" has been left behind in the vast decaying ghettoes of New York, Chicago, and other metropolitan centers. During the 1970s the proportion of whites with incomes below the official poverty line living in the inner city fell 5%; the proportion of impoverished blacks rose 21%. In 1977, 31% of blacks, 7.7 million people, lived in poverty, as compared with 9% of whites. The proportion of black families headed by a female has soared from 18% in 1940 to 39% in 1978. Some argue that it is ethnocentric to consider such family patterns "deviant" and "pathological," and that they are a functional adaptation to ghetto circumstances. Whatever the merits of that argument, the association between residing in a household without a male head and poverty is quite strong. Fifty-one percent of such black families were poor in 1977, but only 14% of two-parent households. And it is clear that children who grow up in poverty are scarred by the experience in a variety of ways that limit their opportunity to lead a happy and productive life. Just what should be done to attack this problem is debatable. In the political climate of today, alas, it is not even being discussed.

REFERENCES

(1) Alba, R.D. 1981. The twilight of ethnicity among American Catholics of European ancestry. Ann. Am. Poli. **454**: 86–87.

(2) Gans, H. 1979. Symbolic ethnicity: the future of ethnic groups and cultures in America. Ethn. Racial **2**: 1–20.

(3) Novak, M. 1972. The Rise of the Unmeltable Ethnics. New York: Macmillan.

(4) Thernstrom, S., ed. 1980. The Harvard Encyclopedia of American Ethnic Groups. Cambridge: Harvard University Press.

(5) Thernstrom, S. 1981. Ethnic groups in American history. **In** Ethnic Relations in America, ed. L. Liebman, pp. 3–27. New York: Prentice-Hall.

(6) Van den Berghe, P. 1967. Race and Racism: A Comparative Perspective. New York: John Wiley and Sons.

(7) Wilson, W.J. 1981. The black community in the 1980's: questions of race, class, and public policy. Ann. Am. Poli. **454**: 26–41.

Minorities: Community and Identity, ed. C. Fried, pp. 255–268
Dahlem Konferenzen 1983. Berlin, Heidelberg, New York, Tokyo: Springer-Verlag.

IMMIGRATION AND CITIZENSHIP IN POSTWAR BRITAIN

R. Sondhi
Asian Resource Center
Handsworth, Birmingham B19 INH, England

Abstract. The purpose of this paper is restricted to recording the rapidly changing immigration and citizenship policies in postwar Britain. In a discussion about the humane options that modern states may consider in the formulation of general immigration and citizenship policies and practices, the British example must stand out as one of the least attractive and most complicated. Immigration rules and the concept of British citizenship have been dictated not by the consideration of universal principles, such as the protection of fundamental human rights, or even international obligations, particularly to the ex-colonies of its empire, but by the overriding need to deny entry and settlement to those with non-European ancestory. As such, British policy carries with it a charge of being racially discriminatory and effectively blocks the emergence of an effective race relations policy.

In the last twenty years the primary aim of British immigration and nationality legislation has been to systematically restrict the entry of black people (1) — as opposed to white — and through an increasingly sophisticated system of internal control, progressively reduce the number of black people settled in the UK. Immigration policy and practice is no longer concerned with the regulation of people on the basis of manpower needs. There has been a clear shift from stopping the flow of "primary" black workers to limiting the eligibility of their dependents to enter the UK. At the same time, changes in nationality law, by "constitutionalizing" discriminatory definitions in immigration rules, have both undermined the status of black British citizens born abroad and sealed off future sources of black immigration. Immigration and nationality are therefore no longer independent issues, but like the whole fraught area of race, largely an expression of how the global economic crisis is felt at every level of British society. Historic distinctions drawn between British subjects, regardless of color, and aliens have been progressively modified to discriminate between black and white.

The genesis of this pernicious development lay in the spread of the British Empire when millions of people all over the world became "natural born" subjects and enjoyed common civic rights – whether they were in the UK or abroad. One of these rights was to be able to enter and settle in any British territory, having obtained permits where necessary. Aliens had no such rights. Their entry and settlement was controlled, first by Royal prerogative and finally by the Aliens Act of 1905 which was passed in a fit of xenophobia and aimed at stopping the entry of a handful of poor Jews fleeing persecution in Eastern Europe.

This distinction was, however, ignored in practice. Throughout history, white aliens have been able to settle through naturalization whilst black British subjects have been denied their rights. As long ago as 1773, the *London Chronicle* mirrored the British obsession with race in the following words:

> "It is therefore humbly hoped that Parliament will provide such remedies as may be adaquate to the occasion by expelling the negroes now here, who are not made free by their owners, and by prohibiting the introduction of them in this kingdom for the future; and save the natural beauty of Britons from the Morisco tint, and remove the envy of our native servants, who have some reason to complain that the negroes enjoy all the happiness of ease in domestic life, while many of those starve for want of places." (2)

Almost exactly two centuries later, these sentiments were to find a striking resonance in the imperialist rhetoric of Prime Minister Margaret Thatcher's rationale of immigration control, broadcast to the nation on a TV program in January 1978. Referring to an estimate that there would be four million people "of the New Commonwealth and Pakistan" in Britain by the end of the century, she said:

> "Now that is an awful lot, and I think it means that this country might be rather swamped by people with a different culture and you know the British

> character has done so much for democracy, and law and done so much throughout the world that if there is any fear that it might be swamped, people are going to react and be rather hostile to those coming in. So if you want good race relations you have got to allay people's fears on numbers."

Immediately after the Second World War as the Empire gave way to the Commonwealth, self-governing Dominions such as Canada and Australia, ruled by settlers of British origin, instituted their own citizenship laws and continued with their racially discriminatory policies of preventing black British subjects (mostly Asians and Chinese) from entering and settling in their territories. Britain, therefore, needed a new law to resolve this contradiction in the free movement of its subjects. The 1948 British Nationality Act was passed. This established a UK and Colonies citizenship acquired by birth or descent. But more significantly, it preserved the concept of British subject and the right of people from the ex-colonial territories to come and live in Britain. This provided the mechanism by which labor could be imported from the Caribbean and Asian native reserves into the industrial heart of England as and when required by the economy of the 1950s.

During the war thousands of young able-bodied men and women from the colonies had assisted Britain in its war efforts. Now in the atmosphere of the economic optimism of postwar reconstruction they were to return, at the invitation of British industry, to fill the unskilled and semiskilled sectors of the labor market. In the early fifties, therefore, the entry of "coloured workers" (as they were popularly known in both parliamentary and popular speech) was regulated by the laws of supply and demand, the numbers entering bearing a direct relationship to numbers of jobs available. But by the mid-fifties the economy had begun to taper off, and the occasional calls for control of "coloured" workers found popular expression in the rhetoric of "race preservation" societies that had sprung up all over Britain. In 1958 Britain's slump-boom economy hit a depression and race riots flared up in London's Notting Hill and Nottingham.

As the numbers of imported black workers grew, such racial disturbances became a public focus for the underlying unease which had been growing throughout the fifties. The initial Government response was to hang on to the principle of free entry from the Commonwealth countries and turn instead to control at the source. India and Pakistan, out of deference to Britain, had assisted by denying passports to their own citizens, and it was hoped that the newly emerging West Indian Federation would be persuaded to do the same.

But the strong anti-immigration lobby which had voiced fears that "the coloured races will exceed the white races in a few years by no less than five to one" (4) had revived public fears about the "evils" of uncontrolled immigration. As soon as the Tories had won the 1961 election with the slogan "You've never had it so good" — the economy had been given a temporary boost — immigration once again became a national preoccupation. Having prevaricated on the issue in the pre-election years, the Conservative Government passed the Commonwealth Immigrants Act of 1962.

This eliminated the right of Commonwealth citizens (British subjects) and those with passports issued by or on behalf of Colonial governments to enter the United Kingdom unless they held a work voucher issued by the British authorities in their own countries. The actual scheme of control was not included in the Act, allowing it to be modified from time to time without reference to Parliament. Immigration officers were given wide discretionary powers which allowed them to treat white people seeking entry more favorably than black people. The Act also rendered Commonwealth (and Colonial) citizens liable to conditions of stay, to deportation and detention without charge. It became a criminal offence to break entry conditions and to harbor entrants guilty of this offence.

The political debates leading up to the 1962 Act created fears among those intending to emigrate to Britain that the doors were about to be closed permanently, with the result that immigration of economically active males rose substantially, particularly from the Indian subcontinent, in the first half of 1962 before the Act came into force. Thereafter it fell back dramatically into its normal pattern reflecting the specific demands for skilled and unskilled workers in Britain's productive and service industries. Substantial numbers of work vouchers were issued by the Ministry of Labor in the early sixties, not all of which were taken up, favoring those who had served in the British forces during the Second World War and who, it could rightly be assumed, would regard an invitation to work for Britain as a call to duty. But the stage had been set for the use of immigration as a powerful election issue so that as the deeply ingrained weaknesses of the British economy gave way once again to recession, and the Labor Party gradually returned to popularity, it had publicly abandoned any previously held principles of unconditional entry for Commonwealth citizens.

The result was that the Labor Party Manifesto of 1964, aimed specifically at the white working classes, stated unequivocally: "The number of immigrants entering this country must be limited." Roy Hattersley, the labor MP for Sparkbrook, Birmingham, had formulated what was to become the raison d'etre of both major political parties for black immigration control when he said: "Without integration limitation is inexcusable, without limitation, integration is impossible."

Labor's victory at the 1964 election was largely engineered by Harold Wilson's promise of introducing the "white heat" of new technology to burn away "13 years of Tory misrule." The economic logic of the times meant that if new technology were introduced, as it had been in the postwar reconstruction of Germany, then a contract labor system was required. But the '62 Act had, of course, forced people to settle in Britain or lose their right to return for work.

However, the Labor Party was unable to formulate an immigration policy based on future economic needs because of immediate pressure for unskilled labor. Many black workers can recall how they were offered bribes by employers in heavy industry to recruit fellow workers from abroad as there was still a demand for labor. The Government used, instead, unsubstantiated allegations that "evasions" were rendering the 1962 Act ineffective to bring in the 1965 White Paper on Immigration from the Commonwealth.

This paper increased the immigration officers' powers and required the entrant to register with the police as a condition of entry. It gave the Home Secretary the power to deport a Commonwealth citizen, lowered the age from 18 to 16 of dependent children who could enter to join their parents settled in the UK, and restricted the quota of work vouchers to 8,500 per year. It was the beginning of the numbers game subsequently to be played by both parties in years to come. R. Moore observed that the White Paper "laid the ideological basis for subsequent policy in this area and as a result the argument that the numbers of immigrants was the essence of the problem." It reduced immigration of economically active males, the source of what the government refers to as "primary immigrants," to a virtual trickle. Thereafter, the number of people entering the UK were primarily the dependent wives and children and some elderly relatives of those settled.

The late sixties then saw immigrants attempting to complete the family-building process and the consolidation of immigrant "colonies" in the inner-city areas where they were, by a combination of circumstances, confined to live. Inevitably, black people became visible not just as workers on the shop floor, or on the buses, or inside hospitals, but in neighborhoods, schools, parks, and swimming baths. Within the framework of a wider political and social crisis — both national and international — white society's fears were once again orchestrated through the self-appointed public spokesman on race and immigration, Enoch Powell. When British citizens of Asian origin living in East Africa were faced with the prospects of having to leave their countries to set up home in England, Powell resuscitated the question of numbers and pronounced:

> "Hundreds of people in Kenya, who never dreamt they belong to this country started to belong to it like you and me... It is monstrous that a loophole in legislation should be able to add another quarter million to that score (of Commonwealth immigrants)."

There was no loophole, there was no quarter of a million, there were only the scenarios of "tidal waves" of black immigrants "flooding" into the country. So when the Conservative Party demanded the tightening of immigration controls, the removal of illegal immigrants, a register of dependents, and repatriation grants, the Labor Government went much further and, within the space of a week, passed the 1968 Commonwealth Immigration Act.

This Act distinguished between citizens of the UK and colonies who had an ancestral connection and those who did not. It specified that holders of UK passports issued outside the British Isles would be subject to immigration controls unless the holders, or one of their parents or grandparents, had been born, naturalized, or adopted in the UK. The effect of the Act was to subject British citizens of Asian origin living in East Africa to immigration control, while in the same instant giving the freedom of entry to white Commonwealth citizens who had previously been excluded by the 1962 Act. A system of special vouchers was set up to cut down the entry of East African Asians to 1,500 a year. There were also further restrictions on the ages of dependents wishing to join those Commonwealth citizens settled here.

In a move unprecedented in the history of immigration legislation, Britain had taken away the rights of its own citizens to freely enter the UK (4). Even so, the British public's fears were not allayed, and once again the Conservative Party demanded that the numbers of immigrants be severely curtailed, that the unconditional right of Commonwealth citizens to settle in the UK after entry be removed, and that those wishing to return should be assisted by public funds. In 1968 Heath, then leader of the Tories, called upon the Government to stop all immigration. The Wilson administration responded immediately by barring male Commonwealth fiancés from entering the country "unless there were compassionate circumstances" (white male commonwealth fiancés qualified for entry in their own right by virtue of their parents'/grandparents' birth in the UK) and a few months later by presenting the Immigration Appeals Act.

This Act, although establishing the right of an immigrant to appeal against a Government decision, also carried with it a requirement — as a result of a last minute amendment — that dependents of Commonwealth citizens settled in the UK must obtain entry clearance from British posts abroad. Immigration control was now instituted as much through administrative control as Parliamentary legislation by creating complex procedures to establish claimed relationships. Through deliberately long delays and unjust refusals, the entry of dependents was drastically reduced in precisely the manner which the Conservatives had wanted (5).

Through the next decade these procedures became increasingly sophisticated and grotesquely cumbersome. Interviews were delayed by anything up to two years — requests were made for documents virtually impossible to obtain — and when obtained, they were disregarded. Applicants were subjected to long and harassing interviews on peripheral details of family history. Medical examinations and X-rays were performed, often by unqualified staff, and sometimes on pregnant women. This growth of the entry clearance officers' discretionary power culminated in the notorious "virginity tests" performed on Asian women entering the UK as fiancées. Those refused entry under these circumstances had a right of appeal, but this took more than a year to prepare and was finally held in the UK where the applicant could not attend. The vast majority of these appeals failed.

Powell made further speeches on race and immigration in 1969 and 1970. It was to be the articulation of the developing belief that the solution to the problems of Britain, whatever they were, were now concerned with not only stemming the "tide" of black people into the country, but reducing the number of those who were "within the gates." In 1969, not the principle but the administrative cost of repatriation was being voiced. In 1970, the speeches were about "high migrant birthrates," black children overcrowding schools, and about the black threat to society generally. The Tories came into power in the same year, and in the next passed the 1971 Immigration Act.

The 1971 Act formalized the 1968 idea of ancestral connection into a new concept of patriality and made this the whole basis of immigration control by giving the right to enter and settle in the UK only to patrials. Patrials were mainly citizens of the UK and colonies who were born, registered, naturalized, or adopted in the UK or who had a parent or grandparent who was patrial. Commonwealth citizens who had a parent born in the UK were also patrial.

When the Government failed in its attempt, despite a powerful white lobby, to include as patrials Commonwealth citizens with a grandparent born there, they managed instead to secure their right to enter and settle in Britain under the immigration rules. The patriality clause was undeniably a racial device to control black immigration since all aliens and all nonwhite Commonwealth and Colonial citizens were non-patrial, while all white British people and several million white Commonwealth citizens became patrial. Having placed blacks and aliens in the same category, legislation then proceeded to remove from EEC aliens most of the constraints it had placed on black Commonwealth workers and their dependents. The ultimate irony was that the 1973 Act came into force the day Britain entered the Common Market. Thereafter, only a handful of blacks were allowed in with work vouchers while white Europeans from the EEC could come in freely without prior entry clearance and obtain admission to seek employment at the point of entry. The settled Commonwealth male had to wait for several years for his wife, children under 18, or his dependent parents to convince entry clearance officers of their family relationships. EEC workers — male or female — could bring at the same time their spouse, children under 21, their parents and grandparents to live with them without having to prove relationships. The Government explained that the Commonwealth connection was becoming less "meaningful and relevant" as the nation prepared to join the white European club.

But apart from laying down a specifically racial basis for immigration control, the Act removed the right to settlement from those Commonwealth citizens who subsequently entered on work vouchers. They now occupied exactly the same position as aliens and reserved only the privilege of the right to vote, stand for Parliament, and work in the public services. The Act also provided for the removal of non-patrials, removed the right of Commonwealth citizens to register, provided public funds for repatriation, and continued the right of appeal against certain immigration decisions. In subsequent years, the appeals system was to become another instrument of control by interpreting the immigration rules so as to make entry and settlement more difficult. For example, in the case of elderly relatives, the condition that they had to be wholly or mainly dependent was extended through appeal decisions into a dependency of "necessity" and not of "choice."

After the 1974 "Winter of Discontent" the Conservative Government was ousted by the industrial action of the miners. The incoming Labor Government, despite its 1973 declaration that "racial intolerance has been aggravated by the structure of our citizenship and immigration laws," did not repeal the 1971 Act. They adopted it as the basis of their policy while making certain cosmetic changes to the rules. For example, they did not enforce the retrospective provisions of the 1971 Act by declaring a dubious amnesty (6) for illegal entrants; removed the restriction on husbands and male fiancés; increased the quota of UK passport holders in East Africa; and through the individual efforts of a liberal Minister Lyon speeded up applications from dependents abroad.

On the other hand, they stage-managed the extension of the definition of illegal immigrant — for whom there is no right of appeal in the UK — by a series of judicial decisions which progressively embraced more people who thought they had beome settled in the UK. Illegal entrants were now not only people who entered immigration control but also those who entered without passing through by deception, wittingly or unwittingly, and who had not revealed information which could later be held to have become material to their eligibility to enter (7). They made increasing use of the power to arrest alleged illegal immigrants without warrants and detain them without time limits, despite the provision of habeus corpus. The basic democratic right of not being imprisoned without a trial has become meaningless in immigration law. Later in 1977 they changed the rules to impose a 12-month "probationary period" on men entering as husbands and financés. This caused grossly objectionable intrusions into matrimonial affairs (8), and although introduced originally to discourage marriages of convenience, was subsequently used to remove husbands in cases of genuine matrimonial breakdown. They did not respond to the plight of the UK passport holders waiting in a four-year queue in India, and, with the sacking of Lyon in 1977, allowed the waiting time and refusal rates of dependents on the Indian subcontinent to escalate dramatically.

This shift in the Labor Party's attitude towards race and immigration was one symptom of the general decline of Britain into what Hall describes as the "law and order society." He writes about "the fining and focussing down of the problem of race into its concrete conditions in the inner city" and adds:

> "In these areas, the programmes of urban aid have failed to stem the tide of poverty and decay. The cycles of unemployment and the fears of recession are beginning to bite. Young blacks are increasingly unemployed — drifting, as every unemployed section of the working class historically has,

into petty crime and pilfering. The colony areas are the incipient basis for an increasingly restless and alienated population. This is where the crisis bites. Practically, these areas have to be policed with increasing strictness. But, also, the crisis has to be explained. Ideologically, it has to be dealt with, contained and managed. Blacks become the bearers, the signifiers, of the crisis of British society in the 70s; racism is its 'final solution'."

In such a climate the Labor Party produced the 1977 Green Paper on Nationality. This completely disregarded the racially diverse character of British citizens — some of whom enjoyed substantially less rights than others. It also failed to specify the future rights of Commonwealth citizens settled in the UK but promised to end sex discrimination in nationality law (10). This Labor Green Paper later became the basis of the Conservative proposals for a new Nationality Act, reflecting again the identical stance taken by both parties on this issue. Sivandan says: "What Powell says today, the Tories say tomorrow, and Labor legislates on the day after."(11)

This unholy alliance is graphically illustrated in the structure and recommendations made by the all-party Select Committee on Race Relations and Immigration which reported in 1978. Although their brief was to investigate the Government's general assumptions on race and immigration, they concentrated only on the restriction of immigration especially from the Indian subcontinent. Some of their recommendations were explicitly racist. For example, they proposed that a quota of dependents be established from the Indian subcontinent alone; that the special voucher system from East Africa should end; and that "the position of arranged marriages" be kept under review with fiancés given low priority in the queue. More disturbingly, they openly suggested ways of instituting the popular demand for tighter internal control. They proposed more resources for the police and the Immigration Service Intelligence Unit, more checks by the D.H.S.S., and tougher sanctions against illegal employment.

By 1978 the numbers of unemployed rose to two million, and the Conservatives sought power on the slogan "Labor isn't working." Thatcher raised the specter of numbers and the shadow Home Secretary, Whitelaw, outlined in a policy speech the means by which they could be reduced to "the absolute minimum." As the Conservatives formed a new government, these measures were laid out in the Change in Statement of Immigration Rules 1979.

These rules were redesigned especially to curtail the consolidation and propagation of family life of specific sections of the black community. They limited entry of husbands and fiancés to those who were sponsored by UK passport holders who were born here or had one parent born in the UK, thereby excluding large numbers of black women from sponsoring their husbands and fiancés from abroad, while at the same time ensuring that white women were not so affected. They also made the entry of elderly dependent relatives particularly from the Asian subcontinent virtually impossible by stipulating that they should have no other relative in their own country to turn to (12). The rules also subjected various categories of people coming for temporary or permanent stay to a requirement that they could not have recourse to public funds. This further restricted black workers from sponsoring their families for settlement or friends and relatives for visits if they happened to be unemployed. The rules also prevented visitors from staying in the UK for more than one year, students from changing courses, and temporary categories from applying for settlement. Control on the entry of black people continued also, as it had always done in the past, through the administration of the rules both abroad and in the UK. For instance, visitors from Bangladesh were fifty times more likely to be refused entry than Canadian visitors at Heathrow airport in 1980. And as the number of blacks entering the UK diminished, resources were diverted into developing internal controls. The number of illegal entrants discovered rose in 1980 following the introduction of an immigration control computer and extensive raids on work-places by police and immigration authorities. There was closer scrutiny of applications for citizenship with police frequently testing language ability of applicants. Marriage Registrars and hospital administrators began to report people whose immigrant status they doubted, the recommendation to deport became more frequently attached to various crimes committed by non-patrials, and police prepared special reports for the Home Office on those convicted of certain offences. Even people stopped in connection with motoring offences had their immigration status checked as a matter of course through the immigration computer. Demands for passports increased both as proof of identity and immigrant status in banks, building societies, tax offices, D.H.S.S., housing departments, schools, hospitals, and labor exchanges. Gradually Britain had, without the general awareness of the white majority, come to resemble a pass-law society.

With unemployment reaching an unprecedented high official figure of 3,500,000, Britain now has a large and increasing labor reserve of which blacks form a proportionately larger part. In these conditions, as growing numbers of black men, women, and young people find their economic functions surplus to the requirements of British industry, the constraints placed upon them by stricter immigration and nationality laws create fear, uncertainty, and resentment. The British Nationality Act of 1981 which comes into force on January 1, 1983, fuelled these fears. Concerned primarily with redefining British citizenship on the concept of patriality, it fails to address itself to the constitutional issues of nationality, the notion of allegiance to a nation, and the rights or obligations of the citizens of that nation. The Act creates six different categories of citizens only one of which carries any right of abode in the UK. The right to citizenship by birth is lost and restricted to those who are born with a parent "settled" here, the overwhelming majority of which are white. The concept of British subject is eradicated, thus opening up the possibility of redefining the rights of Commonwealth citizens (British subjects) settled in the UK. In fact, the Government has already called for an investigation into their continuing right to vote. The right to register as UK citizens will be replaced by discretionary naturalization, the conditions for which include proof of good conduct, fluency in the English language, and the presence of the spouse in this country (13). There is and never has been any right of appeal against the refusal to grant citizenship, and the Home Secretary does not have to state any reasons for refusal. The 1981 Nationality Act is not the end of the story. Immigration rules are soon to be published to conform to the definitions laid out in the Act. Though they may contain, from time to time, apparently liberal changes to appease international opinion (14), the main thrust of immigration and nationality legislation will be, as it has always been, to limit the pool of black workers in the labor market to a size that the rest of the country is prepared to tolerate. This tolerance will in turn be directly related to the way in which Britain feels the economic crisis.

Chapter four of the Conservative Party Manifesto contains the final truth behind black immigration control. Entitled "The Rule of Law," it lists the following subsections: "the fight against crime"; "deterring the criminal"; "immigration and race relations"; "the supremacy of Parliament"; and "Northern Ireland." Couched among phrases such as "a short sharp shock for young criminals"; "the power of strike committees and pickets"; "the defeat of terrorism"; and the "restoration of law and order," immigration and race relations appear tied together in a marriage of political convenience to prescribe a series of proposals, each one of which deals with the reduction of black people in the UK. That the Government has managed to fulfil these proposals, with the acquiescence of the Opposition and the silent white majority, is a measure of the degree to which racism has become, at every level, a respectable English ideology and repatriation has evolved as the logical and acceptable solution to the problem of race and immigration.

NOTES

(1) The term "black" refers to those people who originated in Asia, Africa, and Latin America, and who share a common history of colonization that determines their specific economic, political, and social relationships to Western society.

(2) This article appeared in the *London Chronicle* following the case of Somersett, a freed black slave who had, after a long legal battle, established his right not to be removed from Britain.

(3) Frank Tomney, Labor MP; quoted in Hiro, D. 1971. Black British, White British. London: Eyre and Spottiswoode.

(4) Years later the European Commission of Human Rights was discovered to have pronounced the following judgement in relation to this Act: "The commission finds it established that the 1968 Act had racial motives and that it covered a racial group ... constitutes an interference with human dignity which ... amounted to 'degrading treatment' in the sense of Article 3 of the Convention."

(5) Runnymede Trust Report, "Investigation into Immigration Control Procedures in the Indian Subcontinent," which looked at 58 applicants refused entry and found that only two were fraudulent, one inconclusive, and 55 had a genuine right of entry to the UK.

(6) The amnesty, in fact, applied only to specific categories of illegal immigrants and resulted in the removal of many people, e.g., deserter seamen, who mistakenly thought they qualified.

(7) Many women from the Phillipines who had been recruited through agencies to serve as domestic servants are being removed as illegal entrants years after their entry, when it was disclosed that they had children which their agents had neglected to mention.

(8) Immigration officers have asked husbands and wives separately about their sex life, color of bedsheets, etc.

(9) Hall, S. 1978. Five Views of Multi-Racial Britain. London: CRE Publications.

(10) Under the present Nationality Law, women can obtain citizenship through marriage to UK citizens, but the reverse is not possible.

(11) Sivanandan, A. 1982. A Different Kind of Hunger. London: Pluto Press.

(12) The phrase "to turn to" has been interpreted by court decisions to imply not simply financial but also spiritual, moral, and other assistance. In any extended family of the Asian type, this virtually prohibits entry.

(13) Unemployed Commonwealth Pakistani workers waiting for their wives to join them would not qualify for citizenship. Alternatively, their wives could not enter unless they fulfilled the requirement to accommodate and support or become British.

(15) The Foreign Husbands ruling may be changed following an application to the European Court of Human Rights by British wives not born in the UK to bring their husbands for settlement.

Minorities: Community and Identity, ed. C. Fried, pp. 269–281
Dahlem Konferenzen 1983. Berlin, Heidelberg, New York, Tokyo: Springer-Verlag.

THE POLITICAL, SOCIAL, AND LEGAL STATUS OF ALIENS AND REFUGEES IN THE FEDERAL REPUBLIC OF GERMANY

R. Marx
Amnesty International
6050 Offenbach, F. R. Germany

Abstract. Aliens in general and refugees in particular are marginal social groups affected by problems of a political, social, and legal nature. In the following, the situation of aliens and refugees in the Federal Republic of Germany (FRG) will be outlined. For a better understanding of the subject, the first step will be to elaborate the different groups of foreigners. An analysis of the political and social conditions with respect to aliens follows as well as a portrayal of the corresponding legal position.

THE VARIOUS GROUPS OF ALIENS IN THE FEDERAL REPUBLIC OF GERMANY

Total Size of the Alien Population

On September 30, 1981, there were 4.63 million aliens living in the Federal Republic of Germany. This is the highest number recorded since the foundation of the Federal Republic in 1949. The percentage of aliens in the total population is thus 7.7% (total population of the FRG is 61.56 million). The proportion of aliens in other Western countries is in some cases higher: Switzerland 14%; Belgium 9%; and France, also 7.7%. As a result of deliberate recruitment policies of the Federal and provincial authorities and the corresponding bilateral recruitment agreements between the Federal Government and the respective countries of origin, around 4 million aliens were admitted to the FRG in the period 1955-1974. Beginning in 1955, immigrant workers were recruited on the basis of agreements with Turkey, Yugoslavia, Italy, Greece, Spain, Portugal, Morocco, and Tunisia and were permitted to enter the FRG. The aim of this policy was to remedy the shortage of labor in the national economy. When further recruitment was stopped in 1973, the immigrant workers numbered circa 4 million. The relatively large increase in this group since 1977 is due to family reunification. In 1973, the Federal Government stopped further recruitment, and, as a result, the total number of aliens dropped to just under 4 million by 1977. However, since 1977 there has been a steady increase: 1977, circa 3.95 million; 1978, circa 3.98 million; 1979, circa 4.14 million; and 1980, circa 4.45 million persons.

The largest single group of foreigners is the Turks, of which there are over 1.5 million in the FRG. Out of a total of 4.63 million aliens, 1.06 million are juveniles. This group is also dominated by the Turkish citizens: 47.6% of the young foreigners are of Turkish nationality. Furthermore, 630,000 children of foreign parents now living in the FRG were born there.

The high proportion of Turkish citizens, on the one hand, and a large group of juveniles and children, on the other, are characteristic of the general situation of aliens in the FRG. This directly affects the framework of political and social conditions for foreign citizens. A precise analysis of these, however, must be preceded by a differentiation between the various groups of foreigners, for their political and social problems are correspondingly different.

Aliens from the European Economic Community (EEC) Countries

Foreign citizens from other EEC countries, France, Italy, Great Britain, Netherlands, Belgium, Luxemburg, Denmark, and Ireland, enjoy complete freedom as regards the right to work and the right to settle in the FRG. Thus no social problems of note exist for this group, and they may be excluded from consideration here.

However, political problems arise from the expansion of the EEC to include Spain, Portugal, Greece, and Turkey. As of January 1, 1981, Greek citizens now have limited freedom of movement and right of abode in the FRG, but in 1988 these restrictions will be lifted. No serious problems are expected regarding Greek, Spanish, or Portuguese nationals. But political circles do foresee serious problems if unrestricted right of entry is granted to Turkish nationals. For this reason, the Federal Government is attempting to hinder with the

EEC context the introduction of complete freedom of movement and abode for Turkish citizens.

Foreign Refugees

The term "foreign refugees" is used to differentiate between refugees who are aliens and refugees from the German Democratic Republic (GDR) and others with German citizenship: "evacuees." As the latter cannot properly be regarded as foreigners, they will not be considered here.

It is somewhat unclear which groups may be termed foreign refugees. The authorities consider this group as consisting of refugees who have been recognized as such before entering the FRG, in accordance with §22 of the Aliens Act. This is a very small group. A terminological distinction must be made between these refugees and persons granted the right of asylum in the FRG, as a special procedure exists for granting the right of asylum. Since the foundation of the Federal Republic, around 76,000 people have been granted asylum under this procedure. However, after asylum has been granted there is no difference either socially or legally within the group of foreign refugees. All are accorded the legal status provided by the international convention on refugees in 1951.

This legal status is also accorded to the so-called "quota refugees." These are persons who were admitted by the Federal Government on so-called humanitarian grounds. However, these refugees were not recognized under a juridical procedure such as that applicable to the right of asylum. They were admitted on the basis of a political decision made by the Federal Government in consultation with the provincial authorities (Bundesländer). The issue of a visa by the Federal German embassy in the respective countries automatically recognizes the recipient as a refugee.

Up to 1980, this group was also required to go through the procedure of application for asylum *after* entering the FRG, as were the other applicants for asylum described above. In 1980, though, an act was passed which waived this laborious procedure. To date, approximately 23,000 Vietnamese and 40 Argentinian refugees have benefited under this act.

There are considerable social, political, and legal problems regarding "holders of the right of asylum." As in the case of foreign refugees, holders of the right of asylum and quota refugees are accorded the full rights laid down by the International Convention on Refugees in 1951. Holders of the right of asylum suffer, however, from an individual sense of alienation. By comparison, foreign refugees and quota refugees are, in the main, estranged from their homeland and are thus able to be integrated to a great extent.

Asylum Seekers

These are aliens who apply for asylum, which is guaranteed under the constitution of the FRG as a basic right (Article 16, §2, Clause 2 of Federal German Constitutional Law). The application is considered under a special procedure to determine the status of the applicant (§6 ff., Right of Asylum Procedure Act, 1982; before this time, §29 ff. of the Aliens Act, 1965). Since the foundation of the Federal Republic, around 76,000 applications for asylum have been granted under this instrument. Until a final decision on an application has been reached, the foreign subject concerned is termed an asylum seeker.

Statistics regarding the size of this group, and particularly the growth in numbers of applicants for asylum, are revealing. Up to 1973, about 5,000 persons applied for asylum each year. Only 1969/70 was exceptional, when around 11,000 applicants were recorded due to the invasion of Czechoslovakia by the Warsaw Pact countries. From 1974, numbers rose slightly at first, then considerably, as can be seen in Table 1.

As a result of drastic measures taken by the Federal Government, however, only 49,391 persons applied for asylum in 1981, which represents a decrease of 54.2% compared to 1980. On December 31,1981, a total of 103,338 applications for asylum were still being considered.

The increase in the number of applications was accompanied by a change in the origin of the refugees: until 1973, the majority of applicants were from Eastern European countries, e.g.:1968, 89.4%; 1969, 92.7%; 1971, 64.7%; 1972, 58.4%; and 1973, 51.3%. Since 1974, applicants from the Third World have predominated. At first there were very large groups from Pakistan and India, and they became the subject of political debates in 1976 and 1977. Since 1978, Turkish citizens in particular have applied for asylum in the FRG. In 1980,

TABLE 1 — Increase in number of applications for asylum, 1974–1980.

Year	Number of Applications for Asylum
1974	9,424
1975	9,624
1976	11,125
1977	16,410
1978	33,136
1979	51,493
1980	107,818

about 58,000 or the 107,818 applicants were Turkish nationals. In 1981 there were only 6,302 Turkish applicants, a decrease of about 90% compared to the year before. In 1981 there was an increase in the number of Polish refugees: 1980, 2,090; 1981, 9,901.

With the increase in applications since 1974, there has been a corresponding decrease in the number of persons granted asylum, as shown in Table 2.

Rotation or Integration

Rotation means the enforced termination of stay after 2–5 years and replacement by new persons. The recruitment agreements, made from 1955 onwards, were based on the explicit assumption that the foreign workers would only stay for a limited period of time. However, this assumption proved to be an illusion. Industry objected to releasing workers they had trained after a relatively short period of time. Thus, politically, the model of rotation could not be realized for economic reasons. Yet the rotation model continues to play a relatively important part in political discussions.

The ban on further recruitment in 1973 had a decisive effect. It was a reaction to the oil crisis and the resultant structural crisis in the economy. The Federal Government adhered emphatically to this policy up to the present day, although some sections of industry are demanding a partial or complete lifting of the ban on further recruitment. For in spite of a high level of unemployment, a labor shortage exists in some sectors of the economy such as catering, mining, and agriculture. The demand for further recruitment is usually linked with the demand for the introduction of rotation or even the recruitment of

TABLE 2 — Decrease in number of persons granted asylum (in %), 1953-1981.

Year	Successful Applications for Asylum
1953-1968	20-50.00%
1969	84.80%
1970	76.40%
1971	70.20%
1972	53.30%
1973	52.80%
1974	42.80%
1975	33.00%
1976	24.30%
1977	15.60%
1978	13.40%
1979	14.50%
1980	11.63%
1981	11.01%

seasonal workers. Unlike rotation workers, seasonal workers would only be given a limited work and residence permit for the duration of the season, e.g., harvest time, height of season for restaurants.

Politically, these conceptions do not stand a chance of being put into practice at present. Nevertheless, models such as these are still put forward, particularly in conservative political circles. The ban on further recruitment and the rotation model are political opposites: the ban on recruitment made integration possible for the foreign citizens already in the FRG because planning became possible and long-range solutions could be implemented. The reintroduction of the rotation model would remove the political pressures for integration. A majority of both the Federal and provincial governments have decided in favor of integration.

Hostility towards Foreigners in the FRG

Since 1980, increasing open hostility towards foreigners has been apparent, though it obviously existed in a latent form before this date and also erupted openly on earlier occasions. Neo-fascist elements base their political strategies exclusively on hostility towards foreigners. There have been several bomb-attacks on refugees' hostels attributed to these circles. There is a widespread feeling in the population that the limits of the Federal Republic's capacity have been reached. This attitude has allowed the development of fascist ideology.

No contrary arguments are put forward in the conservative camp, where attention is centered primarily on strengthening existing resentment and the assertion that the FRG's limits of absorption have already been reached (1).

Subjective fears of over-infiltration by aliens and the objective fact of unemployment together produce a dangerous political atmosphere. The former Minister of the Interior of the Federal Government, G.R. Baum, who as such was responsible for policies concerning aliens, gave the following seven reasons responsible for the growing animosity towards foreigners:

1. the 10% increase in 1980;
2. the exceptionally high number of applicants for asylum in 1980;
3. the increased numbers of foreign juveniles joining their parents who have no knowledge of German or adequate elementary education;
4. difficulties concerning integration;
5. increase in the crime rate among foreign citizens;
6. fears of competition on the part of Germans arising from unemployment; and
7. aggravation of the existing housing shortage by the competition of foreigners (2).

The crime rate among aliens has, in fact, risen. In 1980 the proportion of persons of foreign nationality suspected of crimes was 15% as compared with a total of 6.9% of foreign nationals of the whole population. This problem particularly touches foreign juveniles who lack prospects for the future and are affected by a high rate of unemployment. It should be pointed out that these statistics also include offenses specific to foreigners — illegal entry and residence. And, of course, when social crises deepen, the tendency arises to impute blame for criminality to lower marginal groups. However, to the extent that these emotions and reactions are grounded in objective facts, it is high unemployment which is the real background to increasing fears of over-infiltration and animosity towards foreigners.

Unemployment in the FRG

At present there are nearly two million unemployed in the Federal Republic, and the figure is rising. On the other hand, there are 2.01 million foreigners employed. The equation is simple and often used politically: foreigners out = end of unemployment. Politically effective countermeasures to inform and enlighten the public as to the structural complexity of the unemployment problem are virtually nonexistent. In this way a convenient outlet has been created for very real fears about personal material existence.

This problem is aggravated further by an above-average rate of unemployment among foreign workers (in 1981 the unemployment rate as a whole was 5.2%; that of the immigrant workers was 8.1%): this strengthens the already existing prejudice that foreigners from southern countries are lazy and work-shy.

Fears of Over-infiltration by Aliens

The limits of tolerance depend on a variety of social factors and therefore are subject to fluctuation. Even with a foreign population of 7.7%, it cannot be assumed that these limits have been passed. On the other hand, the disproportionate number of foreigners in urban centers and the resultant formation of ghettos must not be forgotten. In Stuttgart, for example, the proportion of foreign citizens is 17%, and the situation in Berlin and Frankfurt (24%) is even higher.

The territorial concentration of foreigners in the urban areas is primarily due to economic structural reasons. However, the emergence of ghettos is also influenced by the conflict of aims which exists for the foreign citizens concerned: those who do not relinquish the idea of returning to their home country and thus will make do with cheap, uncomfortable accommodation that other people do not want.

GENERAL SOCIAL CONDITIONS

Recently, some politicians have accused foreign nationals of not being willing to integrate into society, despite the fact that for over twenty years there was no policy to promote integration. For this reason, not only could no feeling of trust develop among the foreign citizens, but it was also made clear to them that in the long term they were unwanted. In the meantime, the children of the first generation of immigrants have reached adult age, and there is now also a third generation. On the one hand, over 4 million people belonging to ethnic minorities have become de facto part of the populaton. On the other, because of the absence of a policy of integration — particularly in the housing and settlement questions — this group remains predominantly isolated. This politically created social conflict is exacerbated by two factors: the absence of political measures for integration and the increase in hostility toward foreigners, caused originally by politically determined isolation and subsequently carried out by the foreign nationals themselves.

The Conflict Between Willingness to Return to the Country of Origin and the Process of Integration

A recent survey conducted in West Germany showed that 65% of the foreign citizens interviewed have stayed longer in the FRG than they originally planned. Seventy-five percent of those interviewed stated that they wished to return home. Although in 1975, 600,000 of the 4 million foreign residents returned to their home country, the worsening of the economic situation in the countries of origin has since meant that plans to return have become increasingly illusory. In reality, the majority of foreign nationals have accommodated themselves to an indefinite stay. In September 1980, e.g., 41% of the foreign nationals in Baden-Württemberg had been there for ten years or longer. However, even today immigrants are returning to their home countries: in 1979 and 1980, 400,000 people left the FRG to return home. However, it cannot be assumed as a real possibility that the majority of foreign nationals residing in the FRG will return to their countries of origin.

The conflict of aims that exists between the reality of immigration and the subjective unwillingness to give up the idea of returning is particularly predominant among first generation immigrants and has thus a sustained influence on the opportunities and perspectives of subsequent generations. It is often the case that the desire for integration of the younger generations is rebuffed with reference to the wish to return. This obviously has consequences for the social process of integration.

The Concept of Deterrence Regarding Asylum Seekers

The social situation of applicants for asylum is completely different from that of foreign workers. In the case of asylum, integration is consciously and expressly deterred. Thus, in 1980 the Federal authorities introduced a regulation prohibiting applicants for asylum from working for one year after entering the country; this was extended in 1981 to two years. In Baden-Württemberg this ban continued during the whole period of the application procedure. Only in the case of asylum applicants from Eastern Europe is it limited to one year. Increasingly, it is standard practice to keep asylum applicants together in processing camps where they receive social assistance only in the form of payments in kind and a very small amount of pocket money. In addition, their freedom of movement is severely limited. They are not allowed outside of the town limits — in some cases, not even outside of the camps — without permission. The act prescribing the procedure for granting right of asylum which became law in 1982 guarantees these measures of control and deterrence by providing penalties of up to one year for infringement.

Whereas preeminently political grounds govern the social conditions of the immigrant workers, the social conditions of the asylum applicants are determined first and foremost by legal considerations. Whether this form of control is, in fact, permissible depends on the interpretation of the basic right of asylum as stated in the constitution.

THE FRAMEWORK OF LEGAL CONDITIONS

With respect to immigrant workers, the political aim is to consolidate their status as residents. A particularly important legal problem in this connection is that of family reunification.

Right of Residence for Immigrant Workers

The Aliens Act of 1965, which is currently in force, was passed at a time when concepts such as the rotation model and similar ideas were prevalent. The decisive clauses of this act, on which the right of residence depends, are to be determined according to "the interests of the Federal Republic of Germany" (§2, Section 1). It is evident that legal clauses phrased in this way cannot create the basis for trust of foreign citizens in the security of their rights and in the will to political integration.

Recognizing this fact, the Federal Government in 1978 attempted to amend the Aliens Act by means of administrative regulations. According to these, after five years of residence an indefinite residence permit may now be obtained, and after eight years this may be followed by right of residence, which affords greater protection against expulsion. The legal requirements for this are an adequate knowledge of German and suitable living quarters.

Thus, from the legal point of view, the situation of the immigrant workers has improved. After five years in residence they also have the legal right to a work permit. A draft of a new Aliens Act was planned by the previous Federal Government, and this was intended as a political demonstration of integration intentions by legally consolidating the right of residence. It was also planned to replace the "interests" clause with the stipulation of a legally binding set of factual findings. However, it is unlikely that this new draft law and the aims it pursues will be adopted by the new conservative government.

Family Reunification

In the past, there have been considerations which absolutely prevented spouses and children from joining the members of their families living in the FRG. However, this is in total contravention of Article 6 of the Basic Law, which states that the family enjoys the particular protection of the state. In December 1981 the Federal Government passed recommendations which in the meantime have been implemented: minors under 16 years of age may be admitted. If one of the parents is still living abroad, however, minors will not be admitted. As a general principle, spouses may join their partners. However, in the case of foreign citizens who were admitted to the FRG as children of immigrants or who were born there, their spouses living abroad are only permitted to enter if the foreign citizen concerned has lived in the Federal Republic without a break for eight years, is over 18 years of age, and has been married for at least a year. The new government will only allow children under six to be admitted.

Under a law which came into effect on August 3, 1982, the waiting period regulation which had been in practice since 1974 became legally binding. According to this, spouses may not take up employment for four years and children for two after entering the FRG. After this period, they receive work permits, but German applicants and foreign workers of equal status take priority on the labor market. These drastic measures are obviously designed to limit the immigration of relatives from abroad, for the increase in the total number of alien citizens entering the FRG since 1977 is due primarily to family reunification.

Guidelines for Naturalization

In December 1981 the Federal Government proposed a draft law, which still has to be debated, designed to assure the integration of the second and third generations of immigrants. This draft law grants minors the right of naturalization when they have reached the age of 18. The conditions are that they have lived in the Federal Republic for an uninterrupted period of at least eight years and that they surrender their former nationality. After the age of 21, the right of naturalization expires.

The political intention of the authorities is thus assimilation. At the same time it is being considered whether to promote the will to naturalization by introducing sanctions. Such current thoughts are, for example, to make the reunification of families conditional on the readiness to become naturalized and to conscript into the army those foreign citizens who wish to remain in the FRG but who do not take advantage of their right to naturalization (3). This assimilation concept will undoubtedly split a great number of immigrant worker families.

The Legal Conditions of Asylum Seekers

Although applicants for asylum are technically aliens, there is a special legal aspect of their situation related to the constitution: Article 16, Section 2, Paragraph 2 of the Basic Law guarantees politically persecuted foreigners a basic right of admission. No exception is made in the case of aliens who seek asylum at the frontier; they cannot be turned away. The application for asylum must first be considered under constitutional procedures before a decision can be made terminating residency.

In contrast to the USA, which admits a yearly quota of 50,000 refugees, the FRG cannot place a limitation on the numbers of applicants for asylum because of these constitutional grounds. In addition, Article 19, Section 4 of the Basic Law states that *anyone* can appeal to ordinary courts of law against decisions of the administration in the Administrative Courts. Whereas in other countries applications for asylum are only considered through administrative authorities, the procedure is protracted in the Federal Republic by this legal control for constitutional reasons.

It should be considered in this connection that because of the temporary right of stay and lengthy legal procedure involved, the number of foreign citizens who apply for asylum for reasons other than political persecution has increased. Whereas the state has a legitimate interest in hindering the admittance of aliens who are not politically persecuted by hastening the legal procedure, a limitation is placed on designs to speed up and streamline this by the precept that the examination procedure must conform to the arrangements as laid down by the constitution.

It is thus of decisive importance for the constitutionality of political deterrence whether it is possible to conclude this procedure in conformation with constitutional law within a period of time that allows the deterrence concept to remain effective. It was with this aim in mind that a new law was passed on August 1, 1982, to change the application for asylum procedure. However, since the applicants for asylum are forced to remain in processing centers (Sammellager) after rejection of the first application for asylum, for the duration of the application procedure, and since they are not allowed to work and are still under the jurisdiction of the controlling measures during this period, this aim has not been reached. For even under the new law the application procedure can take up to two years.

THE CONNECTION BETWEEN IMMIGRANT LABOR AND POLITICAL REFUGEES

Taking the Federal Republic as an example, two connected factors can be illustrated. On the one hand, if the basic constitutional right of asylum were to be abolished, the legal protection of those who are really politically persecuted would be severely diminished. On the other hand, it is not possible to prevent fully the abuse of the right of asylum by immigrant labor. As the North-South conflict sharpens and as the situation in the countries concerned worsens, threatening the material existence of its inhabitants, the migratory movements will increase further. The European countries will not be bypassed by these movements, but they will attempt to hold them at bay, and this will inevitably hit hard the politically persecuted among the immigrants.

Thus refugee and immigration movements are not unconnected. On the basis of the historical experience of fascism, the FRG has attempted to develop a process making the appropriate distinctions between immigrants and refugees with safeguards provided for under constitutional law. This attempt is now threatened with failure. One of the reasons is that other countries have hesitated to follow the legal example of the Federal Republic. Another is the worldwide economic imbalance that is leading to even greater migratory processes, thus making individual selection procedures in the interest of political refugees more and more difficult.

NOTES

(1) Herzog, R. 1981. Die Ursachen des Ausländerzustroms und die Möglichkeiten einer ausländerrechtlichen Steuerung as der Sicht des Landes Baden-Württemberg. Zeitschrift für Ausländerrecht und Ausländerpolitik **1(1)**: 17f.

(2) Ibid., p. 7.

(3) Schiffer, E. 1981. Ausländeraufnahme — Politik mit Perspektive. Zeitschrift für Ausländerrecht und Ausländerpolitik **1(4)**: 170.

Standing, left to right:
Michael Smith, Henry Shue, Volker Eichener, Paschalis Kitromilides, Ranjit Sondhi, Reinhard Marx.

Seated, left to right:
Hans Rau, Steve Thernstrom, Steven Uran, Michael Walzer, Ari Zolberg.

Minorities: Community and Identity, ed. C. Fried, pp. 283–298
Dahlem Konferenzen 1983. Berlin, Heidelberg, New York, Tokyo: Springer-Verlag.

HUMANE INCORPORATION: THE SHAPE OF ACCEPTABLE OPTIONS FOR RELATIONS BETWEEN MAJORITIES AND MINORITIES

Group Report

H. Shue, Rapporteur
V. Eichener, P.M. Kitromilides, R. Marx, H. Rau, M.G. Smith, R. Sondhi, S. Thernstrom, S. Uran, M. Walzer, A.R. Zolberg

INTRODUCTION

To our surprise, we arrived at a broad consensus about the general shape of acceptable options for relations between minorities and the larger society. Within this overall shape a variety of alternative sets of particular policies may be filled in. Our consensus was thus purchased partly at the price of indeterminacy on subsidiary but important issues. Some of the indeterminacy we regret as a result of a lack of time or information, but much of the indeterminacy we regard as a virtue of our recommendation. For we assume that no single option for relations between minorities and the larger society is an ideal to which all countries ought to aspire. On the contrary, the history and self-understanding of each country set limits upon the options that are possible for that country — no one solution is best, or even possible, for all. But while our consensus is congenial to a variety of specific policies and institutions, it is not hospitable to all solutions. We believe that some constraints apply generally and that some options can therefore be excluded as clearly unacceptable.

Acceptable options should, we concluded, provide minorities with genuine opportunities for humane incorporation: *incorporation* in that members of every minority should be fully able, if they wish, to enjoy the shared status of equal citizens available to all without discrimination; *humane* primarily in that members of each minority should be fully able, if they wish, to enjoy also the special identity of membership in their own community. This standard of humane incorporation requires a practical synthesis of pairs of principles that pull in opposite directions: incorporation into the majority culture and preservation of the minority culture, a universality across the whole society and a particularity within constituent groups. A distinction among occasions makes the synthesis possible: of relevant pairs of principles in tension with each other, sometimes one and sometimes the other holds sway. We will indicate more fully below how this distinction can be maintained.

The standard of humane incorporation is offered as a guide for social policy, not a guide for individual behavior. No suggestion is made, for example, that each member of a minority ought to devote some portion of his or her energies to participation in the activities of the larger society and some other portion to participation in the undertakings of the minority group. Individuals may choose to immerse themselves in either one and completely ignore the other or to divide themselves between the two in any proportions whatsoever — the principle of humane incorporation is silent about what use individuals should make of the opportunities available. What humane incorporation does require is that the society be arranged so that both kinds of options — equal participation in the larger society and full participation in the minority group — are genuinely open and fully protected. Humane incorporation indicates which kinds of opportunities ought to be protected, not which kinds of opportunities individuals ought to choose.

This standard of humane incorporation is directly applicable to naturalization policy: how those already legally inside the gates are to relate to each other. But naturalization policies are only half the story — the other half is immigration policy: who is to be allowed to come inside the gates. Immigration policies must answer two questions: How many? and, Which kinds? An immigration policy must specific the *total number* of people who are to be admitted and the *criteria* for the *selection* of which specific ones from among all who apply are to be admitted.

Our standard of humane incorporation provides only slight guidance on the question of numbers but significantly more on the question of criteria. Concerning numbers, the standard indicates merely that it is acceptable to admit no more immigrants than can be incorporated in accord with the standard's dual requirement of protected opportunity for political equality and protected opportunity for social diversity.

Concerning criteria for admission, the goal of having a nation that humanely incorporates the minorities it admits from outside does strongly constrain immigration procedures in at least two significant respects. The two together constitute what might be termed a prohibition against consignments to limbo. First, apart from persons who are explicitly admitted for only very brief periods, people who are allowed into a country should be permitted to begin, if they wish, a steady, even if gradual, process toward permanent residence and then citizenship. They ought not to be left indefinitely stranded in a shadowy status that allows them neither to participate as equals, if they wish, in the politics and economics of the territory in which they live, nor to participate securely, if they wish, in their original traditions. Second, the end point of the optional process toward full membership in the host polity should consist of a set of clear, achievable, and relevant criteria, the actual satisfaction of which guarantees citizenship. People ought not to be left unable, no matter how lengthy the duration of their residence inside the country, to claim as a right, if they wish, humane incorporation into the nation; and the criteria for admission ought to be no more stringent than necessary for the achievement of humane incorporation.

We also agreed upon an additional criterion for the selection of immigrants to be admitted, which is not entirely derivable from humane incorporation, namely, that some weight should be given both to the affinity of the applicants with the people and institutions of the potential receiving society and to the desperation of the situation of the applicants, especially when the potential receiving society has played a significant role in the creation of the desperate situation. The clearest cases of affinity are close relatives, and the clearest cases of desperation are political refugees.

In order to explain and justify our consensus on humane incorporation, we would like to recount (a) our analysis of the feasible options, acceptable and unacceptable, and (b) our analysis of the appropriate criteria for assessing these options. We will then provide (c) an example of the application of the criteria to some options and (d) an example of the choices unresolved by anything upon which we find it reasonable to agree.

FEASIBLE OPTIONS

Domestic Groups

An internally organized aggregate such as a minority can be incorporated into a modern state in one of three basic ways or, as in the humane incorporation we ultimately recommend, some combination of these basic options. The three basic ways may be called uniform incorporation, segmental consociation, and differential incorporation. Through the process of uniform incorporation, persons are admitted into the state individually, equally, and immediately. Each individual enters on his or her own but gains a status that is universal in the sense of being the same for everyone.

Segmental consociation admits people as members of previously constituted groups, each of which gains equal status. Individuals then participate in the state through membership in the constituent group, as a person participates in the Swiss polity through participation in the affairs of a canton. Differential incorporation similarly admits people as members of prior groups, but is distinctive in that the groups do not gain equal status. De facto combinations of the basic modes are not uncommon. White racist societies, for example, may incorporate whites uniformly but admit blacks only through segmental or differential consociation.

The humane incorporation that we are suggesting is intended as a very specific type of combination of some features of uniform incorporation and some features of segmental consociation. The combination is most certainly not of a type consisting of uniform incorporation for some people and segmental consociation for other people. On the contrary, it is an attempt to provide the universally shared status of uniform incorporation for every individual, along with the equal status of segmental consociation for every group (minority and majority). Our reasons for this choice of a conception of humanely consociational individual incorporation are given in succeeding sections.

Foreign Groups

The main alternatives among conventional approaches to foreign minorities include the following five: rotation, assimilation, mutual acculturation, stable pluralism, and regional autonomy. The policy of *rotation* (a) keeps the foreign group segregated from the broader society during its stay, (b) subsidizes or coerces the departure of the group after a fixed period of time, and (c) is ordinarily used to supply labor that is not available domestically at current wages. Neither as individuals nor as a group do people who are subject to rotation ordinarily enjoy equal membership in the society in which they labor.

The policy of *assimilation* requires the foreigners to relinquish to a significant degree elements of their identity as a group such as language, traditional practices, even values and religious beliefs. Obviously assimilation admits of degrees, but in any very strong form it constitutes the attainment of equal status for individuals at the price of damage to defining characteristics of the group. *Mutual acculturation* involves sacrifices to some extent of distinctive features of both majority and minority (or minorities). Compared to assimilation, it has the appeal of a certain sort of evenhandedness, but equality for the groups concerned is attained by similarly lowering the status of all.

Stable pluralism, according to which attempts are made to prevent the loss of distinctive group characteristics, compares favorably to mutual acculturation in that equality of status for groups is attained by protecting all instead of partially sacrificing all. At the extreme this approach would constitute a pure case of segmental consociation. What is missing from this approach is effective provision for any universally shared participation: group status is gained at the cost of any comprehensive community in which individuals may share the same status. *Regional autonomy* solves the difficulties about relating two groups by, in effect, abandoning the effort to relate them and allowing each to form an undifferentiated community on its own. Cases in which disassociation (e.g., partition or secession) would be preferable to objectionable forms of association arise, but only when the separation can be by negotiation, not force.

APPROPRIATE CRITERIA

Neither the three ways of admitting domestic groups nor the five approaches toward foreign groups strike thoughtful people as being equally acceptable. But even though some are immediately less appealing than others, it can be fruitful to articulate the implicit standards that underlie our intuitive reactions. We sought guidance for the formulation of plausible criteria in historical conceptions of citizenship and philosophical conceptions of community.

Historical Conceptions of Citizenship

Naturally it would have been impossible for us to have explored fully the complex subtleties of any one of the several rich conceptions available in the national histories with which we are familiar. Purely for our own heuristic purposes, then, we concentrated on selective tendencies among conceptions of citizenship, and none of the brief allusions to the conceptions of various nationalities which follow pretends to represent a considered comprehensive judgment of that conception.

In addition to explicitly dominating thought about naturalization, a state's conception of citizenship will directly affect its attitude toward immigration and its treatment of domestic minorities. Naturally, the influence goes both ways, and changing positions on immigrants or on minorities may well lead to modifications in received conceptions of citizenship. We began from the ideal of citizenship that was embodied in the ideology of the American revolution (without pursuing questions about the extent to which the ideal was observed in practice, because we were at this point making a catalogue of conceptions), since the American revolutionary ideal, along with the French revolutionary ideal, represented an intellectual watershed. American citizenship was conceived by contrast with British subjecthood, and it was conceived as universal in two critical respects. Citizenship was to be readily available to anyone who sought it, with no barriers except a five-year waiting period. And citizenship was to admit of no gradations or degrees: citizenship would be an equal status for all who held it.

Indeed, the ideology of the American revolution was an ideology of equality, and the basis of the equality was to lie in shared ideas. On this self-understanding, being an American would be a state of mind, not a matter of blood. Racist practices have been a violation of this self-understanding. Otherwise the function of the ideology was to include people: America would, Thomas Paine suggested in *Common Sense*, be an amalgam of the peoples of the world. This inclusiveness and the equality of status were its two forms of universality. But the ideology did also serve to exclude those who rejected its own content: Tories were not welcome because they did not share the vision. Hence the ideology of the American revolution did not represent the ultimate in inclusiveness; that may have been constituted by the Brazilian era in which immigration was entirely open and the government simply declared at one point that everyone within its territory who did not protest was a Brazilian, irrespective even of beliefs and of awareness of having become a citizen.

The same two elements of universality — broad inclusiveness and equality of status — are found in the ideology of the French revolution as well. For our purposes there is little to distinguish between revolutionary French and revolutionary American conceptions on the dimension of equal status. With regard to inclusiveness, however, one might say that the price had been raised (many would also say that the rewards were at least equally greater as well). Anyone who wishes may become a French citizen, but only by becoming French — French in culture. In the extreme Brazilian case, citizenship was granted and nothing more than acquiescence was asked. According to the

American ideology, nothing was to be asked but allegiance to the American ideal. It is plausible to think that on its own terms the French conception asks (and offers) a more complex and far-reaching change. But the conception is still inclusive, because what it asks anyone who chooses may do. Citizenship remains a state of mind, not a matter of blood, at the level of the ideal. And if the person becomes French in the relevant respects, he or she has then a right to citizenship.

Of the instances briefly considered, some of the tendencies in the current German and British conceptions of citizenship offered the most provocative contrasts with elements of the revolutionary American and French conceptions. (Whether there is any good reason why the British, German, or other conceptions should be more rather than less like the revolutionary American and French conceptions is, of course, the kind of normative judgment toward which we are still working. And, once again, the extent to which current American and French practices are betrayals of their own ideologies is a separate and important topic — but not *our* topic.) Naturally, there was no conception of German citizenship until 1871, nearly a century after the revolutionary American and French articulations. It is our understanding that the current German conception differs significantly on the dimension of inclusiveness, being positioned, as it were, considerably farther from the extreme American conception than the French conception is. Just as one must become French in significant respects in order to become a French citizen, one must become German in order to become a German citizen. But becoming German is still not enough — it merely constitutes eligibility for citizenship, not qualification in any definitive sense. The granting of citizenship, except in very special cases such as marriage to a German citizen, remains a matter of administrative discretion, subject to appeal only upon procedural grounds. Except in special cases, no outsider is recognized on the German conception to have a claim of right upon German citizenship. One cannot deserve it if the German state does not choose to grant it.

The current British conception of citizenship struck us as by far the most rapidly changing and complicated of those to which we referred. Significant new legislation took effect on January 1, 1983. The conception of British subjecthood is, of course, ancient — it was against it that the revolutionary American conception was shaped — but the conception of British citizenship has developed since World War II. The striking feature of the evolving British conception of citizenship is the centrality to it of the notion of patriality. Patriality is a matter of descent from someone who was born in, or otherwise entitled to permanent residence in, the United Kingdom. Reliance on patriality represents a radical narrowing of who is British, compared to the criteria of who is British entailed by the traditional conception of a British subject. It is also less inclusive than the revolutionary conceptions we are using for contrast in that it makes citizenship dependent on an historical fact of birth, not upon current behavior or beliefs that one might be able to choose to modify. Insofar as patrials tend to be white, it prefers whites who have never lived in the United Kingdom but had a grandparent who did, over nonwhites who earlier counted as British subjects. While there is nothing one can do to guarantee that one will be granted German citizenship, there are things one can do in order to qualify. For British citizenship as it is now evolving, there appears to be nothing that many long-time British subjects can do even to qualify, if they were born of the wrong parents. This appears to be approaching the other pole on the dimension of inclusiveness, moving toward making citizenship a matter of blood.

Philosophical Conceptions of Community

Whatever the differences among the various conceptions of citizenship available in our history, except perhaps for short-lived limiting cases like the Brazilian instance mentioned above, all states have had "admissions policies" in the sense of immigration policies, naturalization policies, or — usually — both. But what legitimate values, if any, do state admissions policies serve? Two seem particularly important.

What is perhaps the lesser of the two is the ideal of social democracy, or the welfare state, to which many people in the modern world are deeply committed. Governments are considered to have major responsibilities for the social and economic security of their citizens. But the resources under the control of each government are limited. If a given government is to provide the requisite level of economic and social security, then it must limit the number of people toward whom it bears these responsibilities. This entails ultimate control over immigration.

The more important argument rests upon the value of the self-determination of communities. A political community often embodies a way of life — it is an expression, and a continuing working-out, of how life ought to be organized. Without closure, no character. The right of a political community to shape its own character is perhaps the key ingredient of sovereignty. This right of self-determination is one of the most widely and deeply held values of our time.

This right of closure for the sake of self-definition is naturally not absolute. Two bases for qualifying it are affinity and necessity. One cannot appeal to self-determination as a ground for excluding those who share the dream, the heritage, or whatever is essential to the community's identity. Thus those with affinity have a strong claim, unless the limits specified by the argument from social democracy have been reached. Then one might exclude even one's fellows rather than overburden the society.

Also limiting the appeal to self-definition, and possibly the appeal to inability to fulfill the responsibilities of a social democracy, and in some cases competing even with affinity, is necessity, or desperation: the refugee who will pay in death or torture if he cannot flee his own state and who must have somewhere to go. Perhaps economic destitution as well as political persecution sometimes creates such a necessity for somewhere else to go that simple decency requires that the gates not be shut by those with more than enough of what is needed. Where the desperation is clearly in major part the making of the potential country of refuge, as the need for asylum by political refugees from Vietnam was in part the making of the United States, the responsibility for granting refuge is especially strong and clear.

ATTEMPTED RESOLUTIONS

No one assumed that either the preceding conceptions of citizenship or the conceptions of community offered any simple key to the best choices of options for relations between majorities and minorities. Against any claim to move directly from the value of self-determining communities to any conclusion about national immigration policies, for example, two points were raised. First, settling the boundaries of the unit with the right to exercise self-determination is a major additional problem. Subnational and transnational communities have their own claims to self-determination that conflict with the self-determination of existing national states; which claims should triumph requires further argument.

Second, granting that there is a right of closure is not the same as determining how tight, so to speak, the closure ought to be. Even given that a particular nation has a right to closure at some point, where would it be best for that nation to draw the circle? It might, for example, prefer to be rather open even though it would have been within its rights to have excluded people it actually chose to admit. It is possible to have one's reasons for choosing of one's own accord to be an immigration country.

Nevertheless, it did appear to us that policies toward minorities, domestic and foreign, with a certain general shape are (a) perfectly feasible and practical politically and economically and (b) manage to serve to a very considerable degree several of the most compelling values in our history and philosophy. That general shape was already introduced above as "humane incorporation." Using the materials in the sections on **FEASIBLE OPTIONS** and **APPROPRIATE CRITERIA,** we would like now to explain briefly how it is conceived and how it is justified, as well as its very real limitations.

The construction of the notion has already been foreshadowed. For domestic minorities we rejected the option of differential incorporation, leaving uniform incorporation and segmental consociation (on the assumption that the original three exhausted the significant alternatives). We refused to choose between uniform incorporation and segmental consociation — we chose both, and we chose both for everyone, sometimes. Why? Why precisely these choices?

Differential incorporation systematically divides the members of a common society into two or more radically unequal sections and requires that individuals belong to one or another of these in order to participate in the wider society. If the inequality among units is at all significant (and there is nothing to prevent further consolidations of power by the initially more powerful), the units inferior in power stand little chance of being self-determining; thus this important value is likely to be little enjoyed by the members of inferior units. Indeed, the way is left open for severe domination by the majority. And there is absolutely no provision for equal participation in self-government at the state level; thus the value of equal citizenship, much prized historically, is also served not at all. Perhaps some traditions (like the American and French revolutionary ones) have overrated the value of equal participation at the state level, but it would be difficult to defend assigning it a very low value.

This leaves uniform incorporation of individuals, which appears to serve equal participation directly and effectively, and segmental consociation, which seems well designed to allow for the self-determination and self-definition of each of the constituent groups granted equal status at the group level. The choice between the two modes would be a difficult choice between the two values — equal participation for individuals and self-determination for groups — that the two modes respectively serve. Better to cling to both if it can be done.

And it appears that it can be done if some workable line can be maintained between the one statewide arena and the several narrower arenas. Equal participation in self-government requires agreement on more values than just the value of equal participation in self-government — how many more, political sociologists' research tries to determine and we did not discuss. But a great variety of religions, races, customs, attitudes, etc., is compatible with sufficient sharing of attitudes and values for the functioning of a self-governing arena. And if the state arena is not allowed to encompass too much of community life — as it seems less likely to do if the individuals who share in it belong to smaller communities that they also value highly — diverse communities may be able to flourish as well.

This ideal of humane incorporation, which we suggest is a practical ideal, will become clearer if we use it to work through a concrete problem. So will its limitations. Consider now the assessment of the five approaches to foreign groups that constitute major conventional alternatives in immigration policy: rotation, assimilation, mutual acculturation, stable pluralism, and regional autonomy.

Clearly, a sixth approach that was the embodiment of humane incorporation would, provided it is workable, be superior to all five conventional approaches. Assimilation trades self-determination for equal participation, stable pluralism and regional autonomy trade equal participation for self-determination, and mutual acculturation buys equality with a piece out of every group's self-determination. That leaves rotation, which is important, controversial, and intellectually elusive.

Rotation has been a popular idea. It seems to be one of Britain's preferred methods for dealing with noncitizen subjects, one of Germany's disappointed hopes for its Turks, and one America's bright "new" ideas for reducing illegal immigration from Mexico. Rotation appears to be the fulfillment of every economist's dream: a Pareto optimal program that provides jobs for the unemployed in the sending country, cheap labor (with no job security or union militancy) for the receiving country, an improvement on the balance of payments (through repatriated wages) for the sending country, and a brake on inflation for the receiving country (through lowest available production costs) — everyone benefits and no one loses. (In fact, the worst-off workers of the receiving country lose significantly through wage depression, but the dimensions of the losses are controversial.) How could anything this good economically be bad?

Our standard of humane incorporation suggests that we look at the provisions for equal participation at the state level and for self-determination at the group level. It is difficult, although not impossible, for these workers to participate in the statewide arena of the sending country when they are away from it for extended periods of years, and no provision is made under a rotation program for them to participate in the statewide arena of the receiving country. And self-determination as a group? The first question here is: What is the group? Consider, for example, Turkish citizens in the Federal Republic of Gemany. If their relevant group is the Turkish polity, self-determination for *their* group is the same as equal participation at the state level; and it is difficult, as before, although not entirely impossible, at a great distance for an extended period. But if, as seems more plausible, the relevant group is "the Turkish minority in the Federal Republic" (or some subgroup of that), there will not be much more self-determination for the group than the Federal Republic makes provision for.

But why should the Federal Republic be the state to make provision for self-determination of the Turkish minority within the Federal Republic — not to mention provision for equal participation in government at the state level for individual Turks — when the whole point of a rotation program is, in Max Frisch's much-quoted phrase, to get workers, not human beings? The Federal Republic did not offer the Turks group self-determination and equal individual participation (any more than the proposed U.S. temporary worker programs offer them to Mexicans); it only offered them jobs.

The implications of humane incorporation for the rotation approach are less obvious than we would have liked. But we would judge that the implications include what in the **INTRODUCTION** was called a prohibition against consignment to limbo. Maybe the Federal Republic was only interested in the labor of the Turkish workers, just as many of them were doubtless interested only in the Germans' wages. Maybe both sides sincerely believed the arrangement was temporary, although a fixed date of departure was certainly not one of the terms of the mature version of the bilateral treaty (after the 1964 exchange of diplomatic notes). But as the lives of the Turks who continue to live in the Federal Republic pass, it increasingly appears that the only political game in which they are ever likely to be able to play is the one in the Federal Republic. The burden of argument seems to shift: what costs for the Federal Republic of citizenship for the resident Turks would be so great as to outweigh the values for the Turks, especially if the particular plan had the general shape of humane incorporation?

But what if the guestworker program had remained an authentic rotation plan with the maximum term of two years in the original bilateral recruitment agreement of October 1961? What if the Americans build a date of departure into the terms of a temporary worker program for Mexicans? Is it acceptable if the terms are explicit? On this, the standard of humane incorporation is not determinative, because the fundamental issue is, of course, whether there must be any sort of incorporation at all. If the period of temporary work is not so long as to constitute a consignment to limbo, our group was split on whether the rotation program was unacceptable. Deciding would involve appeal to other values we did not discuss thoroughly. These other values include the avoidance of the sheer exploitation of some human beings by others, which some members of the group believed would rule out any rotation plan even when its terms were explicit and agreed to in advance. Agreement to the terms might well be the forced product of sheer economic desperation and thus not carry the ethical weight of free consent. The rotation plan would use the foreign workers merely and strictly as means to the advancement of the goals of their employers (and the employers' community) and make no provision at all for the workers to form and pursue human goals of their own.

UNRESOLVED CHOICES

Needless to say, more is left unresolved than is resolved by our suggestions. At the conceptual level the major problem for humane incorporation is drawing the line between the statewide and the narrower arenas.

We have said that satisfying the standard of humane incorporation involves some practical synthesis of pairs of principles that pull against each other, such as the principle that every citizen, irrespective of his or her membership in a minority group, ought to be able to participate equally in the larger society, and the principle that members of every minority ought to be able to participate fully in the distinctive enterprises that contribute to their identity. When considerations are in tension with each other, to say merely that they should somehow be balanced is to offer very little guidance. We will try to be more specific — mostly by excluding what we take to be unacceptable arrangements — while recognizing that it is impossible for any general principle to determine specific decisions fully.

Our initial inclination was to suggest that the line between the sphere of participation in the larger society on an equal footing with all other citizens and the sphere of participation in the minority's distinctive activities should be the line between the public and private. This would yield a standard that is extremely neat — too neat, as it turns out — by allotting equal participation to public affairs and shared identity with one's minority to private affairs: the same in public, different in private. In public all are, say, Germans; in private, some are Christians, some are Jews, and some are Muslims. In public all are Americans; in private, some are whites, some are blacks, and some are Hispanics.

The serious difficulty with this tidy thought is that Muslims or blacks may not wish to be Muslims or blacks merely in private: they may wish to act on behalf of their minority in the public sphere. This does not mean that the public sphere must legitimate or authenticate their minority — on the contrary, members of other minorities, too, may wish to participate as members of those minorities in the public sphere. No group can expect favoritism in the public arena, of course, but its members cannot be expected to shed their identity as members of that minority as the price of admission to the public arena. Black citizens may want to take action as citizens, to be sure, but also as black ones on behalf of the rights, needs, or wishes of blacks. Hispanics, then, may choose to act as Hispanics (and either to join a coalition with blacks on a particular issue or to ally with whites) or not to pursue any position in the name of Hispanics.

But to *require* that, say, Hispanics never act publicly and politically as Hispanics is to make loss of identity the price of citizenship. This would constitute citizenship as the lowest common denominator, and we saw no reason for such an anemic form of citizenship. We saw no reason to relegate all expression of group identity to the private realm — and good reason not to relegate it so, namely, that minorities as such would thereby become politically impotent. The unavailability of public defense, like political action, must sooner or later mean that the private sphere can be threatened with impunity.

But avoiding the relegation of minority identity to the private sphere by no means implies swinging toward the other extreme of expecting public recognition for minority interests or rights in any fashion that would grant a special advantage to the minority in question. Muslims in Germany, for instance, should not be forced to be Muslims only in private, but they also should not expect Islam to be taught in the public schools in any way in which other religions are not also taught or private Muslim schools to have any advantages that private Catholic schools do not have.

For this reason we remained cautious about the endorsement by some participants of "mediating structures." In the context of discussion about the place of minority groups within a state, we take mediating structures to be institutions and associations that are neither state nor minority but are somehow intermediate — and therefore able to mediate — between the two. Thus a business or labor organization that is open to individuals from all minorities but is not a state organization is one kind of mediating structure.

We appreciate that a German Muslim, for example, might often prefer to act neither simply as a German citizen nor specifically as a Muslim but instead as a worker belonging to a certain trade union. Often what he or she sought would be no different from the objectives of a non-Muslim worker in the same union. Individuals have multiple identities: citizenship and minority-group membership are not the only identities people may value, and they may not value either as highly as other identities.

It is important, however, that these "mediating structures" not take on any official or quasi-official status if the balance is not to tip too far in the direction of the state. If some mediating structures come to have an official imprimatur, the others lacking that imprimatur will be at a disadvantage and will almost certainly come to be second-class organizations. We definitely think that any political expressions by members of minorities should neither need nor be given any official status beyond the protection of their equal opportunity to be pursued. Some minorities on some occasions may choose "to go public" with some issues, introducing those issues into the statewide political arena. The choice between the public pursuit and the private pursuit of the minority's interests and values should, however, lie with the minority, for whom the state should simply preserve both options.

At the practical level, these conceptual issues translate into difficult choices about the areas of social life that are arguably an inextricable tangle of statewide and minority matters: supremely difficult, perhaps, are education policy and language policy, which are of course intimately related to each other, and close behind in difficulty is housing policy. Time prevented us from seriously testing our standard against these hard cases, so we think the question of its fruitfulness (and of the viability of the attempt to maintain a reasonably clear line between statewide and community matters) simply remains open. We hope others will try to test and refine our admittedly crude suggestion.

To sense the difficulty, sample it in the area of education. How can we possibly say which aspects of education are state concerns and which community? Will the public schools of a liberal state not be secularizing? Will the public schools of a theocracy not be intolerant toward competing faiths? The state must educate its citizens in the skills necessary for effective participation if the equality of participation is to be of much value, but the minority group may wish to protect its members against the erosion of their distinctive traits by the homogenizing forces of majority-run institutions. How can all this actually be compatible?

We did not have time to pursue a case such as education in depth, but we think that the standard of humane incorporation does provide some basic direction. For example, just as it suggests that naturalization requirements should mandate no more assimilation that is necessary for effective participation in the larger polity, educational curricula should promote no more assimilation than is necessary for effective participation. On the other side, no minority group should expect any public subsidies for any of its educational endeavors that it is not willing to see go equally to the similar endeavors of other groups (majority or minority), since each group is equally entitled to self-determination and self-development.

We think, at least, that such approaches to selecting policy options are worth serious exploration.

Minorities: Community and Identity, ed. C. Fried, pp. 299–313
Dahlem Konferenzen 1983. Berlin, Heidelberg, New York, Tokyo: Springer-Verlag.

POLICIES FOR POLICE-MINORITY RELATIONS

M.P. Banton
Dept. of Sociology, University of Bristol
Bristol BS8 1UQ, England

INTRODUCTION

The criminal law, with very few exceptions, imposes the same obligations and confers the same rights upon all persons living within the state's boundaries. In practice, apparently respectable, rich, or powerful persons may be treated better when offenses are being investigated, but such differences arise in the implementation of the rules and are not authorized by them. Only in restricted circumstances are special rights conferred upon minorities. In England and Wales, special instructions are issued to the police concerning procedures for interviewing children and young persons, mentally handicapped persons, deaf persons, and those unable to speak English. Failure to comply with the directions may lead to the resulting evidence being declared inadmissible and excluded from the trial process, so members of these minorities have special rights. Not surprisingly, judges sometimes have difficulty determining whether the police have complied with these directions. I suggest, therefore, that in our discussions we should concentrate upon matters arising from the implementation of laws phrased in universalistic terms.

At present, particular concern is expressed about the position of ethnic minorities, especially those whose members can be identified from their outward appearance. Ethnic minorities will often be composed of immigrants or the descendants of immigrants and will come from larger ethnic groups with distinctive cultural traditions. Their relations with the police may be different because (a) the minority has its own culture and social structure; (b) it occupies a special position in the structure of the larger society; (c) police attitudes towards minoriity members are different. Once an ethnic minority secures a niche in the economic and social structure of an industrial society, relations between majority and minority can be expected to develop in certain predictable ways. This essay will start from some observations about the Turkish minority in Berlin as an example of the sequence in its early stages. It will then go on to discuss possible future developments and their implications for the future.

THE TURKISH MINORITY IN BERLIN

Out of a population of nearly two million in West Berlin, 246,000 (or nearly one in eight) are aliens, and over 118,000 (or more than one in seventeen) are Turks. Most of this increase (and especially that of the Turkish minority) has occurred since 1965. Since any opinions we express may concern policies to be implemented by the police, it would be well to begin by taking note of how the police themselves perceive the problems.

In November 1981, Hübner, the West Berlin police chief, delivered an address to the Bundeskriminalamt on this very topic (8). He began by remarking that in the contemporary debate it was presupposed that Germany was not, and could not become, a country of immigration; that the old assumption that Germany could utilize a migrant labor force of variable size consisting of workers who returned to their countries of origin had been proved invalid; and that the saturation point had been reached with respect to the demands the foreign population was making upon Germany's cultural and social structure. Integration could not be forced upon people. Those immigrants who came from Turkish cities had relevant work skills; they were both capable of being integrated and ready for it, but those who came from the peasant areas of Anatolia were neither. For them Islam provided not only a religion but a distinctive set of laws for daily living. Having come to Germany with a dream of returning as rich people to their homeland, they had seen no need to learn German and had created ethnic colonies in which they sought to live according to the norms of their homeland. In 1973 further recruitment was stopped, yet the immigrant population had grown as children and other family members came to join relatives in Berlin and more children were born. The social consequences of this should surprise no one. Whereas 15.6 percent of the German population were in the age group 0-17 and 23.1 percent were over 65 years old, the corresponding proportions for foreigners were 32.3 and 1.8 percent, respectively. The proportion of foreigners who were in the work force had declined substantially. In the Kreuzberg district of Berlin one half of all the children were of foreign origin, 77 percent of them Turkish. There were whole classes of Turkish pupils and the schools could make little contribution to social integration. As a result, new immigrants might not be permitted to take up residence in certain districts of the city. Until recently, little attention seems to have been paid to the social and political status of the children, and, as I understand it, they are growing up without any prospect of obtaining German citizenship or the right to vote.

Nor will it surprise many people that the incidence of juvenile delinquency amongst minority youngsters should be higher and is increasing. As Hübner pointed out, the group dynamics in which minority children have been involved have frequently pointed them towards criminal actions and have often provided them with accomplices. Their offenses were, again, those which might have been expected: theft from shops, unlawful taking of motor vehicles, handbag-snatching, causing bodily harm, theft from dwelling houses, attacking other children and threatening them with knives. The police experienced great difficulty in communicating with these children, even when they had the assistance of interpreters, and regretted that they did not have sufficient resources to grapple with the problems they perceived.

For the Berlin police the Turkish, Yugoslav, Italian, and Greek minorities posed one set of problems. Another set stemmed from the administration of the laws governing the right to political asylum. In 1981, 13,942 foreigners sought asylum: 5,064 from Arab countries, including Palestinians; 2,682 from Pakistan; 2,883 from Sri Lanka; and 1,028 from Ghana. Many Arab and Pakistani applicants for asylum entered from East Berlin, apparently assisted by the PLO office in that portion of the city. Hübner said that at times groups of Pakistanis were to be seen in the street outside the immigration office, each with his stencilled application for asylum which he regarded as an application for a work permit. They had already paid excessive sums to get to Berlin and were only too likely to be exploited if they were allowed to take up residence. Other immigrants strengthened their claims to residence by contracting marriages of convenience, and indeed, an organization existed to facilitate such marriages. In six months during 1981, 149 such marriages were identified. There was also a significant number of Ghanaian women who had married German men outside the country and were engaged in prostitution in Berlin. Palestinians who held Lebanese passports posed special problems, both as applicants for asylum and, if given entry, for their criminal associations. Whereas Lebanese constituted 1.5 percent of the foreign population, they accounted for 11 percent of the convicted foreign population; indeed, the figures showed that over 55 percent of the Lebansese population had been convicted, 9.2

percent of them for narcotics offenses. Hübner gave details of a man born in 1951 who sought asylum in Berlin in 1972 and on three occasions since had been sentenced to a total of six years' imprisonment for trading in heroin. Another man who was born in Beirut and came to Berlin in 1977 at the age of 15 had by the beginning of 1981 been convicted of 17 varied offenses including theft, attempted murder, robbery with violence, unlawful possession of firearms, and trading heroin with a street-value of 90,000 DM. A 1981 Berlin police report stated that hostility towards foreigners had been growing among the German population. It is surely dangerous to the interests of the law-abiding foreign workers if they are identified with the activities of a small minority who abuse the right to asylum that has been accorded them.

Hübner also drew attention to problems for the police which arose from conflicts within or between ethnic minorities. Turkish trade unionists tended to have leftist connections and to be opposed to groups associated with the mosques. Recently, there had been mass fights, mostly with wooden cudgels, though one Turkish teacher died from a knife wound received in such a confrontation. Representatives of an Islamic organization had entered schools and insisted, with threats, that female Turkish pupils wear headdresses. A leftist group had occupied the Turkish consulate, injuring the consul and members of his staff. Iranian political groups had fought one another, as had Lebanese and Pakistanis. Yugoslav groups had attacked the Jugoslav diplomatic representative in Berlin, and Yugoslav shopkeepers in Germany had been financing the Croat liberation movement. There had also been fierce fighting between left- and right-wing Vietnamese students.

Hübner referred to the measures necessary to combat special forms of crime, such as those associated with narcotics, but stressed the general issues. The police, he said, had little trouble with foreigners who had a similar outlook and were able to live together with Germans according to commonly accepted norms. It was much more difficult if groups with different outlooks lived in cultural islands and produced a reaction from those around them. So, independently of political decisions about the social position to be accorded minority members, there was "a practical requirement that foreigners must be integrated in the sense that they are not marginal groups, not problem groups, but participants in our society." The contribution which the police can make to such policies is limited, but they arrange special training courses to help selected police officers acquire a working knowledge of the Turkish language and of the culture of the Anatolian village community.

A GENERAL MODEL

Anatolian immigrants in Berlin appear to constitute what sociologists call an incapsulating minority, i.e., a group that attempts to preserve in a new environment the way of life to which it was earlier accustomed. In that new environment their values and goals gradually change, so that after about twenty years they have often created a new culture intermediate between those of what may conveniently be called the sending and receiving societies. Very few minorities maintain a distinctive culture after three generations, the usual course being one of assimilation. But assimilation is not a simple process of unidirectional change. This section will outline some of the processes of change to be expected within such a minority as a result of their changing relations with the majority.

In peasant regions the family is usually the landowning unit; families compete with one another for power and prestige, much of the competition centering on the negotiation of favorable marriages. To obtain the most valued things in life, an individual usually has to depend upon his family and upon its head who controls the allocation of resources. This is the economic basis of the social relationships most important to the maintenance of any distinct culture: those between men and women and between parents and children. The man manages the land and the labor outside the home, the woman manages the domestic sphere. Children grow up observing their parents' work and increasingly participating in it. When such a family transfers itself to an industrialized society, the members are no longer economically dependent upon one another in the old ways. The parents can no longer speak with authority on how their children should behave outside the home, indeed, the children may become the interpreters and a source of information about the receiving society's expectations. For role models children look to their peers and sometimes to teachers who can assist them toward scholarships and occupational advance. When the wife obtains paid employment, she expects a say in decisions about how her earnings are to be spent. The general pattern is one of people seizing new opportunities which shift the balance of power and reciprocity, forcing everyone to change their earlier ideas as to what each expects from the other. There is a conflict between role expectations of two cultures which can sweep in so fast as to cause great emotional strain and contribute to breaches of the criminal law. Girls who have disgraced the family honor have been murdered or ejected from the home. Wives have been beaten or killed. Men have attacked one another. The incidence of mental disorder has increased. Children have run off or taken to stealing.

Immigrant family life in the receiving society may change more slowly if it is still tied to the economic structure of the homeland. Sometimes migrants send money back to be invested in land, housing, or property. They may seek brides or bridegrooms for their children from the homeland. They will seek to resolve any disputes within the family or by negotiation between two families. So long as such informal controls are effective, immigrants less frequently invoke action by the police or the social agencies of the receiving society. At times, though, the internal controls are ineffective, as when the dispute is between family groups with few shared interests. Then either or both parties may attempt to involve the police in such a way as to punish their opponents. Using their normal methods of inquiry, the police may be unable to get a sufficiently clear view of the dispute to determine what response is appropriate according to the law of the receiving society. Their ignorance of minority culture can be a serious worry when they have a dead body on their hands.

Life in industrial societies promotes much more intense competition between individuals. Children are brought up from an early age so as to prepare them for such competition, though much depends upon the kind of advantages they inherit from their parents. In Sowell's expression, social mobility is an intergenerational relay race. The first generation of immigrants is self-selected (except when they are refugees) and consists of people with above-average ambition who often respond heroically to the challenges and disappointments of the transition. Because of their own problems it is difficult for them to give their children as much support as parents in the native population, so the children grow up at a disadvantage and feel dissatisfied (as do many working-class native children) by the restriction in the opportunities available to those who start the race with a handicap. Between the first immigrant generation (born overseas) and the second generation (born in the new country) there is what I call the sesque generation (from the Latin for one-and-a-half), consisting of overseas born individuals brought as children to the new country whose socialization is split in two. It may be expectd that this generation will be subjected to the most stress.

In general, crime rates in the first generation will be well below those in the native population; in the sesque and second generations they may be higher (depending on the strength of the immigrant family system), but there will be much variation from one section of the minority population to another and according to the type of offense. Any higher incidence of minority criminality is soon noticed by the police and easily promotes stereotyping. If a large proportion of, e.g., car thefts is committed by minority youths, it is rational for police to investigate whenever they see a smart car driven by a minority youth since there is a greater likelihood of discovering an offender this way than by random checks. It is rational in the same sense as it is rational for a manufacturer to discharge a noxious chemical into a nearby river when there is no law against environmental pollution. In securing his own ends more economically, the individual derives a private benefit while creating a public cost. Unless the policeman has some reason additional to minority status for stopping the minority driver, he will contribute, first, to an unwillingness to cooperate with the police among that minority, and second, to an amplification of the initial identification. Because minority drivers are stopped more frequently, more offenses will be discovered among them relative to their proportion of the total population, even if there is no difference in the frequency with which they commit the offense in question. The initial belief will have turned into a self-fulfilling prophecy (3).

An immigrant minority can usually catch up economically with the native working class within twenty years, but it is unlikely to attain the same proportion of highly paid occupations within three generations. Its very presence, therefore, tends to draw attention to the barriers to upward mobility which affect everyone in that society. Wherever that society's institutions function badly, they will function worse for the minority, and this is likely to influence the latter's position in the political structure. Which native political parties the minority supports will depend largely upon their occupational distribution. Some majority political groups will seek support from the majority by displaying hostility towards the minority. At the other extreme, some majority groups wishing to change the political structure will attempt to channel minority political activity so as to further their ends.

EXPERIENCE IN BRITAIN

From my inquiries it would seem that the British police have done as much as, if not more than, the police of any other European country to identify the problems of police-minority relations and to develop policies for responding to them. This may excuse my drawing disproportionately upon the British experience. It should be added that since British immigration control is concentrated upon checks at ports of entry, the police are less involved than those of some countries in the enforcement of immigration and labor legislation. (This is also the place to note that while there is a substantial literature on police-black relations in the United States, there seems to be nothing on police-Hispanic relations, which could provide a more relevant analogy for us in

Europe. In general it would seem that the main focus of the former is now on police-black relations *within* police departments (9).)

Great Britain has, especially since 1948, received a substantial number of immigrants from the West Indies, Africa, and Asia. They and their descendants constituted in 1981 about 3.3 percent of a population of roughly fifty million. Not until the end of the 1980s will the proportion of UK-born members of these minorities exceed the proportion of overseas-born. By the year 2000 they will constitute 5.9 percent of the population, after which little further growth is expected since further immigration has been virtually stopped (apart from family reunification) and the immigrant birthrate is expected soon to resemble the native rate. Most members of these minorities have from the beginning possessed British citizenship; others (apart from illegal immigrants) will be eligible for it. The Asian minorities can be classed as incapsulating, particularly the Mirpuris and Bengalis, but also, to some extent, the Gujaratis and Sikhs. People of West Indian origin do not constitute an incapsulating minority; they immigrated with an assimilationist orientation but were rebuffed; their community controls have been weaker than the Asians' and the processes of change have hit them harder.

In 1967 the Home Secretary issued to all police forces in England and Wales (of which there are now 43) some general information and advice on means of ensuring good relations between the police and colored people. This led to the designation of selected policemen as Community Liaison Officers and to additional training. It was also about this time that (belatedly, as can now be seen) more attention was paid to the recruitment of police officers from the minorities. A government working party on Police Training in Race Relations reported in 1971.

About this time I addressed a variety of police training classes. The dominant opinion expressed in these classes was that such training was unnecessary and misconceived. Policemen asserted that their job was to enforce English law and that they did so without discrimination. It had not been demonstrated that, in general, police performance had been deficient. It was the immigrants who needed training in English law and customs. If the immigrants were to live in England they must adjust to the English. For the police to be given training about their expectations was to give the immigrants a specially favored status which they did not deserve. The Home Office working party took a different view. They noted that when a policeman comes from the same background as the people of the neighborhood he serves, he can draw upon an understanding of them acquired in his childhood while his off-duty associations serve to keep him in touch. A policeman who moves to another part of the country or who has to deal with citizens of a different social class may have to learn how to get on with people with whom he cannot so easily identify. Extra training about minority groups for police officers from the majority population was to complement a kind of training about the majority group which they had received before joining the police.

Four principles for police training in this field were stated. The first was that it should relate to the general duties of citizenship and not be seen as something special or separate. The second principle concerned the minorities and listed the features of their lives that needed to be understood. The third acknowledged the relevance of majority attitudes, including evidence on white prejudice and discrimination. The fourth principle recognized that the police themselves were caught up in the attitudes of the white majority and needed more insight into the roles they were called upon to play. New training courses have been introduced which give increased attention to the social and psychological aspects of police work at the expense of tuition concerning laws with which the ordinary policeman rarely has to concern himself.

In the early 1970s the police were agreed that, on the whole, immigrant crime rates were slightly lower than those of the general population. There were some minor exceptions. The illlegal sale of liquor in "shebeens" was said to be a nuisance in several cities, and in Manchester the colored population was said to be more negligent over the licensing, insurance, driving, and maintenance of motor vehicles. In London black youths were increasingly participating in nonviolent street crime, particularly in the snatching of women's handbags. An analysis of the London figures for 1975 found that, after correcting for the age distributions of the populations in question, Asians were no more likely to be arrested than whites, but blacks were more than twice as likely to be arrested for indictable crimes.

Hübner remarked at one point in his address that in connection with immigrant behavior patterns there is "yet more trouble" to be expected. Many British police officers would say that one source of such trouble lies in the mass media and the educational system. Inspired by the black power movement in the United States, many liberals came to feel so guilty about the whole history of black-white relations that they were ready to make excuses for black misconduct. The July 1982 issue of the monthly magazine of the Police Federation (representing all ranks below superintendent) carries an article,"Who taught the rioters?" by a former London teacher who writes that many teachers in the immigrant areas had studied in the atmosphere of "post 1968 radicalism" and sought to sensitize pupils to "the wickedness of late capitalism." West Indian parents brought with them ideas about how to discipline children which horrified liberal English social workers. A mother is quoted as saying "The biggest culprits are the welfare officers who ... can't wait to take black kids from their homes to put them with nice white aunties and uncles, where they are allowed to run riot." Teachers were afraid that they would be accused of racism if they took energetic action to restrain misbehavior by black children or if they assigned them to special schools. Many policemen would argue that the liberal intelligentsia has behaved in a manner subversive to the interests of both the state and the minorities. They have provided a rationale and a legitimation for resistance to lawful authority. Since members of the minorities must in the end conform to the expectations of the society in which they live, their adjustment has been only hindered by the elaboration of historical grievances. Minority members suffer more from crime than majority members of comparable socioeconomic status. They have reason to be anxious about the reputation of their communities. Reponsible minority representatives are concerned about law enforcement, but white political activists have made it more difficult for them to give a lead.

Police spokesmen do not speak in these terms. They list various things they have done to improve training, too build community contracts, and to recruit policemen from the minorities (the Metropolitan Police have had to struggle hard to lift the number of minority police officers to 138 in a force of over 25,000), and they say that they are doing at least as well as many other national institutions, all of which are finding "racial equality" a challenging objective. Spokesmen, of course, seek to present their own organizations favorably. Older and more senior police officers usually want as quiet a life as possible. Younger ones, especially the men, want excitement. (For a discussion of how the occupational culture of the lower ranks can affect the policing of a mixed neighborhood, see (7), pp. 66-82.) In some important respects these efforts have been quite unsuccessful. A Home Office study in 1981 discovered that the incidence of racial attacks on Asians was 50 times greater, and that on blacks over 36 times greater, than upon white people. The Home Secretary acknowledged that such attacks were more common than had been realized and that they seemed to be on the increase. The attacks were not orchestrated by extremist organizations, a conclusion that makes the evidence more frightening. The police are dependent upon public cooperation to solve crimes of this kind; it would seem that only limited cooperation had been forthcoming and that the police had given the investigation of such attacks low priority.

Nineteen eighty-one, of course, was the year of the riots, described in Lord Scarman's report as "communal disturbances arising from a complex political, social and economic situation ... there was a strong racial element ... essentially an outburst of anger and resentment by young black people against the police." The first serious riot was triggered off by two young policemen whose conduct, though not unlawful, lacked "the discretion and judgement which maturer years might have brought. Perhaps they had become inured by their experience of the hostility which police action could arouse in Brixton to the point where they failed to recognize real danger signals or to strike the correct balance between enforcing the law and keeping the peace." Lord Scarman recommended improved ethnic minority recruitment into the police; the screening of recruits for racial prejudice; expanded training for recruits and compulsory in-service training in community relations for officers up to and including the rank of superintendent; improved supervision; dismissal for racially prejudiced behavior; random checks by lay visitors on the interrogation and detention of suspects in police stations; reform of the procedure for considering complaints against the police; restrictions upon processions in racially sensitive neighborhoods; statutory provision for local consultation about policing; and better coordination of, and public participation in, measures to deal with inner-city problems, including those of housing, education, and employment.

The British have been so concerned about keeping issues of policing outside the arena of party politics that they have been unwilling to recognize that the allocation of resources for law enforcement is of real and increasing concern to citizens; it is a political matter even when it is not party-political. According to recent reports, the incidence of serious crimes per caput is officially expected to increase by three hundred percent in the next ten years, while the pressure on the prison service had already reached the crisis point two years ago. Recent years have seen a public debate about "community policing" without previous parallel, while the riots have helped to bring police methods into the arena of public debate as never before in the country's history. Any policy for police-minority relations, therefore, has to form part of a larger policy for maintaining the peace within the nation. I have argued that the problems of determining the optimum use of the criminal justice budget and of assessing the criminogenic implications of official and unofficial policies will lead to creation of a body of specialist analysts, and that the problem for the police will be whether they wish to participate in staffing the decision-making processes or not (2). The British police tradition has comprehended both law enforcement and crime prevention, but crime prevention in contemporary circumstances requires new initiatives.

THE NATURE OF THE OPTIONS

By way of conclusion, it may be helpful to try to identify the options open in regard to police-minority relations, even if the generalizations cannot be supported by many empirical studies. Police officers are apt to claim that the range of options is narrow because their powers and obligations are laid down in the law. Sometimes, indeed, the law severely constrains police decision-making, particularly in dealing with individual offenders, but in matters of internal organization much is left to the discretion of police administrators. The power of such officials independently to determine the allocation of police resources in preventing and detecting crimes varies from one country to another, but it is usually considerable. These two areas will be considered in turn.

Three aspects of internal police organization and practice which bear closely upon minority relations are those of recruitment, training, and supervision. As to the first, it would be widely agreed that the composition of the police should reflect that of the population, and that minorities should be adequately represented in the police. Applicants or serving police officers who display prejudice towards minority members should be considered unsuitable for police employment. The problems that arise, however, are of a different order. Is it desirable to reduce recruitment standards for minority members in order to increase their representation? The general view is firmly to the contrary. The effectiveness of minority police officers would be lowered if it were thought that they had not been able to meet the same tests as others. It is better to provide additional pre-entry tuition for minority applicants. Similarly, the issue concerning the ineligibilty of prejudiced people is not that of principle but of the effectiveness of tests for screening applicants.

The options with reference to training are again practical ones: a) how much (given that training is expensive)? b) at what stages (recruits, in-service, supervisors)? and c) what kind (the imparting of objective information, or sensitivity training; by police or non-police tutors)? To decide such matters information is needed about the effectiveness of particular programs for given kinds of trainees in varied local circumstances. The evidence from studies in the United States and Britain has not so far offered any encouragement at all (see **BIBLIOGRAPHY**). But while it has not been possible to demonstrate that any kind of training in this field has improved either police attitudes or performance, police commanders believe that the position would deteriorate were such training discontinued. The problem calls for systematic experimentation and research. The same conclusion might be urged in connection with the effect of different kinds of supervision, were that not an even more difficult topic to study. If first-line supervisors are convinced of the need to take disciplinary action against police officers who treat minority members badly, this has positive effects upon minority relations generally. The effectiveness of actions by more senior officers to see that first-line supervisors implement policies is equally crucial.

Questions of minority relations do not so much raise new issues regarding the allocation of resources as provide variations upon old issues. The police regularly tolerate lower standards of conduct in "red light" and "skid row" areas of the inner city, knowing that if they were to enforce the law there more strictly, they would only drive the offenders to districts where it would be more difficult to supervise them. The police also recognize that they can often achieve their objectives more easily by utilizing the informal social controls that form part of the community structure (as by seeking the assistance of a priest or a head teacher). It is, therefore, nothing new if they underenforce some laws in immigrant areas (e.g., the laws against the use of cannabis — or marijuana — in some parts of British cities frequented by black youths) or if they allow an immigrant minority to settle informally disputes which among the majority might go before the courts. The central issue is that of the accountability of police chiefs for their decisions as to the relative priority to be attached to the investigation and prevention of particular kinds of crime. Countries vary more in the means by which they ensure accountability than in the ends themselves (4). Police chiefs may be accountable in varying circumstances and degrees to a central government ministry, to a mayor or prefect, and to local committees representative of the public. Responsibility for criminal investigation, for preventive policing, and for registration of aliens may be separate or combined. What matters is that minority interests should be adequately represented in the bodies or institutions to which police chiefs are accountable; that offenses by minority members are not pursued with disproportionate vigor; and that minority demands for services from the police evoke an equitable response. These are problems at the local level as much as matters of central policy. Frequently, social agencies and institutions are insufficiently well informed about the tensions within ethnic minorities and in their relations with members of the majority, and therefore they are unaware of their shortcomings. Administrative improvements to such matters as recruitment and training will never achieve their goals so long as minority interests are inadequately represented in the policy-making process.

BIBLIOGRAPHY

(1) Banton, M. 1973. Police-Community Relations. London: Collins.

(2) Banton, M. 1978. Crime prevention in the context of criminal policy. Police Studies **1**: 3–9.

(3) Banton, M. 1983. Categorical and statistical discrimination. Ethn. Racial **6(3)**:269–283.

(4) Bayley,D.H. 1982. Accountabilty and Control of Police: Lessons for Britain. Paper presented at Cropwood Conference, Cambridge, England, December 1982.

(5) Butler, A.J.P. 1982. An Examination of the Influences of Training and Experience on the Attitudes and Perceptions of Police Constables. West Midlands Police, Birmingham.

(6) Field, S., and Southgate, P. 1982. Public Disorder: a review of research and a study in one inner city area. Home Office Research Study No. 72. London: Home Office.

(7) Holdaway, S., ed. 1979. The British Police. London: Arnold.

(8) Hübner, K. 1982. Integrationsprobleme der Ausländer und ihre Auswirkungen auf die Polizei. Kriminalist.**(Mai)**: 283–290.

(9) Jacobs, J.B., and Cohen, J. 1978. The Impact of Racial Integration on the Police. J. Polic. Sci. **6**: 168–183.

(10) Kettle, M., and Hodges, L. 1982. Uprising: The Police, the People and the Riots in Britain's Cities. London: Pan Books.

(11) Scarman, Lord. 1981. The Brixton Disorders 10-12 April 1981. Cmnd 8427. London.

(12) Southgate, P. 1982. Police Probationer Training in Race Relations. Home Office Research and Planning Unit Paper No. 8. London: Home Office.

(13) Teahan, J.E.; Adams, M.; and Podany, C. 1980. A comparison of the value structure of British and U.S. police. Int. J. Soc. P. **26**: 246–254.

(14) Tuck, M., and Southgate, P. 1981. Ethnic Minorities, Crime and Policing: a survey of the experiences of West Indians and whites. Home Office Research Study No. 70. London: Home Office.

Minorities: Community and Identity, ed. C. Fried, pp. 315–327
Dahlem Konferenzen 1983. Berlin, Heidelberg, New York, Tokyo: Springer-Verlag.

WRITING A CONSTITUTION TO PROTECT MINORITIES: THE CANADIAN EXPERIENCE

P.C. Weiler
Harvard Law School
Cambridge, MA 02138, USA

INTRODUCTION

It is commonplace to say that equality is the predominant value in modern political discourse, whether for good or ill. One of the central attractions of this notion is its usefulness in the fight against discriminatory treatment of discrete and historically-oppressed minorities. And it is only natural that *constitutional* arrangements should be seen as the crucial instrument through which a minority can be guaranteed protection against unfair treatment at the hands of the majority.

It is impossible in a short paper to say anything about this subject which would be both broadly applicable and tangibly significant. Instead I shall treat in some detail a single case, the recent Canadian attempt to grapple with its minority problem by writing a new constitution. At the end of the paper I shall briefly note some of the general lessons from this specific experience.

THE SPECIAL CHARACTER OF THE CANADIAN MINORITY ISSUE

It goes without saying that in fashioning a constitutional provision to protect a minority group, much depends on the characteristic features of this minority. In Canada, as we shall see, the perennial concern has been the condition of the French-Canadian. But the modern impetus for writing constitutional guarantees of rights actually came from ill-treatment in the Forties of the Japanese Canadians (on the West coast) and the Jehovah Witnesses (in

Quebec) (14). The claim made by each of these groups was simply to be treated under the law as individual Canadian citizens, entitled to the same rights as anyone else, and not to be singled out for invidious treatment on account of their respective race or religion.

Language, at least in the Canadian context, presents a much different, more ticklish challenge.* Language seems to present a peculiar combination of race and religion in its divisive and explosive potential. Like religion, a person's language shapes his intellectual and cultural worldview†. Modern philosophy teaches us that the way we talk profoundly affects the way we think. But, like race, language is also an inescapable, external badge of identity. As soon as one begins to speak, everyone knows what you are, with unhappy results for the French at some times and in some parts of Canada.

There is an even more important legal complication from language. One cannot speak intelligently about a purely individual, negative right respecting language (as one can, for example, about the individual's right not to be discriminated against on account of race or religion). Language is inherently a social activity. If one wants to speak it, one needs a listener. If one wants, as do the French Canadians, to be part of a linguistic and cultural group which survives, which flourishes, a considerably more delicate problem is posed for the constitution-writer.

Aptly illustrating this point is one of the crucial turning points of Canadian history: the Manitoba School Question of the 1890s (2). The issue was whether the English Protestant majority in Manitoba would grant the French Catholic minority the educational rights the latter claimed. The right asserted by the French Canadian minority was to "separate but equal" schooling, funded from the public coffers to which they contributed as taxpayers. This was the same "right," then being forced on the black minority in the United States, in the now notorious *Plessy v. Ferguson* decision. As is currently becoming clear in the United States, as well, equal rights, expecially in education, imply very different things when one deals with a linguistic instead of a racial minority.

* I realize, of course, that the rights sought by racial groups can be much more complicated as well, as the American debate of the last fifteen years testifies to vividly.

† In Canada, at least, the linguistic division between French and English historically overlaps the religious division between Catholic and Protestant.

THE CANADIAN CONSTITUTIONAL TRADITION

Notwithstanding its British heritage, Canada has always had a written (and judicially enforceable) constitution. The British North America Act of 1867 was needed to knit together the colonies into a nation with a *federal* system, while allocating specific responsibilities to each level of government. The only fundamental right protected from governmental intrusion concerned the linguistic minority (although to some extent these guarantees were expressed in terms of religion). It was to these constitutional rights that French Catholics in Manitoba appealed. The then-ultimate arbiter of the Canadian constitution – the Privy Council – true to the prevailing British view of the judicial power, gave a very narrow reading to these constitutional and judicially-enforceable limits on parliamentary sovereignty (overruling the contrary holding of the Supreme Court of Canada) (12).

One reason why the Privy Council felt this to be a legitimate construction was that the original BNA Act afforded another *political* avenue for securing these minority rights: through a protective jurisdiction of the central government against such intrusions by a province. Unfortunately, the government in Ottawa was unwilling to avail itself of the opportunity left open by the Privy Council to force the Manitoba government to take corrective action. This result would not be found surprising to those in the dominant American tradition of constitutionalism: who would surmise that even the most democratic of governments, responsive as they are to the majority, cannot be relied on to protect the minority interest where there is a deeply-felt conflict between the two. Only entrenched constitutional rights enforced by an independent judiciary are up to the task.

There is a problem in this explanation of the Canadian case. The 1896 national election in Canada was fought over the Manitoba School Question, between the Conservatives, led by Tupper, whose main constituency was in Ontario, and the Liberals, led by Laurier, whose main constituency was in Quebec. But it was the Conservatives who wanted to force the hand of the Manitoba government, and it was Laurier's Liberals who defeated them on a platform of provincial autonomy: in particular, the freedom of a province to set its own education policy, come what may. Laurier's constitutional rationale was that this basic principle of federalism likewise protected the authority of the Government of Quebec, the one government in North America whose constituency was predominantly French.

THE MODERN CONSTITUTIONAL DEBATE IN CANADA

The seeds of the modern Canadian dilemma and debate were sown in that Manitoba School Question nearly a century ago. French Canada, effectively frozen out of the West, the major growth area in twentieth century Canada, has declined from approximately one-third of the population to slightly over one quarter now and is predicted to be nearer 20% by the turn of this century (8). These unfavorable demographic trends have made French Canadians even more insistent that their language and culture not just be guaranteed survival in Canada but be treated on a footing of full equality with the English, as one of the two founding peoples in this country. Elite groups in English Canada largely suscribe to this view (reflected in all-party acceptance of the Federal Official Languages Act enacted a decade ago). No one disputes that the French Canadians have a fundamental right to protection against assimilation into the numerically-dominant English world: i.e., against the Louisiana solution to the problem of linguistic and cultural dualism. But for the past twenty years there has been a deep division of opinion about what is the best means — in particular, the best constitutional means — of achieving this end.

Intriguingly, this debate has largely been conducted within the French Canadian community itself. One position is primarily associated with Prime Minister Trudeau, the chief spokesman of French Canada in the national government in Ottawa (4, 15). Under Trudeau's vision of the language issue, each and every Canadian has a fundamental right to speak, use, be educated in either (or both) of the country's two official languages, French and English. The legal mechanism for nurturing and realizing this individual right is to be a system of *institutional bilingualism*, under which a wide variety of services, particularly public services, are made available in either of these two languages (at least where numbers make this feasible). Trudeau's goal is that all Canadians should feel confident in their ability to move to and live in all regions of the country without facing a threat to the linguistic heritage of their family. To a considerable extent the national government has been able to implement this principle through a statute, the Official Language Act, governing the availability of federal positions and services (e.g., radio and television) in both languages throughout the country. But the Trudeau objective has always been to entrench this right in the constitution, not only to protect the current federal policies from change in the political winds (of which the air traffic control dispute in 1976 was an unhappy harbinger) (10), but, more important, to require provincial governments to adhere to the same principle, particularly in the crucial sphere of education (and also in the administration of criminal justice).

One way of capsulizing this overall philosophy is that Toronto must be made as hospitable to the French language as Montreal has been to the English.

In the last decade the sharpest opposition to this approach has come not from English Canada, from Toronto, for instance, but from Quebec City, from Premier René Levesque, leader of the Government of Quebec. Levesque and his co-workers believe that the French language is effectively doomed outside the Province of Quebec, because legal rights of the individual cannot withstand the sociological pressure to assimilation to the dominant language in an urban industrialized world (1, 3). But there remains an important struggle to protect the viability of the French language inside Quebec from the attraction and the threat of the numerically-small but economically-powerful Anglophone community. The Quebec City view of the appropriate language policy is now embodied in its own statute,, the Charter of the French Language (5, 16), based on a philosophy of *territorial unilingualism.* The province has an administrative Office of the French Language which has embarked on an ambitious program to try to make Montreal as French as Toronto has always been English — in the language of government, work, education, even the street-scape. The constitutional principle which underlies this policy is political nationalism, not individual rights.* The legal authority and the effective power of the government in Quebec City must continually be buttressed and expanded, because this is the one government in North America which answers to a substantial French minority. Within this nationalist strain, there is a vigorous debate about whether Quebec should be a province with a special status in a heavily-decentralized Canadian federalism or a politically sovereign nation within an economic association with English Canada (9). These differ more as points along the same spectrum than as starkly opposed alternatives. Neither could abide the development of new constitutional arrangements which would subject Quebec City to external limits on its authority, expecially in a sensitive area such as education, even though the intended beneficiary of any such constitutional right might be the French language itself.

THE CONSTITUTIONAL RESPONSE (7, 11, 13)

This was the constitutional dilemma which had to be confronted, especially after the November, 1976 election of the independentiste Parti Quebecois as the

* As one scholarly defender of this persuasion candidly admitted, the price of this approach to nurturing the French language in Quebec is one which will have to be paid by the English inside Quebec and the French outside Quebec (6).

Government of Quebec. The PQ conducted a referendum in May, 1980, asking for a mandate to begin negotiations which ultimately would lead to political sovereignty for Quebec. Trudeau occupied a prominent role in the campaign against this proposal. He promised the Quebec people a "renewed Canadian Federalism" which would be a third option between independence and the status quo (and he was seconded in this by a number of key English-Canadian provincial premiers). While the referendum verdict was "Non," by a 60-40 margin, nearly 50% of French-speaking Quebecers voted "Oui." As far as Trudeau was concerned, the time had come for constitutional reform in Canada, which would make good on his undertaking.

The constitutional package he unveiled in October of 1980 — basically the patriation of the Canadian constitution with a domestic amending formula together with an entrenched Charter of Rights, including language rights — was clearly designed to cement into law his vision of Canada as a single nation uniformly guaranteeing the rights of individual Canadians across the country. The proposals evoked vigorous objections from eight of the ten provincial capitals, led by Quebec and the West (and supported by the Conservative opposition in Ottowa), not just because of what the package contained, but also because it ignored their aspiration for greater provincial autonomy to protect the distinctive interests of the Canadian regions. Through a complicated sequence of political and legal manoeuvers, the debate over the Trudeau package was conducted in two forums: in the national Parliament and also in federal-provincial conferences of First Ministers. This separation of forum sharply accentuated the schizophrenic quality often found in discussion about the constitutionalizing of rights.

Inside Parliament, both in the Joint Commons-Senate Committee deliberations and in debates on the floor of each House, the focus was on the scope and definition of the "rights" themselves. The draft Charter contained the general array of such rights: in Canadian parlance, the "democratic" (the equal right to vote or run for a regularly-elected Parliament), "fundamental" (religion, speech, press, etc.), "legal" (due process protections, especially in the criminal justice system), and "egalitarian." The focus of the debate in this forum was how to buttress and expand the legal force of these rights which everyone found terribly difficult to question in the abstract.*

* The one exception was with regard to abortion. Here the groups interested in an explicit constitutional guarantee of free choice ran up against groups just as intensely interested in constitutionalizing the right to life. Eventually the politicians found caution to be the better part of valor and left the relevant language of the Charter oblique as to this issue.

A few examples make this trend clear. The initial draft of the "legal" rights regarding arrest and bail, search and seizure, and bail limited the protection to grounds and procedures "established by law." To vigorous objections that it was insufficient to require the legislature just to specify the grounds and establish the procedures for such intrusions on individual freedom, the government responded with new language guaranteeing the individual against any "arbitrary" arrest, any "unreasonable" search or seizure. More pertinent to this topic, the original "egalitarian" right was actually labeled a "Nondiscrimination Right" and provided:

> S.15(1) "Everyone has the right to equality before the law and to the equal protection of the law without discrimination because of race, national or ethnic origin, color, religion, age or sex."

Again this was attacked as inadequate and eventually the government offered a draft of what were now called "equality rights":

> S.15(1) "Every individual is equal before and under the law and has the right under the equal protection and equal benefit of the law without discrimination and, in particular, without discrimination based on race, national or ethnic origin, color, religion, sex or age."

Equally significant were the changes in the general interpretation and enforcement clauses of the Charter. Unlike the American Founding Fathers, the modern-day writers of constitutions (in Europe, for example) are under no illusion that fundamental individual rights, even if entrenched, can ever be absolute. They have to be limited in their scope and application in light of a variety of competing values (including other fundamental rights). Article I of the Canadian Charter was written in that vein:

> S.1 "the Canadian Charter Rights and Freedoms guarantees the rights and freedoms set out in it subject only to such reasonable limits as are generally accepted in a free and democratic society with a parliamentary system of government."

This language evoked widespread protest, led by the Canadian Civil Liberties Association, which feared that it would incline Canadian courts towards far too much deference to the limits actually established by the Canadian parliament (admittedly freely-and-democratically-elected). Once more the reaction in Ottawa was favorable, as the ultimate text of Article I testifies:

> S.1 "the Canadian Charter...guarantees the rights and freedoms...subject to only such reasonable limits prescribed by law as can be demonstrably justified in a free and democratic society".

Added was a new "enforcement" provision:

> S.24 "Anyone whose rights and freedoms, as guaranteed by this Charter, had been infringed or denied may apply to a court of competent jurisdiction to obtain such remedy as the court considers appropriate and just in the circumstances."

The meaning of these and the other changes I have recounted was unmistakable. Canadian courts were to play the central role in protecting the fundamental rights and freedoms of individual Canadians.

The debate taking place at the same time in the federal-provincial forum was on an entirely different plane: concerned not with the *substantive* character of these fundamental rights (except for language, as to which more later), but rather with the *institutional* question of who was to have the final say about what these rights meant — the legislature or the courts. The opposition to entrenching these rights in the Constitution was led by Allen Blakeney of Saskatchewan, an Oxford-educated lawyer, a social democrat, and (along with Trudeau) the most sophisticated Canadian leader of the past decade. His objection did not rest on the abstract principle of democratic rule. He recognized that it was of the very nature of these rights that a majority, no matter how large, could not legitimately ignore them. It stemmed from his distrust of the capacity of Canadian courts, of the Canadian legal culture generally, to make the delicate practical judgements about the scope and limit of these rights in ambiguous real-life conflicts. Needless to say, the direction of the rewording of the Charter's language inside the Parliamentary Committee added to the uneasiness of the Premiers.

Remarkably enough, protracted negotiations finally produced a solution to this conundrum: a typical Canadian compromise between pure parliamentary sovereignty (a la the United Kingdom) and judicial constitutional supremacy (a la the United States). The basic rights of Canadians would be entrenched in a new Canadian constitution and thus enforceable in the courts as against any offending legislative or official action. But a new Section 33 was added to the Charter, one which empowered

> "Parliament or the legislature of a province [to] expressly declare in an Act of Parliament or the legislature...that the Act or a provision thereof shall operate *notwithstanding* a provision...of the Charter."

Section 33(2) imposed a "sunset" limitation on any such non obstante declaration, deeming that it would expire every five years unless it was reenacted.

The rationale of this provision was quite simple.* These important rights were to be formally entrenched in a written constitution wrapped in this mantle of symbolic authority. Individuals aggrieved about intrustions on their rights were offered a judicial forum in which they could challenge official action and legislative enactment in a comparatively reasoned, nonpartisan atmosphere and win a verdict on the merits of the arguments advanced, irrespective of the number of votes commanded. The central core of democratic rights — the right to vote in a free and regular election — was excluded from the non obstante provision and thus untouchable by the legislature which happens to be sitting at any one time. But as regards the other fundamental rights — to speech, religion, fair procedure, equal protection, etc. — while the courts were invited to subject any intrusion to a first and searching scrutiny, the legislature would have the last word. If a government believed that the court was simply wrong in its judgement about what "reasonable limits could be demonstrably justified in a free and democratic society," if it felt strongly enough about the issue that it was willing to face the political flak of visibly overriding the very popular Charter, then the political branch was to be trusted with the final say.

THE CHARTER AND THE LANGUAGE ISSUE

On the face of it, this seemed to be a happy solution to the constitutional conundrum. The initial reaction of most observers was that the non obstante power was one which any Canadian government would find it terribly difficult to use. Indeed, a number of provincial premiers — including a couple who opposed the Charter — quickly vowed that they would never override the now-constitutional rights of their citizens. But the fact that an escape valve was still there, permitting an ultimate political judgment about hotly-contested issues of public policy, was sufficient to satisfy those who were uneasy about full-blown, once-and-for-all constitutional entrenchment.

In actual fact, the denouement was not so happy, because of the perennial problem of language. Trudeau was adamant that his treasured language rights would not be subject to the non obstante provision, fearing, quite rightly, that

* I appreciate that it is often illusory to impute too high a degree of rationality to a political compromise, especially one such as this which was eventually arrived at in an Ottawa kitchen at 4 a.m. I take this liberty here, if only because this particular proposal for breaking the logjam in the Canadian constitutional debate was first advanced in a paper of mine published the year before (17) and then circulated among and discussed with some of the key actors just before the last crucial conference.

the Levesque government would quickly use this power to insulate its own language policy. In the end the English Canadian premiers conceded this point to Trudeau, since language rights were no longer a major political issue, even in the West. But Levesque indignantly refused to sign or accept the final accord.

Since then, the new constitution has been enveloped in legal and political controversy inside Quebec. When Parliament passed the revised Resolution, Levesque had the Quebec flag flown at half mast. Quebec City challenged the propriety of amending the Canadian constitution without the consent of the Province of Quebec, a challenge which has just been rejected by the Supreme Court of Canada. In the meantime, faced with a court challenge by English-speaking parents of the key educational provisions of its language legislation, the government has tried to defend them on the basis that they were "reasonable limits...demonstrably justified in a free and democratic society" (S.1). Chief Justice Deschenes of the Quebec Superior Court resoundingly rejected that argument. But the Quebec legislature did purport to use its non obstante power through a blanket (and legally dubious) override of all the other constitutional rights of Quebecers — fundamental, legal, and egalitarian — even though it had never advanced any objection to *their* value and scope. More ominously, Parti Quebecois has hammered home the theme that the new Canadian constitution was imposed on Quebec through a deal struck between Ottawa and the English premiers, that its main visible result so far has been to strike down a central feature of the popular Charter of the French Language, and thus the only way in which Quebecers can free themselves from this alien document is through political sovereignty.

CONCLUSION

What are the general lessons to be seen in the details of this Canadian experiment in writing a constitution to protect a minority:

A. Fundamental rights present issues which are ambiguous and entangled. The popular drive to enshrine them in enduring legal form is fueled by memories of historic oppression and abuse. But by the time a polity is ready to grant the right (whether in writing or interpreting its constitution), the rights and wrongs in current real-life cases are nowhere near so clear-cut. It is not enough to take a stand in favor of equality and against discrimination in the abstract. One must commit oneself to a view of what these mean in the concrete, a subject upon which people can have serious and legitimate differences, even within the minority community itself.

B. The decision about whether to constitutionalize a right implies, then, not simply a judgement about the value of the right in question but, even more important, an institutional judgment about who should have the final say regarding its scope and limitation — the frankly political or the judicial branches of government (with their respective participants, constituencies, styles of analysis, and so on).

C. But even along this institutional dimension, the constitution-writer is not locked into an either/or choice between legislature or court. The non obstante device in Canada was conceived as one arrangement for eliciting a more sophisticated dialogue between the two bodies, one which would draw on the respective strengths of each. By putting the courts on the front lines of this struggle, Canadians would get just about everything of value that comes from a judicial forum for the elaboration and enforcement of individual rights against unwarranted governmental intrusion. But the ultimate political authority was to be kept in reserve to be used — albeit with considerable difficulty — when it was believed that the judges simply have pursued legal principles too far. Needless to say, the jury will be out for some considerable time on this Canadian experiment.

EPILOGUE

There is another, older minority group in Canada which only recently has emerged into the constitutional limelight, the native peoples of Canada — the Indians, Inuit, Metis, etc. The path taken by their rights tracked that of the Charter itself, towards a similarly murky conclusion (12). The original Trudeau initiative purported only to save, not to entrench, any rights hitherto enjoyed by the native Canadians. But the parliamentary process in Ottawa produced a characteristically strong declaration to the effect that "the aboriginal and treaty rights of the aboriginal peoples of Canada are hereby recognized and affirmed." This statement was attacked as inadequate by most native groups who carried the battle to governments and courts in the United Kingdom, even in Western Europe. But it was viewed with an equally jaundiced eye by the provincial governments (even Ontario, which otherwise supported the Charter) who would primarily be responsible for settling land claims and other such "rights," and whose flexibility and leverage would now be hampered by the possibility of judicial scrutiny. In the November Accord, the "aboriginal rights" clause was dropped altogether in view of the fact that neither its beneficiaries nor its targets seemed to appreciate it.

This omission produced a considerable outcry from native groups and their sympathizers (although much less so than another change affecting women's rights). In the result, the Premiers agreed to amend the Accord to the extent of recognizing and affirming just "*existing* aboriginal and treaty rights." Sufficient uneasiness was felt about the open-ended character of even this step that the Constitution scheduled a First Ministers Conference within one year, which was to have on its agenda "constitutional matters that directly affect the aboriginal peoples of Canada, including the identification and definition of the rights of these peoples to be included in the Constitution." As I write this, the Governments and native groups are in the final stages of preparation for this fateful meeting.

REFERENCES

(1) Beaujot, R. 1979. A Demographic View of Canadian Language Policy. Canadian Public Policy **V:16**.

(2) Berger, T. 1981. Fragile Freedoms, Ch. 3. Toronto: Clarke and Irwin.

(3) Castonguay, C. 1979. Why Hide the Facts? The Federalist Approach to the Language Crisis in Canada. Canadian Public Policy **V:4**.

(4) Government of Canada. 1977. A National Understanding. White Paper. Ottawa: Government of Canada.

(5) Government of Quebec. 1977. Quebec's Policy on the French Language. White Paper. Quebec City: Government of Quebec.

(6) Guindon, H. 1978. The Modernization of Quebec. **In** Glenday, D.; Guindon, H.; and Turowetz, A. Modernization and the Canadian State, p. 244. Toronto: MacMillan of Canada.

(7) Hogg, P. 1982. Canada Act 1982 Annot. Toronto: Carswell.

(8) Lachapelle, R., and Henripin, J. 1982. The Demolinguistic Situation in Canada. Montreal: Institute for Research in Public Policy.

(9) McRoberts, K., and Postgate, D. 1980. Quebec: Social Change and Political Crisis, revised ed. Toronto: McClelland and Stewart.

(10) McWhinney, E. 1979. Quebec and the Constitution: 1960-78, pp. 105–107. Toronto: University of Toronto Press.

(11) McWhinney, E. 1982. Canada and the Constitution. Toronto: University of Toronto Press.

(12) Schmeiser, D. 1964. Civil Liberties in Canada. London: Oxford University Press.

(13) Sheppard, R., and Valpy, M. 1982. The National Deal. Toronto: Fleet Books.

(14) Tarnopolsky, W. 1975. The Canadian Bill of Rights, 2nd ed., ch. 1. Toronto: McClelland and Stewart.

(15) Trudeau, P.E. 1968. Federalism and the French Canadians. Toronto: MacMillan of Canada.

(16) Vaillancourt, F. 1978. La Charte de la Langue Francaise du Quebec. Canadian Public Policy **IV**: 284.

(17) Weiler, P. 1979. Of Judges and Rights: Or Should Canada Have a Constitutional Bill of Rights? Dalhous. Rev. 205.

Minorities: Community and Identity, ed. C. Fried, pp. 329–339
Dahlem Konferenzen 1983. Berlin, Heidelberg, New York, Tokyo: Springer-Verlag.

THE RIGHT OF ETHNIC MINORITIES TO POLITICAL REPRESENTATION

A.M. Thernstrom
The Twentieth Century Fund
New York, NY 10021, USA

INTRODUCTION

In the past few years a new issue has surfaced in American politics: What is the meaning of the right to vote? Are blacks and other historically disadvantaged groups entitled to electoral arrangements which will facilitate the election of their candidates to office? Is there a right, in short, to political representation? And if so, under what circumstances?

In concrete terms, the question takes many forms. For example:

> A state has redrawn its Congressional districts to comply with the findings of the 1980 census. A different plan would increase the concentration of blacks in one more district and make the election of one more minority Congressman more likely. Under what circumstances is the state obligated to adopt the plan that is more favorable to blacks?
>
> A city which is one-third black elects its three city commissioners at large. No black in this century has been elected to city government, but if voting were by wards, a majority black district could be drawn. When are blacks entitled to an electoral system in which black candidates are protected from the competition of whites?
>
> A city which is forty-five percent black is governed by a council of eight, elected at large. Three councilmen are black, but they won only as part

of a white mayor's slate. Blacks (and whites) who run as independents lose. Under what conditions are black voters entitled to single-member districts which would eliminate the slating and probably result in the election of one more black councilman?

A rural county is sixty-five percent black, but black registration and turnout are low and blacks are therefore a minority of the voting population. If the at-large method of election were abolished and ward voting substituted, the low level of minority participation would have little impact. When are blacks entitled to electoral arrangements which, in effect, compensate for their failure to register and vote?

Until 1965 the majority of blacks in the South in the United States were disfranchised. The fifteenth amendment, passed in 1870, guaranteed that "[t]he right of citizens of the United States to vote shall not be denied or abridged by the United States or by any State on account of race, color, or previous condition of servitude," but fraudulent literacy tests, combined with poll taxes and the ever-present potential for violence, kept most Southern blacks from registering to vote. Beginning in 1957 a series of Congressional civil rights acts were passed, each of which gave some measure of protection to black voting rights, but it was not until 1965, when an actual Voting Rights Act was passed, that Southern whites were definitively stripped of their power to keep blacks from the polls.

The result was that, for the first time since Reconstruction, blacks registered in large numbers. By 1969 the proportion of blacks registered in Mississippi, for instance, had jumped from 7 percent (in 1965) to approximately 60 percent. In other Southern states the change was less dramatic but impressive. In all, by 1980 two million blacks had been added to the registration rolls and the consequence was, of course, a radical shift in the rules of the political game. The old political guard was forced to change its tune in order to retain office, and in many places the new voters swept new faces (both black and white) into office. There has never been even an approximate count of the number of Southern whites who owe their political position to black support, but the number of blacks in office in the South has jumped from under 100 in 1965 to over 2000 by 1980.

The change has been striking, but seventeen years after the passage of the Voting Rights Act, the number of blacks holding public office in most Southern jurisdictions is not proportionate to the black population. In some places the disproportionately low number of minority officeholders is the consequence of low voter participation — which is, in turn, partially a legacy of the history of disfranchisement. In other jurisdictions, the structure of the electoral system, combined with continuing white opposition to black political participation, has had the effect of reducing the number of blacks elected to office. And in still other places, minority candidates have lost elections for a variety of normal reasons: membership in the wrong political party, poorly run campaigns, inadequate credentials, lack of popularity even among minority voters, etc. Questions only arise, of course, when the defeat on election day appears connected in some way to the candidate's racial or ethnic background.

In 1965, the right to vote meant the right to cast a ballot and have that ballot properly counted. In 1982, few would advocate such a narrow definition. Indeed, from the national media, it would appear that a radically different view had been embraced. In *The New York Times, The Washington Post, Newsweek, Time,* and *The New Republic,* among others, the assumption is that blacks, Hispanics, and perhaps other ethnic groups are entitled to protection from electoral arrangements which dilute the impact of their vote — which deprive them, as *The New York Times* put it, "of the representation their numbers would allow." In fact, the views which many in the press appear to hold do not necessarily represent a national consensus on the right of racial and ethnic minorities to representation in proportion to their numbers. But on the proposition that the promise of the Fifteenth Amendment and the Voting Rights Act has not necessarily been fulfilled wherever blacks can vote, there is widespread agreement.

The shift in perspective requires little explanation. It occurred in the context of a heightened sensitivity to the inadequacy of traditional notions of equal opportunity for groups whose extended experience with discrimination left them unable to compete against those historically more advantaged. Just as, by the 1970s, it no longer seemed sufficient to allocate seats in elementary and secondary schools, to admit students to institutions of higher education, to hire police officers, or to award public contracts without regard to race or ethnic background, so it became widely regarded as inadequate to allow minority voters inexperienced in American politics and still the object of considerable hostility and suspicion to win or lose in the electoral game as conventionally played. In other words, as the commitment to affirmative action to rapidly alter the status of historically disadvantaged racial and ethnic minorities spread, it was logical that the principles that increasingly governed educational and employment policy, most notably, would likewise be applied to questions of political representation. There are few universities in America without some sort of affirmative action program to ensure the representation of blacks, Hispanics, and sometimes Asian Americans, and there came to be relatively little dissent from the proposition that, at least under some circumstances, affirmative action was necessary to ensure the representation of minorities in legislative bodies — on city councils, county commissions, in state legislatures, and in the United States Congress.

There are some, of course, who continue to argue that blacks and other ethnic groups are everywhere guaranteed their full political rights as long as there are no legal obstacles to registering and casting a ballot, but that view is difficult to maintain in the face of evidence that there certainly remain places in the South in which blacks, on election day, might as well stay home. To take the extreme case, where blacks are a distinct political interest group and in the minority, where there is no competition for black votes, and where no white will vote for a candidate publicly linked with that vote, black ballots are worthless and blacks are without representation. Some sort of protection to ensure a voice in the political process would seem warranted. It is obviously offensive to basic notions of justice and democracy to permit the continued exclusion of a historically disfranchised group, and it is politically unwise to court the social instability that is the likely result of such exclusion. Moreover, even where blacks gain nothing tangible from the election of one or two city councilmen to a council of nine, there is a symbolic and psychological significance which is hard to dismiss.

Nevertheless, the widespread consensus on the inadequacy of the traditional emphasis on access to the ballot disguises the fact that, in reality, the issue of the right of minorities to representation is far from settled. The extreme cases have precipitated agreement on a broad principle, but the principle embraced is so general as to be close to useless. By contrast, even freedom of speech and other rights embodied in the first ten amendments to the Constitution, famous for their susceptibility to varying interpretations, seem precise. What constitutes a legislative voice or adequate participation in the political process? If, as it is often said, blacks, Hispanics, and other groups (American Indians, Native Alaskans, and Asian Americans are specifically mentioned in the Voting Rights Act) are guaranteed "equal access" to the electoral process, or "an equal opportunity to elect candidates of their choice," or "an equally effective vote," what do these seductive but evasive (if not deliberately obscure) phrases mean?

Situations of unmistakable electoral discrimination eliminate the necessity of defining equal electoral opportunity; where opportunities are manifestly *un*-equal, there is no need to specify precisely the point at which absolute or true equality is reached. But there are now few jurisdictions in which ballots cast by minority voters are consistently, predictably worthless, and while the arguments of Justice Department and minority plaintiffs' attorneys, as well as those of many lower court judges, tend to obscure that fact, it has become increasingly apparent that the task of defining equal access to the political process must be tackled. In the wake of the 1980 census, in jurisdictions across the nation, reapportionment plans are being scrutinized for their impact on black and Hispanic voting strength, and yet the principles which underlie the standards by which improper dilution of the minority vote is measured in most of the cases have never been made clear.

THE EXISTING FRAMEWORK OF DEBATE

The standards themselves have been the object of considerable discussion. The central question has been whether the test for electoral discrimination should be the impact that a particular method of voting has, or, alternatively, the intentions that either lay behind the adoption of that particular method or continue to sustain it. Do results or purpose make discrimination evident, in other words?

The debate has produced much heat and little light, for, in fact, whichever side wins, the central question of what the "right to vote" now means remains unanswered. Effect may be the measure of discrimination, but the question of what constitutes a discriminatory effect must still be addressed.The right to vote may mean the right to cast a ballot in an electoral system, the impact of which is not ethnically or racially biased, but that definition still leaves us in the dark unless the concept of bias is also clarified. With considerable justification, it has been said that the right implicit in the effects test is clearly that of legislative seats in proportion to minority population. Discriminatory effect is nothing more than a disproportionately low number of minority officeholders. And, indeed, an entitlement to proportional racial or ethnic representation is often explicitly or inadvertently acknowledged. But such acknowledgments are inevitable, since it is hard to imagine what other "results" advocates of an effects test might have in mind. Those who disclaim a belief in simply looking at election returns and counting suggest the alternative of assessing "equal access" or "equal opportunity" to elect candidates of one's choice, but that is a solution which only serves to reintroduce the initial problem. What does "equal access" or "equal opportunity" mean?

The objection to simply looking at election results and counting the number of minority officeholders should be self-evident. Minority candidates lose elections for a variety of racially neutral reasons. Numbers alone are insufficient evidence of discrimination. It is this obvious insight that gives weight to the arguments of those who advocate an intent test. A particular method of election may disadvantage an ethnic group, such advocates point out, but that alone should not condemn it. In majority white communities ward voting is undoubtedly advantageous to blacks. One or more majority black districts can be drawn and the election of black officeholders greatly facilitated. But since no group is entitled to representation, none has a right to an optimal electoral environment. They are entitled, however, to protection from a system, the impact of which is designedly adverse. That candidates affiliated with a particular group do not win is normal, everyday politics. That they do not win because they belong to an ethnic or racial minority and the deck has consequently been deliberately stacked against them is cause for alarm.

The intent test contains an implicit definition of the right to vote: the right to cast a ballot in a system that has not been deliberately instituted and maintained to perpetuate white rule. But this definition, like that which inheres in the effects test, raises more problems than it solves. To begin with, does the right to vote, then, mean only the right to be free from government sponsored electoral discrimination?

In American law, discriminatory intent refers to action by the state itself — the unequal treatment of individuals by government. And yet voting rights cases are odd. State action is almost never the root of the problem. Methods of voting were often established many decades ago, when blacks could not vote; keeping them out of office was not a plausible aim. And the relevance of the motives of legislators in the late nineteenth and early twentieth centuries is not, in any case, clear. "Past discrimination cannot, in the manner of original sin, condemn governmental action that is not itself unlawful," U.S. Supreme Court Justice Stewart has admonished, and yet the Court iself (in adopting an intent test for a certain kind of case) has encouraged condemning methods of voting precisely on the ground of original sin.

Even when an electoral system — the placement of district lines, for instance — is of recent vintage, discrimination cannot usually be proven on the basis of the intentions of public officials. Apparent motives will be mixed and hidden ones elusive. And, in any case, most voting discrimination is the consequence not of state action per se, but of the uses to which that action is put by white voters who would exclude blacks (or Hispanics) from political participation. A district may be gerrymandered to reduce a minority candidate's chance of success, but, in the end, whether he wins or loses depends on the voters themselves.

Is it the motives of the voters themselves, then, that should be examined? Does the right to vote mean the right to cast a ballot in a color-blind context — in an environment in which whites are receptive to the inclusion of blacks in government? If so, then in the absence of such an environment, minority voters would be entitled to judicial or administrative intervention to restructure the electoral system to minimize the impact of white hostility.

Switching the focus from the intentions of the state to the intentions of voters is not only legally unorthodox but raises other problems. Often it is not white hostility which works to exclude blacks (or Hispanics) from the political process, but simply white indifference in a setting in which that indifference perpetuates unbroken white rule. Moreover, even if hostility were the sole test, evidence of such opposition would normally be exceedingly difficult to obtain. And even where it can be obtained, there remain problems of assessment. How much hostility is unacceptable? At what point does a political environment become sufficiently "color-blind"?

There are two customary sources from which evidence as to white attitudes toward minority political participation is obtained: returns on a referendum on the method of voting and returns in elections in which black candidates have run. Neither is satisfactory.

Evidence of discriminatory intent would perhaps seem easiest to establish in cases in which minority voters can point to a recent referendum on at-large voting in which whites defeated a proposed change. If a city council has been all white (or nearly so) and voters have been specifically asked to choose between an at-large system and single-member districts, and if it is public knowledge that the election of blacks would be the consequence of change, the decision to maintain the status quo would seem to indicate opposition to minority participation in the political process.

But referenda are defeated for a variety of reasons. The change to ward voting has consequences other than the greater likelihood of minorities in office, and those consequences may be unacceptable to the majority. Citizens may prefer to stick with the familiar. The wording of the proposal may be poor. The campaign on its behalf may be badly run. The support of the minority community itself may be lukewarm. In fact, even where it can be shown that racism did play some part in the defeat of a proposed change, the question of its significance would remain. How should one motive be weighed against another? And if there are three racists in town, can it be said that white racism kept blacks out of office? Perhaps the majority of whites were not eager to keep blacks out of office but found the disproportionately low number of minority officeholders acceptable. How should their intentions be judged?

Similar questions can be raised with reference to the returns in electoral contests. Where 98 percent of the whites in a community are consistently voting against black candidates (or against white candidates allied with black voters), the hostility to black political representation is clear. But such solid white bloc voting is now rare. And where even 15 percent of the whites have demonstrated a willingness to support black candidates, the picture alters. That 15 percent is often sufficient to get those minority candidates elected.

Can whites be said to be thwarting black political hopes when those hopes are realized despite considerable continuing white racism? Does hostility shade into receptivity at the point at which black candidates can win? If so, then implicitly blacks have a right to win. Electoral discrimination is present in every defeat. But what if minority candidates lose for reasons other than race? And how are racial responses to be distinguished from those which rest on considerations of ideology or personality?

Space limitations obviously forbid an extended discussion of the problem of assessing the significance of white bloc voting, but, in fact, once white solidarity substantially dissolves, there is no satisfactory way of using election data to demonstrate that ill-intentioned whites are keeping blacks from office. If intent is the test for electoral discrimination, there is no obvious way to measure it. And even if there were, there would remain the insurmountable problem of identifying the point at which an environment becomes properly colorblind — properly receptive to the inclusion of blacks in office.

AN ALTERNATIVE APPROACH

The magic words "intent" and "effect," it should be clear, settle little. The definitions of the right to vote that inhere in them are either unacceptable or incomplete. And the argument is never conclusively settled, in part because both sides have a point and in part because the wrong question is posed. Asking whether intent or effect should be the measure of discrimination is like asking whether Democrats or Republicans are best suited to govern in the United States while neglecting to specify what makes for a well-run nation. In fact, you cannot know what the test of discrimination should be until you know what is discriminatory. Defining the condition of inequality must be the first step: devising a means of testing that inequality is the second. Or to put it another way: the first task is to decide where to draw the line between discrimination and disadvantage — to ascertain where the one leaves off and the other begins. Only then will criteria upon which to properly base judgement be evident.

If a disproportionately low number of minority representatives always signifies a political process unequally open to minorities and whites, then it is obviously legitimate to test the racial neutrality of the method of voting by looking at its impact on black officeholding. At the other extreme, if blacks are on an equal political footing with whites as long as the method of voting has not been designed to keep minorities from public office, then focusing on legislative intent is proper. But if neither is true, then a different test — inevitably a more complicated one — must be devised. There is a way out of every blind alley, but it is always necessary to backtrack. In this case, a retreat to the question of when the right of minority citizens to vote has been circumscribed — when the election process is discriminatory — is the appropriate means of escape.

An election process is discriminatory and minorities are entitled to protection, I would suggest, when (a) it is the desired or accepted outcome that legislative seats are largely or exclusively reserved for whites, and (b) it can be shown that without minority officeholders there is no minority representation. And those conditions hold (c) where the minority group remains a community with distinctive interests, (d) where as a result of white hostility or long-standing callous indifference it is without a means of asserting its interests through the political process, and (e) where there seems little prospect for change without outside intervention. This is a definition which prompts not an inquiry into "intent" or "effect," but a series of considerably more complicated questions. Of course, if racial or ethnic hostility can be clearly inferred — if 98 percent of the whites in a jurisdiction are consistently aligned against any candidate linked with black voters — then electoral discrimination is indisputable. But in the overwhelming majority of cases, the inquiry would have to focus on the political process itself. Are blacks (or other ethnic minorities) without a means of asserting their interests through the normal electoral and political channels, it would have to be asked. Are white candidates indifferent to black voters, black candidates without routine access to white voters, white officials unresponsive to black interests? And where minority candidates are losing, does the responsibility clearly lie with the whites? Did the black community field viable candidates, organize a proper campaign, and get out the vote? Lastly, is this a community in which internally generated change seems unlikely? Will minority voters, in all probability, remain politically excluded unless outside authorities intervene? Are the problems insoluble within the community itself?

Shifting the focus to the election process through which minority candidates attain office and to the political process through which the needs of minority voters are voiced and met certainly leaves many problems. The task of determining when the state should intervene remains complex. The central questions spawn a host of others. For example:

> Are blacks always a true political interest group, and if not, what are the signs of cohesiveness?
>
> When is the process of candidate selection closed? Where blacks and Hispanics are free to run, but no one wins who has not been formally or informally slated by a white-dominated group, is there discrimination?
>
> When are the election contests themselves manifestly unequal? Candidates who are the choice of blacks and Hispanics are often at a disadvantage. At what point does that disadvantage shade into discrimination, such that it becomes legitimate to protect those candidates from white competition? If the local newspaper never supports minority-backed candidates, if white money never flows into minority campaigns, or if clubs and other organizations before which candidates normally appear invite only whites, is the political atmosphere unacceptably inhospitable to those who would represent black and Hispanic interests?
>
> Where votes of minority citizens have an indisputable impact on the outcome of elections but minority candidates themselves have never won and probably cannot, is the electoral process discriminatory? Have legislative seats, in effect, been reserved for whites?

These are questions which are difficult but not impossible to answer. Their difficulty is a reflection of the complexity of politics itself. One of the problems with both the intent and the effect test is their demand that the large and sloppy world of politics conform to procrustean legal beds. If the problem of when the impact of the minority vote is impermissibly diluted is to be seriously attacked, acknowledging the complexity of the political process is an essential first step.

CONCLUSION

The problem of minority voting rights arose in the late 1960s and early 1970s in the context of remarkably widespread agreement that the promissory notes that America had too long been delivering to blacks were past due. No other groups could have triggered a concern that access to the ballot was not enough. None had been comparably disfranchised and none had acquired the vote in so inhospitable an environment. Other groups were clearly on the road to political integration once they could vote. With blacks that was not so clearly true, or true only over an intolerably long haul. By the 1970s it was inevitable that at least those judges, federal administrators, and attorneys who were sensitized to issues of civil rights would conclude that the nation's obligations to blacks remained unmet as long as black citizens were not on an equal political footing with those who were white.

Blacks were the only group that could have sparked the initial concern with "equal political access." But, once formulated, the principle that (under some conditions, at least) blacks were entitled to special political protection was soon extended to encompass other groups — most notably, Hispanic Americans. The assumption was that these groups were "black-like" — that in important ways their experience had been comparable. Whether that was true is the subject of another discussion. Whether, outside the American context, ethnic and language minority groups qualify for the kind of political assistance that we have elected to give blacks and Hispanics is a question best addressed by those more familiar than I with the status of ethnic minorities in England, on the Continent, and in other settings.

Acknowledgement. This paper is an outgrowth of work in progress supported by The Twentieth Century Fund, New York, NY.

Minorities: Community and Identity, ed. C. Fried, pp. 341–351
Dahlem Konferenzen 1983. Berlin, Heidelberg, New York, Tokyo: Springer-Verlag.

POLITICAL COMMUNITY IN PLURAL SOCIETIES

P.M. Kitromilides
University of Athens
Athens, Greece

Abstract. This paper considers the theoretical challenges posed by the problem of minority integration in multinational states and of the preconditions of political community in plural societies, with special reference to the experience of Cyprus.

The question I propose to address in this brief essay concerns the political preconditions of the integration of ethnic minorities in a national community. I should like to suggest that this question constitutes one of the major challenges to contemporary political theory and political practice. The complications arising out of the problem of minorities for political practice are well-known and obvious to everyone who has paid even cursory and incidental attention to efforts to construct national states, cope with social change, and regulate the smooth functioning of political institutions in plural societies: just consider the civil wars attendant upon the dissolution of colonial empires, racial strife in the USA, the nationalities problem in the USSR, and, more recently, the centrifugal tendencies of "peripheral" nationalism in the ancient nation-states of Europe. The challenges posed by the issue of minority integration to political thought are registered in the fact that none of the theories of the state prevalent in modern political philosophy can provide an adequate analytical framework for the examination of the problem. This is due to the fact that nationalism, with its parochial and morally outrageous values, has consistently remained outside the purview of political analysis which is distinctly universalist in its normative premises and ethical claims (1).

The preconditions of minority integration in the larger society are fundamentally identical with the range of problems posed by the phenomenon of nationalism. How can an ethnic minority be made to feel a part of a larger national society at the lowest possible human cost, is a question that raises all the problems, empirical and normative, about national identity and national sentiment. The broad range of issues associated with these forms of human expression have not received the attention they deserve in social theory — with some notable exceptions, of course. Classical political thought has largely ignored nationalism, as becomes obvious if an attempt is made to bring the perspectives of liberal and republican theories of the state to bear on the pertinent problems. This is amply demonstrated if one considers the host of problems that remain unresolved and the critical questions which arise if the issue of minority integration is considered in the context of the "nightwatchman" minimal state of the liberal tradition or, alternatively, in connection with the participatory republic of civic virtue visualized by the radicals. The liberal state which leaves all individuals and groups to their own devices in the struggle for survival can hardly be considered a good place for socially disadvantaged minorities in need of affirmative action programs in order to cope with the structural constraints imposed upon them by an environment of social inequality. At the very least, the liberal state might be a tolerable place for religious and other intellectual minorities, since in principle it will leave them alone to cultivate their faiths as they see fit. The republican state, however, with its tendency to invade the privacy of the citizen with its participatory demands and the civil religion it requires in order to motivate participation and dedication to the common weal, could very well be a quite intolerable environment for a dissenting minority. A dominant national doctrine is bound to be the most usual content of the civil religion, and for the national community to work, those who might have doubts about official nationalism or are suspected of nurturing alternative loyalties might well "be forced to be free."

Thus classical political theory, with some rare exceptions which, however, approach the question of national minorities only to beg it, has generally overlooked the problem. More recent modes of political discourse, most notably some strands in twentieth century Marxism and the empirical theories of modernization, have attempted to come to grips with the issues of nationality and "nation-building" (2). However, the specific problem of creating political community in societies where articulate ethnic minorities are present has not been adequately treated. Marxism, despite the important contributions of East European Marxist thinkers to the analysis of nationality (3), has optimistically sidestepped the problem by delegating its solution to a future socialist society that will achieve the freedom of all its members, including members of national minorities, from the forms of oppression and alienation nurtured by capitalism. Modernization theory, on the other hand, although it has included political "integration" among its major concepts, has delegated the problem of minorities to a secondary place. With some notable exceptions, modernization theory has used the concept of integration to analyze the process of the extension of the power of state institutions and the regimentation of society through unifying policies. The cultivation of national identity and the emergence of nationalism are considered part of this process. This has generally been the analytical approach to the problem of national community in the most influential sources on modernization and political change (4).

Despite the seminal contribution of one of the pioneers of the study of modernization, Rupert Emerson, who has laid out in all its complexity the problem of minority integration in the process of national self-assertion (5), it has provoked only tangential interest in the mainstream of the empirical theory of political change. It is true that studies of "political development" in Africa and Asia in the 1950s and 1960s have not failed to notice the centrality of ethnic conflict in the process of state building. This emphasis, however, was confined to particular case studies and, with rare exceptions, did not provide the focus of theoretical reflection. Only belatedly has the problem of minorities, ethnic relations, and civil conflict been given the attention it deserves in political analysis (6). This can be considered a characteristic instance of the modification of theory and of change in analytical emphasis under the pressure of actual problems and political experience. Overall, however, the issue of minorities has tended to be left to sociologists and social psychologists who have focused on ethnic stratification, race relations, and prejudice (7), or to scholars of international law and relations who have treated minorities as one more nuisance in the regulation of relations among nations and in the orderly functioning of world society. The discussion of the problem in terms of legal norms tends to transpose a degree of formalism into it that obscures the human drama and urgency of the issues involved (8).

What I am trying to convey with the preceding brief survey of approaches and modes of analysis is basically a dissatisfaction with the available ways of conceptualizing the problem. The fundamental issue contingent on the problem of minority integration concerns the preconditions of political community, and this has been explored with a good deal of formalism. I do not propose to try to correct all this through the elaboration of an alternative theory. On the contrary, I suggest that the attempt be made to bring a corrective to theory by proceeding inductively from the failures of practice. In this connection, I propose to put forward some reflections based on a consideration of the experience of Cyprus. This is a classic case of two articulate ethnic communities, one demographic majority and a substantial minority (in a ratio of approximately 4 to 1) having to coexist in a state, with the easy solution of separation precluded from among their choices by the facts of geography, ethnic demography, and economic viability, yet failing to achieve political community and consequently shouldering the appalling costs of conflict, foreign invasion, and violence on a large scale.

The facts of the case have become generally known to the informed observer by recent tragic developments and need not be repeated here. Furthermore, they are covered in a voluminous and easily accessible literature (9). What I should like to do is identify the major problem areas that, in my judgement, caused the failure of minority integration and on the basis of this evidence venture some suggestions on what might be needed to make political community in ethnically plural societies possible. I should like to stress, nevertheless, that my tentative suggestions derive primarily from critical reflection upon past strategies and political options and therefore are not meant as a generalized policy blueprint either for the shaping of Cyprus's future or for the resolution of conflicts in parallel cases such as Lebanon or Northern Ireland. My diagnosis, in short, is much more a criticism of the past than a specific vision for the future.

A first major impediment to political integration might be considered constitutional formalism and rigidity. The institutionalization of communal representation, the distribution of offices on the basis of communal criteria, and the stipulation of ethnic quotas in public services by writ of the constitution make ethnic divisiveness a part of the formal political culture and undermine political integration by stressing ethnic identity over democratic citizenship. Furthermore, the over-detailed specification of institutional arrangements and minority privileges writes rigidity and intransigence into the constitution and undermines the possibility of liberal solutions. The respect of the rights and the equal treatment of the minority should be made the major test of the majority leadership's statesmanship, not an institutional brake on the democratic functioning of the state. The choice of corporatism over democracy precludes piecemeal political solutions and preserves an all encompassing dynamic in the settlement of political problems whereby the whole can collapse over a minor practical issue which might not otherwise put in question any fundamental constitutional principles.

Such had precisely been the experience of Cyprus. When the island became an independent state in 1960, it was endowed with a constitution built upon rigid bicommunalism in order to meet the requirements of the minority and to assure their participation in the republic. In this manner, it incorporated and institutionalized in the new state structure the traditional ethnic communities and identities which had been preserved and politicized by the British colonial administration. The most characteristic ideological expression of this configuration was the recognition by the constitution of membership in one of the ethnic communities, not of the status of individual citizenship, as the primary and decisive basis of the political identity of the subjects of the new state. In this manner, the classic liberal basis of political democracy, the primacy of the individual and his or her rights as a person independent of special characteristics, was delegated to a secondary place while predominant weight was ascribed to ethnic, religious, and racial attributes in the organization of political life. Racial discrimination became, in a way, the price of the minority's agreement to participate in the bicommunal partnership of the republic.

All this meant that the preconditions of liberal political life and democratic change away from the inherited structures of colonialism were excluded constitutionally from available political choices at the inception of the republic. Instead of embarking on a quest for a political and cultural identity of its own, the republic remained a practically inconvenient and symbolically uninspiring bicommunal compromise. By officially preserving traditional ethnic identities, the republic could not capture the emotional allegiances of its subjects and failed to nurture a shared loyalty for the common homeland.

The respective nationalist legacies of the majority and minority communities which remained intact in the new political structure constituted a powerful ideological factor which pushed the republic in a direction opposite to that of an "integrative revolution" (10). The clash of two symbolically antagonistic, mutually exclusive, and highly authoritarian nationalisms provided the ideological content of ethnic conflict. On the eve of her independence in the context of the anticolonial struggle of the 1950s, Cyprus had experienced the crystallization of ethnic confrontation that stretched from the ideological to most other levels of public life. Furthermore, political role inversion between the two ethnic communities bred mutual paranoia: the Turks who had been the master race in the three centuries of Ottoman rule in Cyprus found themselves in the position of a minority under the republic, while the Greeks changed from oppressed subjects to the dominant community in a state whose existence stood in the way of their national aspiration for union with Greece. Each side consequently found itself locked into the fear that willingness to compromise and cooperate might lead to even greater symbolic losses. Thus the bicommunal experiment of the Republic of Cyprus had to face from its inception serious subverting forces in the dynamics of local politics.

The effects of the dialectic of intolerance which divided the two communities were deeply felt in the political culture of independent Cyprus. All timid voices raised against communal isolation and antagonism and evoking the survival of the republic were simply stamped out. On the Greek Cypriot side, the dominant nationalist orthodoxy thriving on the absence of liberal values was ruthless in discrediting as treason all dissenting voices. On the Turkish Cypriot side, the extremists, firmly entrenched in the leadership of the community, did not limit themselves to moral and psychological coercion but went all the way in using their terrorist gangs to achieve the extermination of their critics who advocated interethnic cooperation. Alternative forms of social organization such as professional associations, trade unions, and the units of local government that could bring members of both communities together and had an important record of interethnic cooperation under the colonial administration were blocked from developing any further and undermined under the republic. The fact that these agencies of intercommunal cooperation were consistently under the ideological influence of the Left precipitated the strong reaction of the nationalists of both sides against them.

Besides the constitutional and ideological impediments to the creation of political community in the republic of Cyprus, a structural source of division and minority antagonism toward the majority has been the socioeconomic inequalities between the two communities. Although the Turkish Cypriots had been politically and socially the dominant element in the three centuries of Ottoman rule (1571–1878) and experienced privileged treatment under the British (1878–1960), the form of capitalist development followed by Cypriot society in the twentieth century left them at an economic disadvantage. The Greeks as a more enterprising element, free from the cultural obstacles that hindered the economic development of the Turks, profited from an earlier start on mercantile activities and then reaped the benefits of economic modernization. Ethnic separation in the republic after 1963 and the self-imposed isolation of the Turkish Cypriots deprived the minority community of the fruits of the economic boom and prosperity brought by development planning after independence. These structural inequalities blended with other forms of antagonism to further obstruct the feeling of political community from developing. In the absence of timely gestures of generosity in the minority's favor, and since material condition and political status are inextricably interconnected, the fears and insecurities of the Turkish Cypriots about a precarious future in the republic despite their excessive constitutional privileges were, at least subjectively, not without foundation.

The previous observations point to what could be considered the most critical factor in the attainment of political community. On the evidence of the Cyprus case, the motivations of political leadership appear as the decisive element in the determination of political outcomes. Comparative evidence suggests that the motivation of political leadership constitutes the critical variable in the achievement of conflict regulation in segmented societies (11). On this factor hinges the workability of constitutional formulas, the exploitation or abeyance of ideological tensions, the rectification of structural inequalities through planning and affirmative action, and the alleviation of fears and insecurities in the collective psychology. The motivation of political leadership makes all the difference in the achievement of compromise. This was precisely what Cyprus lacked. The Greeks felt they had sacrificed too much. The Turks feared they had too much to lose. None trusted the motives of the other. The vicious circle of ideological rhetoric never failed to provoke mutual mistrust. Maximalist goals on both sides precluded all serious efforts to make the republic work. Naturally, it broke down with a little help from outside. But I want to stress that the assignment of the major responsibility to "outside interference" which has provided the Cypriots with a convenient scapegoat for their own failures, is no more than comforting self-deception. British colonialism, American imperalism, Turkish expansionism, Greek fascism — all played their role to a catastrophic degree at the expense of an innocent and good-hearted people. But on the evidence of the historical record, especially since 1960, I tend to suspect that all these overpowering forces might have failed to work out their poisonous schemes, had they not found the appropriate conduits in the contradictions of local society. The major channel through which foreign conspiracies were made operational was the lack of motivation on the part of local leadership groups to make the republic work, however difficult this might have been. In this, at least, the majority and the minority leadership, each motivated by their respective maximalist goals, were for once united. The force of this fact turned Cyprus into a classic example of a

Thucydidean political tragedy: self-destructive blindness and violent human passions were let loose to wreck a whole civilization and sink hundreds of thousands of men, women, and children of both races into appalling suffering.

The conclusions that emerge from the foregoing considerations, as far as the prerequisites of political community in plural societies are concerned, suggest basically a counterpoint to the experience of Cyprus. First and foremost, political community might be achieved if democratic citizenship is given precedence over communal membership as the basis of political identity. The vexing issue of the effective protection of the minority would be achieved through the full democratization of society and not by resorting to corporatism which is the certain avenue to the preservation of the infrastructure of conflict. Second, affirmative action programs to redress communally based social inequalities might provide the needed evidence of good intentions on the part of the majority leadership to allay the fears of the minority and to disprove the arguments of its extremists. Furthermore, this might be the only way to cope with the fundamental component of the minority's social experience, objective and subjective exclusion. Third, a serious effort to cultivate loyalty to the common state by stressing shared rather than divisive symbols might act as the catalyst for the emergence of a viable psychological and cultural context within which majority and minority can coexist. This is probably the greatest challenge of all. The argument for the transcendence of group values, mentalities, and stereotypes raises the question of how much people can sacrifice in terms of their emotional attachments without risking large-scale anomic consequences. This problem has not been adequately appreciated by those who think of nationalism as a purely artificial and largely contingent contrivance (12). It is at this point that one of the greatest tests facing the effort to construct political community in plural societies arises. If a national society is to be made viable, both the tyranny of the majority must be avoided and the risk of driving minorities into rebellion should be preempted by generous recognition of their desire for differential treatment. Concurrently, however, the majority's sensitivities must be respected by not yielding "too much and at the wrong time" — something that might encourage separatist forces to shatter the fragile national community while it is still in the making (13).

It might be suggested against the line of argument developed above that what a minority fears most is full democratization itself. Constitutional formalism and corporate recognition might be exactly what minorities are striving for: democratic decisions may turn against affirmative action policies, and the chances of individual mobility in a democratic society may undermine the minority's cohesion by encouraging its members to integrate in the majority's political culture, thus abandoning their heritage and tradition. It is quite conceivable that the minority's claims would focus on the safeguard of precisely these constitutive elements of its identity against the amalgamating pressures of an open society. This, of course, is morally and politically a highly debatable claim in that it poses the issue of individual autonomy and self-determination versus corporate pressures for conformity. I should like to stress that this is the fundamental issue at stake which gets ideologically clouded — not entirely innocently — by appeals for ethnic rights.

A few analytical distinctions might be helpful in placing the problem in perspective. First, I think it must be recognized that so long as we are honestly concerned with basic humane options, the individual's right to belong, but also to modify or change allegiances, should be given priority over all corporate aspirations. Otherwise, the case becomes a straightforward issue of coercion: the preservation of cultural heritages in terms of personal identity should be made an individual option, not a constitutional compulsion. Naturally, if a group of individual citizens wants to preserve and transmit a particular tradition, it should be free and also assisted through state subsidies to its cultural institutions to do so. This, however, as a form of free individual expression, should be limited to what Hegel has defined as civil society and should not be allowed to introduce particularist values into the public domain. State aid to minority institutions should be premised on the ideal of equality and not on the recognition of corporate legitimacy. Should the state, however, attempt to stifle, either positively or negatively, such initiatives in civil society, the minority would be the victim of persecution and tyranny and would have legitimate claims to revolt and secession. If the multiple majorities which control the state want to avoid such eventualities, they have no other option but to respect the minority's sensitivities and make their goodwill felt through affirmative action. It is at this point that the greatest political failures usually occur in plural societies.

Fundamentally, the issue of the creation of a viable community is a question of political justice. Safeguard of basic civil liberties, respect for equal political rights, and the enjoyment of the protection of citizenship by all can answer the question of community. Their absence amounts to tyranny, and community under tyranny is impossible, as ancients and moderns recognized too well. It is precisely over the willingness to strive for political justice that the motivations of leadership can best be appraised in connection with the attainment of community. I conclude with this issue because, beyond the structural and cultural parameters of the problem, it opens the possibility of political criticism.

NOTES

(1) Dunn, J. 1979. Western Political Theory in the Face of the Future, pp. 55–79. Cambridge: Cambridge University Press.

(2) Cf. Hobsbawm, E. 1972. Some reflections on nationalism. **In** Imagination and Precision in the Social Sciences, ed. T.S. Nossiter, pp. 385–406. London: L.S.E. Publishers.

(3) See Davis, H.B. 1967. Nationalism and Socialism. New York: Monthly Review Press. See also the selections from Bauer, O., and Renner, K. 1978. **In** Austromarxism, eds. T. Bottomore and P. Goode, pp. 102–125. Oxford: Clarendon Press.

(4) See Bendix, R. 1977. Nation Building and Citizenship. Berkeley: University of California Press. Also, see Huntington, S.P. 1968. Political Order in Changing Societies. New Haven: Yale University Press.

(5) Emerson, R. 1960. From Empire to Nation. Cambridge, MA: Harvard University Press.

(6) Enloe, C.H. 1973. Ethnic Conflict and Political Development. Boston: Little, Brown. Also, see Young, C.M. 1976. The Politics of Cultural Pluralism. Madison: University of Wisconsin Press.

(7) Cf. the survey by Rose, A.M. 1968. Minorities. Int. Encyc. Soc. Sci. **10**: 365–371.

(8) I should like to note a recent work which treats this aspect of the problem in a remarkably humanist perspective: Ténékidès, G.C. 1980. L'action des Nations Unies contre la discrimination raciale. **In** Academie de Droit International. Recueil des Cours **168**: 271–487.

(9) For a survey of the bibliography see Kitromilides, P.M., and Evriviades, M.L. 1982. World Bibliographical Series 28, Cyprus, pp. 53–98. Oxford: Clio Press.

(10) Cf. Geertz, C. 1963. The integrative revolution. **In** Old Societies and New States, ed. C. Geertz, pp. 105–157. New York: Free Press.

(11) Nordlinger, E.A. 1972. Conflict Regulation in Divided Societies. Cambridge, MA: Harvard University Center for International Affairs.

(12) Gellner, E. 1964. Thought and Change, pp. 147–178. London: Weidenfeld and Nicolson.

(13) Cf. Emerson, op. cit., pp. 332–333.

Minorities: Community and Identity, ed. C. Fried, pp. 353–364
Dahlem Konferenzen 1983. Berlin, Heidelberg, New York, Tokyo: Springer-Verlag.

AFFIRMATIVE ACTION AND THE RIGHTS OF THE MAJORITY

C. Cohen
Dept. of Philosophy, University of Michigan
Ann Arbor, MI 41809, USA

GROUP CONSCIOUSNESS AND GROUP PREFERENCE

Honorable parties, seeking just remedies for social wrongs, often disagree heatedly about the rights of members of minority groups and of the majority. Disagreement is compounded by misunderstanding. One source of such misunderstanding is the ambiguity of the phrase "affirmative action." I aim to expose this ambiguity, to identify some of the errors associated with it, and thus to suggest both the proper functions of affirmative action and some limits imposed upon it by the rights of all citizens.

A less controversial phrase, "ethnic group consciousness," (or "race consciousness") provides a beginning. Great and prolonged injury has been done to many only because they happened to be of a certain race, or religion, or nationality. When the ethnic group of the victim was the ground of injury, consciousness of this group identification may prove essential in providing redress. Ideally, the goddess of justice is blind to color or creed; the historical legacy of group oppression, however, sometimes obliges her to peek through her blindfold in order to know what considerations are fairly put on the scales of justice. Without ethnic group consciousness we are sometimes unable to do what is right now because we are insensitive to what was done wrong before.

Does minority group *consciousness* necessarily involve minority group *preference*? Here the seeds of ambiguity are sown. "Yes and no," one is tempted to reply. We may be obliged (morally or legally) to give preference to a certain class of persons that can only be defined, in part, by ethnic group. In that sense our answer will be yes; group consciousness leads to group preference. On the other hand, every such obligation stems from some identifiable damage done to actual persons; obligatory preference to those damaged is not preference given to them *because* of their ethnic identification. Preference may be an instrument of redress for injury, where the proper recipients of that redress are identified partly by race or nationality because the injury in question was triggered by race or nationality. Legitimate recipients of redress are identifiable more fundamentally, however, as the persons injured in such-and-such ways. Preference may arise as a moral requirement today only because of wrongful injury yesterday; there is and should be no preference because of race or nationality *in themselves.* For example, a worker who has been discriminated against by his employer because of his nationality may be entitled now to preferential treatment by that employer, not because he *is* a member of that minority, but because, and to the extent that, he was discriminated against. In this sense group consciousness is not equivalent to group preference, and certainly does not entail preference *on the basis of* group membership. The common failure to draw and apply this distinction between group consciousness and group preference is the source of much sore confusion about affirmative action, and what may or may not be fairly done under its name. The justice of a present remedy depends critically upon the nature and source of the injury that requires it.

Earnest good will and the anxiety to make up for past injustice lead some to go beyond remedy to a morally impermissible ethnic favoritism. Programs whose original motivation was genuinely compensatory then overshoot their target, giving special advantage to persons who happen to be of the same racial or ethnic group as were others who were injured for that reason in times past. Benefits are then awarded and handicaps imposed *on the basis of* race or ethnicity. This is "group preference" in the strict sense, and the phrase is wisely reserved for cases of this kind, to distinguish them from other cases of more narrowly tailored remedy. The distinction is needed to sort the just from the unjust uses of group consciousness.

To the question asked above (does group consciousness necessarily involve group preference?) the correct answer is, therefore, not "yes and no" but simply "no." Morally appropriate group conscious practices do not entail preference by race in the strict sense — and it is precisely this strict sense of racial preference that is our target when we condemn racial "discrimination."

We do not use this word meaning that racial differences must never be discriminated; clearly they must be in precisely those circumstances in which race appropriately figures in the definition of the class of beneficiaries of justified redress. We do mean by this word that no individual should get less (or more) than he would otherwise get, of goods or opportunities, only because of his group membership.

In sum: Any individual who is disadvantaged (or advantaged) simply because he is in some racial or national or religious category is the victim (or the beneficiary) of precisely the sort of discrimination we rightly condemn. Programs that implement group preference, in the strict sense of that term described above, have precisely that result.

AFFIRMATIVE ACTION

Now we can return to "affirmative action." Clearly, this phrase can be used to refer to practices that do, and to practices that do not, involve racial or group preference in the strict sense. This helps to account for the fact that while affirmative action programs had an honorable genesis, some of them now have a bad aroma.

A long history of racially discriminatory conduct must have consequences that will not soon be eliminated if all consciousness of race is suddenly erased. *Equal treatment* is the right principle. But the refined application of that principle requires the equal treatment *of equals.* Those who have suffered the damage of ethnic discrimination in education or employment are not the equals of those who have not, for the purpose of determining equal treatment in those spheres. We must treat equals equally, to be sure. We must also take deliberate steps, affirmative steps, to insure that unequals are treated unequally, to adjust our practices so that they are fair in the light of the real circumstances of those upon whom the practices bear. We are rightly called upon, therefore, not simply to give equal opportunities to all, regardless of their group or race, but to take *affirmative action* to insure that persons deserving special attention get that attention. Affirmative action to remedy injury done by racial discrimination is *compensatory* in spirit, and justly so. This is exactly the sense in which the phrase "affirmative action" is used in the American Civil Rights Act. That statute carefully provides that a court, after having found some practice to be unlawful discrimination, may enjoin it, and also may "order such affirmative action as may be appropriate" [42 U.S.C. 2000e-5(g) (1976)].

Affirmative steps, deliberate attention, may be needed not only to provide deserved remedy, but also to insure that treatment intended to be fully equal is genuinely so. Where institutional practices — in recruitment, promotion, and the like — have long been invidiously discriminatory, it will not be enough to eliminate the intention to discriminate against minority groups. Institutional machinery originating in that discriminatory ambience must be reexamined and cleansed of elements — procedures, criteria, emphases — that were introduced with discriminatory intent (or with expectation of discriminatory effects) long ago, and whose retention has continuing discriminatory results. This variety of affirmative action is *corrective* in spirit, not compensatory, and is the proper business of affirmative action officers in corporate structures, universities, or any institutions whose long-entrenched ways of doing things may need adjustment to achieve fair treatment for all. No group preference, even in the weaker sense, is entailed by affirmative action of this corrective or prophylactic sort.

Both species of affirmative action, the compensatory and the corrective, have the same target: equal treatment for equals. Both are sensitive to the residual impact of past discrimination and are to that extent race (or group) conscious. Neither involves any group preference in the strict sense. Both are humane and morally right.

AFFIRMATIVE ACTION AND GROUP REPRESENTATION

There is another species of affirmative action — a poisonous species — which is designed to yield certain numerical results in the distribution of racial and ethnic groups in employment, schooling, residence, and other spheres. The difference between group consciousness introduced to achieve genuinely equal treatment regardless of group, and group preference introduced to achieve proportionality (or near proportionality) of group representation, is very great. This latter species is poisonous because, however well-intended, it enforces group preference in the strict sense and thus imposes that very discriminatory inequality of treatment we now strive to eliminate.

Unfortunately, it is this species that is most commonly thought of when reference is made to "affirmative action." It is an instrument so blunt and so ill-designed in its distributive impact as to damage many persons, including some whom the programs were intended to assist. By awarding to some on ethnic grounds what is taken from others on ethnic grounds, such preferential programs do, in the good name of affirmative action, exactly what affirmative action was instituted to undo. When "affirmative action" comes to be equated with numerical group representation, it gets — as it deserves — a bad reputation.

What leads to this degeneration? What brings honorable persons to the defense of outright ethnic preference? They are motivated by a sense of moral urgency. They want justice done, not just talked about. They want results. This far, all may rightly join them. But what *are* results, and how are they to be *measured*?

Preferentialists commonly rely upon a straightforward numerical assessment of results to determine whether justice has been done. To justify that reliance they tacitly invoke a false assumption, one whose plausibility is protected by its tacit use and its consequent insulation from scrutiny. This assumption may be formulated thus: "If no ethnic discrimination had infected the distribution of goods in society, we should now find those goods distributed randomly among the ethnic groups of which society is composed; wherever more and less goods appear in distinct ethnic patterns, that is because and only because of ethnic oppression." Assuming this, preferentialists conclude that no corrective or compensatory remedies can do what must be done. After the repeated application of such remedies, they point out, jobs and housing and educational attainments remain clustered by ethnic category. It must be, they infer, that the discriminatory infection is so deeply set that no normal medicine can uproot it. What can? Only a frontal attack upon the manifestations of discrimination, the ethnic clustering itself. "Never mind (say they in effect) who or what was responsible for present societal imbalances; never mind whether there was wrongful injury done for which redress is justified. Ethnic imbalance is injury in itself; the only satisfactory solution is one that moves toward, and ultimately requires, the establishment of numerical ratios that would have been achieved if the original distribution of goods had been random."

How much justice will be done, for the preferentialist, thus becomes quantitatively measurable. It is measured by the closeness with which the outcome of any distributive policy approximates ethnic proportionality. Perfect proportionality may never be attainable, but it remains (on that view) the ideal objective.

Wiith this vision of society ideally homogenized, the task of uprooting racial discrimination is transformed into the task of achieving certain numerical outcomes — in employment categories, professional schools, juries, everywhere. Some versions of this position explicitly replace the ideal of equal *treatment*, or equal *opportunity*, with the ideal of equal *results* — that is, results proportionately equal in their impact on ethnic groups. More subtle versions do not openly abandon equal opportunity as ideal, but insist that wherever numerical ethnic imbalance appears or reappears we may infer conclusively that equal opportunities had not been given. Equality of group results, because it is supposed a necessary concomitant of genuinely equal opportunity, thus becomes the standard with which equality of treatment is to be measured.

The sociological supposition upon which this approach is grounded — that truly equal opportunities would inevitably yield approximately equal group results — is naive and seriously mistaken. It is not the case that, when ethnic discrimination is eliminated, employment patterns, educational patterns, and the distributive patterns of other important social characteristics will be random across ethnic groups, yielding approximate numerical proportionality in each sphere. Long experience around the globe confirms the falsity of that assumption. Decades of historical and sociological inquiry by leading scholars, who have recognized and carefully discounted actual discriminatory practices, leave the matter in no doubt. The evidence — too voluminous to include here in full — is overwhelming.

Does nondiscrimination yield random distribution? Here follows some of the evidence showing, or tending to show, that it does not.

The "degree of enclosure" of an ethnic group has been shown sharply distinguishable in many contexts from the "relative social status" of that group (7). In examining ethnic stratification, Haug draws a similar distinction, concluding that ethnic stratification cannot be accounted for by any single factor. The real plurality of ethnic groups introduces a "special condition of diversity which varies widely in degrees across societies"(5).

Parsons reports what he calls the "notable confirmation" of an important general feature of ethnic identification: that it is in large degree "optional and voluntary" and, at least in the United States, will often be maintained by a group even when doing so runs *counter* to the striving for upward mobility (6).

Many anthropologists have argued that the causes of ethnic patterns and ethnic differentiation are essentially *economic.* Pursuing this theoretical approach, Haaland accounts for the maintenance of sharp ethnic boundaries between major ethnic groups in the Western Sudan, while at the same time explaining some erosion of these boundaries in response to varying geographical and climatic circumstances. The underlying determinants of the actually resulting pattern, he concludes, are a combination of "specific economic structures" and the given "ecological setting" (4). Siverts takes a similar approach in explaining the relatively rigid retention of ethnic boundaries and hierarchies in southern Mexico. Individuals in the several ethnic groups repeatedly face, for reasons having nothing to do with racial discrimination, "similar dilemmas of allocation of labour and capital to which the repertoire of responses is limited and stereotyped"(8).

The importance of *political* factors in dividing ethnic groups, causing distinct patterns in their activities and reinforcing cohesion within each group, is emphasized by Bell. He concludes that the "very nature of interest-group rivalry, where the plural groups are evidently distinct, makes it certain that the political arena becomes the most salient in the competition for the chief values of the society"(1).

One of the most influential of all studies of ethnic groups, published almost two decades ago by Glazer and Moynihan, explored the patterns and relations among ethnic groups in New York City. This work, *Beyond the Melting Pot: The Negroes, Puerto Ricans, Jews, Italians and Irish of New York City*, demonstrated that elements of society long thought of as marginal and in the process of absorption were solid and central (3). Ethnicity, in the years since, has come to be recognized as a healthy feature of American society, though it results in patterns of distribution very far from that random distribution idealized in "the melting pot." With Glazer and Moynihan's massive recent publication, *Ethnicity: Theory and Experience*, that understanding has deepened. Elsewhere Glazer has compiled masses of evidence to establish beyond reasonable doubt that concentrations of blacks and other minorities in America in housing, education, and jobs is simply not to be accounted for by deliberate discriminatory behavior. The use of racial and ethnic categories to force distribution in accord with an idealized pattern inconsistent with reality, he concludes, arises from

> "... a radical misunderstanding of how we in the United States have attempted to deal with the problems of a multiracial and multiethnic society. The pattern we have developed is not easily summed up in slogans — which is perhaps its defect — for we have decided against both the forcible assimilation of all groups into one mold and the legal recognition of each group for the establishment of a formal parity between them. It is a pattern that has emerged from the complex interplay of constitutional principles, political institutions, [and] American culture..." (2).

One could go on and on. In sum, cultural homogenity is not now a reality, in America or in any other complex society, and is not likely ever to become a reality. When a distinguished American woman once announced to Thomas Carlyle that she accepted the universe, he replied that she had better.

In the real world ethnic groups exhibit characteristics and preferences that, in one way or another, commonly result in distinct patterns marked by clustering in residence, employment, education, and other spheres. Cultural pluralism — to which honorific lip service is widely given — does not arise and is not maintained merely by the oppression of majorities. Anthropological data astutely refined, confirmed by the evidence of everyday experience, establishes the point: the demand for numerical ethnic "balance" in social distribution is the product of a romantic misconception.

In part, of course, ethnic disproportions *are* the result of discrimination, overt and covert. To suppose that invidious maltreatment had no role in creating present ethnic clusters would be as wrong-headed as to insist that clustering is explained by discrimination alone. In matters of such complexity, what could be clumsier, or more obtuse, than social programs built on the assumption of a single causal factor? Yet that is what the preferentialists — meaning to do good, of course — generally promote. They may not see that so crude a view of social causation has been assumed, but it is upon just such an assumption that their arguments essentially depend. Without it they could not plausibly use numerical proportionality as the criterion of justice.

Ironically, the quest for ethnic "balance" has consequences the very reverse of those ultimately sought. Wanting justice, the advocates of group proportionality do injustice; hoping to eliminate ethnic discrimination, they impose it to attain the numerical ratios they believe ideal; seeking to reduce racial disharmony, they exacerbate it.

GROUP RIGHTS AND INDIVIDUAL RIGHTS

The misconception that engenders the ideal of a homogenized society is distinct from, but compounded by, moral mistakes about the ways in which, where there has been ethnic discrimination, redress is justly given.

The most common and most serious moral error in this sphere is that of treating ethnic and racial groups as the bearers of rights and liabilities, and therefore as the appropriate recipients of compensatory satisfaction. This blunder magnifies the damage done by the vision of homogenization. That vision distorts the assessment of injuries actually done, obscuring the real injuries by viewing every ethnic cluster as a sign of evil. Yet there has been much evil in this arena. The moral tangle is enlarged and inflamed by lumping the innocent with the guilty, exacting redress from persons (because of their group identification) to whom no wrong can be sensibly ascribed, and then further enlarged and further inflamed by lumping the undamaged with the damaged, distributing benefits to persons (because of their group identification) without regard to genuine entitlements.

This supposition of group responsibility and group entitlement has contributed greatly to the decline in the reputation of affirmative action. Groups are involved, clearly; but the distinction between appropriate consciousness of group and inappropriate preference by group must be borne in mind. When persons have been injured because they are members of an ethnic group it makes sense to use membership in that group in delineating the class to be compensated, as we have seen. Discriminatory injury has commonly been hidden, or partly hidden, by institutional policies that may not have been explicit, and may not be easy to document. Therefore it is not reasonable to expect that every member of such a group who has been so injured must prove the specific damage that he or she has suffered. If remedy for discriminatory injury can only be achieved by litigation focused exclusively on the circumstances of one individual, most such injuries will not be remediable.

But to affirm that ethnic groupings are rationally introduced is not to affirm that racial grouping is the factor of sole or even primary significance in giving redress. It is not the members of ethnic minorities only who have suffered employment discrimination or educational deprivation, and the like — and not every member of each such minority has been thus damaged. Therefore, it surely does not follow from the fact that a given set of persons were injured because they were blacks, or Jews, or orientals, that blacks generally, or Jews or orientals generally, are entitled to the redress that damage calls for. Precisely that fallacious step, however, is the implicit moral linchpin of arguments for group preference. The injury to be redressed having been attributed to group membership, the remedy is awarded to the group as a whole. Ethnic groups, rather than the individuals who compose them, are treated as moral agents and bearers of rights.

Group-think of this kind arises in part from the failure to see that a class of persons injured must be defined by more than ethnicity. But it also arises, in part, from confusion over the ways in which groups can be injured and may deserve recompense. It is possible for a group to be injured by the injury done to one or a few of its members. This can occur, however, only if that group has distinct corporate interests and is so organized that each of its members shares those interests by virtue of his role in the integrated whole. Thus, a football team may be injured as a team and be entitled to remedy as a team when a few of its members have been unfairly kept from playing. A company of actors may, as a company, be entitled to redress for the damage done to it when some of its members have been unjustly hurt or detained. The legitimacy of such group claims depends essentially upon the organic linkage between the group as a whole and its wrongfully injured members. Not every collection of persons is so organized; most groups are not so organized; racial and ethnic groups certainly are not so organized. To treat a race, religion, or nationality as though it were a corporate unity or team is a grave error.

The error is partly obscured by the fact that injuries done to some of the members of an ethnic group (because of their group identification) do understandably offend other members of that group and do threaten them. When, as a Jew, I learn of injuries done to other Jews just because they are Jews, my indignation will be justifiable and my anger appropriate. But I am surely not entitled, by virtue of being a Jew, to the same redress due those who were injured. To say that all Jews are hurt by such practices is not foolish — but putting it so is a kind of deliberate hyperbole. It is a way of calling attention to the generally offensive character of anti-Semitism and to the danger that it may spread, as it so often has. Yet others who are not Jews may be equally angry and equally pained. Simply being Jewish does not entitle one to compensation for anti-Semitism everywhere. Jews, as a group, hold no rights; it is persons, not ethnic groups, who hold rights, and who may be entitled to compensatory relief.

If circumstances may be imagined in which a social debt was owed to all or almost all members of a large class of persons, that debt could in any case not be discharged by giving special preference to a few of the members of that class. Group-think invites this second, also common moral error: the assumption that the class is compensated when some of its members are compensated. That, too, would require organic relations among the group members, relations known and agreed upon beforehand by all. Nothing like that is true of ethnic groups.

Thus individuals and groups are repeatedly confused with one another, first in one way, then in the other. For injuries suffered by individuals the group is supposed entitled to relief; and for entitlements held by groups, special payments to a few of its members are considered satisfaction. Affirmative action of the preferential species is encouraged by the first mistake and guided in practice by the second.

Moreover, if ever debts are owed to all or almost all members of an ethnic group, surely it would not be the case that they are owed equally to all members of that group. Just recompense would certainly require some appraisal of the nature and degree of the injuries actually suffered by the many persons involved, in order to give redress with instruments that at least *aim* to match the remedy to the wrong. Any public policy that gives special but equal favor to all members of an ethnic group necessarily ignores this proportionality in response. Many public policies, under the name of affirmative action, do worse than ignore this proportionality; they confer preference by ethnic group for jobs or schooling opportunities in contexts requiring (for the preferences to have effect) a rather high minimum level of attainment or promise. It is therefore a common result in the United States that the beneficiaries of preferential programs are those minority group members whose injuries (assuming arguendo that every member had been injured) were the least damaging. Those most gravely hurt, whose ability to perform in school or on the job had been most seriously affected, rarely benefit from programs giving preference to ethnic groups as a whole. Instruments that would apportion redress to those injured with perfect or near perfect proportionality are too much to hope for, of course; expectations must be reasonable. Yet it is a common error, a kind of moral blindness encouraged by group-think in matters ethnic, not merely to ignore the proportionality of relief to injury, but to reverse it.

The moral mistakes of ethnic preference are thus three: it errs in ascribing rights to the group instead of the person; it errs in supposing that when an entire class of persons is entitled to relief, payment to some is partial payment to all; and it errs in failing, not accidentally but necessarily, to fashion the remedy to fit the wrong.

AFFIRMATIVE ACTION AND THE RIGHTS OF CITIZENS

The wrongs done to ethnic minorities across the globe have been grievous and cry for remedy. Affirmative action carefully tailored to give just remedy and to insure that the habits and devices of discrimination are thoroughly eliminated is a moral (and often a legal) duty. In the fulfillment of this duty, however, we must carefully avoid that excess of zeal which infringes upon the right of all citizens, whether of the minority or of the majority, not to suffer loss or damage simply because of ethnic identification.

Some affirmative steps to deal with the problems of ethnic minorities are not only permissible but morally obligatory. Justice to all parties requires that we design, imaginatively but with restraint and with as much precision as circumstances permit, public policies both corrective and compensatory. The moral enthusiasm needed to develop such policies, and the political energy needed to implement them, will be undermined by well-intentioned favoritism insensitive to universal individual rights. Humane solutions to the problems of minorities are not advanced by the careless equation of affirmative action with simple ethnic preference.

REFERENCES

(1) Bell, D. 1975. Ethnicity and social change. **In** Ethnicity: Theory and Experience, eds. N. Glazer and D. Moynihan, pp. 141, 161. Cambridge, MA: Harvard University Press.

(2) Glazer, N. 1975. Affirmative Discrimination: Ethnic Inequality and Public Policy. New York: Basic Books.

(3) Glazer, N., and Moynihan, D. 1963. Beyond the Melting Pot: The Negroes, Puerto Ricans, Jews, Italians and Irish of New York City. Cambridge, MA: MIT Press.

(4) Haaland, G. 1969. Economic determinants in ethnic processes. **In** Ethnic Groups and Boundaries: The Social Organization of Cultural Difference, ed. F. Barth, pp. 58, 71. London: George Allen & Urwin.

(5) Haug, M. 1967. Social and cultural pluralism as a concept in social system analysis. 73 Am. J. Soc. **294**: 303.

(6) Parsons, T. 1975. Some theoretical considerations on the nature and trends of change of ethnicity. **In** Ethnicity: Theory and Experience, eds. N. Glazer and D. Moynihan, pp. 53, 64. Cambridge, MA: Harvard University Press.

(7) Schermerhorn, R. 1970. Comparative Ethnic Relations: A Framework for Theory and Research. New York: Random House.

(8) Siverts, H. 1969. Ethnic Stability and Boundary Dynamics in Southern Mexico. **In** Ethnic Groups and Boundaries: The Social Organization of Cultural Difference, ed. F. Barth, pp. 101, 102. London: George Allen & Urwin.

BIBLIOGRAPHY

Glazer, N. 1975. Affirmative Discriminaton; Ethnic Inequality and Public Policy. New York: Basic Books.

Goldman, A.H. 1979. Justice and Reverse Discrimination. Princeton: Princeton University Press.

Livingston, J.C. 1979. Fair Game? Inequality and Affirmative Action. San Francisco: W.H. Freeman and Co.

Minorities: Community and Identity, ed. C. Fried, pp. 365–373
Dahlem Konferenzen 1983. Berlin, Heidelberg, New York, Tokyo: Springer-Verlag.

POLITICS OF AFFIRMATIVE ACTION: RACE AND PUBLIC POLICY IN AMERICA

M.L. Kilson
Dept. of Government, Harvard University
Cambridge, MA 02138, USA

INTRODUCTION

The policy of Affirmative Action seeks to remedy consequences of a century of both law-induced and custom-induced racial or caste constraints on the social mobility of blacks in American society. It is a color-conscious or race-conscious policy or, in the case of female beneficiaries, a gender-conscious policy. The execution of this policy involves "goals," "quotas," and "preferential treatment" in such areas as employment, education (colleges and professional schools), and government contractual procurement.

The first use of the term "Affirmative Action" occurred in an Executive Order by President John F. Kennedy (Order 10925 of 1961) that forbade racial discrimination by federal contractors. Further Executive Orders by Presidents Johnson and Carter broadened and strengthened Affirmative Action, tilting it ultimately in the direction of goals, quotas, and preferences. A variety of federal statutes aided this evolution — e.g., The Public Works Employment Act of 1977 required a 10% allocation to minority enterprises; the Department of Transportation Act of 1976 required that agency to provide venture capital and bonding facilities for minority entrepreneurs; a variety of federal educational fellowships have been skewed disproportionately to minorities; and other federal statutes and administrative guidelines were evolved with the same Affirmative Action intent. Perhaps the key federal instrument in the development of Affirmative Action is the Equal Employment Opportunity

Commission, created first under Title VII of the Civil Rights Act of 1964 but given wide powers of intervention into job market discrimination by the Equal Employment Opportunity Act of 1972. Employment goals and quotas for blacks and other minorities became more prominent features of Affirmative Action from this point onward. And, of course, legal and political conflict over Affirmative Action intensified from this time as well.

The opponents of Affirmative Action have pursued several lines of argument:

First, the skewing of opportunity for jobs, education, and general social mobility in favor of blacks, women, and other minorities is unprecedented and thus unfair, for no other groups have been *affirmatively assisted* through politics or public policy.

Second, disproportionately aiding blacks over whites in job and educational opportunity, after a century of massive discrimination against blacks, violates the new status of equality-before-the-law that was created by the Civil Rights legislation of the 1960s. Those citizens who are asked to sustain restitution to Afro-Americans by way of Affirmative Action (e.g., Bakke) did not commit injurious or oppressive acts against Afro-Americans in the past.

Third, it is argued that merit, an important value in modern democratic societies, will be made a shambles by Affirmative Action. Some defenders of merit even claim that persons or groups benefiting from Affirmative Action suffer when merit is modified, for they bear a stigma — a stigma of self-doubt and inadequacy.

IS AFFIRMATIVE ACTION UNPRECEDENTED?

Practices akin to Affirmative Action are not without precedent in American politics and public policy. Since the Civil War, politics in American cities, counties, states, and at the federal level have involved the skewing of extensive social resources for long periods on behalf of specific ethnic and interest groups. This process might be called de facto Affirmative Action, a prominent mechanism of which was "patronage politics," often without a legal basis but also with a legal basis. This was the preserve, in different periods, of Irish Americans and other white ethnics in cities and states, or the preserve of Anglo-Protestants in cities or in a whole region such as the South. This de facto Affirmative Action has also been the preserve of all kinds of white interest groups — groups such as veterans, producer groups such as tobacco farmers, cotton, dairy, soybean, wheat farmers, and other interest groups.

In San Francisco between 1870 and 1900, for example, Irish Americans — a small but influential force in the city's politics — received affirmative assistance through control of 30% of all white-collar public jobs, and by 1900 the public sector in San Francisco employed 10% of all Irish white-collar workers. Between 1935 and 1955, millions of white Americans were affirmatively assisted through the Federal Housing Administration's (FHA) allocation of some 10 million housing units almost exclusively to whites, with blacks receiving barely 1% of FHA-provided housing. This affirmative assistance to whites was administratively mandated by FHA guidelines, guidelines that explicitly excluded Afro-Americans as housing recipients. A similar form of affirmative assistance for whites was apparent in the admissions practices of America's state university system from their founding in the late 19th century to the 1960s. No whites ever met the intense and rigid exclusion from as wide a range of state education institutions as Afro-Americans endured for virtually a century.

These instances should suffice to illustrate the point that something akin to Affirmative Action or affirmative assistance characterized important aspects of American politics and public policies well before the evolution of contemporary Affirmative Action policy for blacks, other minorities, and women. While not usually declared as preferential practices for a *given class of persons* (though occupational and educational preferences for war veterans in fact were so declared), the above-mentioned instances of what I call de facto Affirmative Action have the "quota" and "preference" dimensions now associated with official Affirmative Action. And, like the current practices, the de facto ones had a morally and politically workable rationale surrounding them: they served the ends of justice — of "moral and lawful equality," as Rousseau put it — in a society seeking to serve the "general will," the will of society viewed as a single whole.

The dialectic between a *particular will*, always grasping for privilege in its primitive (unsocialized) state, and the *general will* is inevitably messy. This is especially so when the particular will is organized as interest groups, factions, ethnic groups, etc., rendering the democratic society's quest for equality forever problematic. The use of politics and public policy to regulate this dialectic between the general and particular will is not likely to be any less messy — especially if equality is one's purpose and goal.

AFFIRMATIVE ACTION AS REVERSE DISCRIMINATION

Allowing the dialectic between the general will and particular will to evolve free of intervention (regulation) from politics and public policy is, therefore, a recipe for inaction in regard to a substantive equality in American society.

Since the early 1970s the opponents of Affirmative Action have, interestingly enough, characterized it as "affirmative discrimination" or "reverse discrimination." This perspective highlights the doctrine basic to American democracy of "equality-before-the-law," which from the reverse-discrimination perspective is purported to be a substantive reality *now* for both white and black Americans, owing to court decisions and federal legislation promulgated since the 1960s. Thus the practices associated with Affirmative Action, such as skewing job and educational opportunity on behalf of racial and gender groups, are considered unduly discriminatory toward whites, violative of the principle of "equality-before-the-law."

In this perspective, the paradigm for defining the role of politics in regard to the dialectic of the general and particular will is different from the one I delineate. For the opponents of what they call reverse-discrimination, the dialectic of the general and particular will requires no regulation, no politics and public policy. It is, as it were, *self-regulating*, guided by generic properties, such as the American creed of fair play, equity, and justice which now *apply as values* without regard to race, ethnicity, and gender.

Thus the purpose of politics in a democracy is to ensure each citizen's right to dress himself in those values that define equality, free from ascriptive impediments. For those who oppose Affirmative Action, it is not the purpose of politics and public policy to correct or compensate for any short-circuiting of or flaw in the acquisition of parity in equality status — flaws that, as in the case of Afro-Americans, have crucial impact upon the *utility of equality* in the social process.

Alas, the *utility of equality* in the social process is a subordinate value when compared to *parity in equality status*, as far as the reverse-discrimination perspective is concerned. This perspective on equality is, at the very least, ahistorical and amoral. It is doubtful that its proponents can apply it with very much consistency, owing to the objective messiness surrounding the dialectic of the general and particular will in democratic societies.

Grossly self-evident is this perspective's shallow understanding of the awful legacy of white racism in American life — the intricately calibrated institutional constraints upon blacks' competitiveness and the terrible psychic trauma that racist violence (even since the presumptive post-1960s equality-before-the-law era) visited upon blacks individually and collectively. The mandating by the state (courts and legislatures) of a universally applicable American creed is still a fragile accomplishment — in spite of the teeth in the mandate —

in face of this perplexing legacy of racism. Herein lies a crucial moral lacunae — a lapse of moral agility — in the reverse-discrimination perspective. It is a kind of moral stumbling that the German intellectual Grass has so acutely brought home to us in his response of outrage at the human wreckage that fascism visited upon European Jewry. The antidote to the ahistorical and amoral perspective of "reverse-discrimination" was sounded by Grass's response when he first set eyes on Dachau, realizing, he said, "what unbelievable crimes had been done in the name of my... generation and... what guilt, knowingly and unknowingly, our people had brought upon themselves."

Thus the shallow historical and moral structure of the reverse-discrimination perspective produces shallow analysis. The conservatives who propagate the reverse-discrimination viewpoint would surely agree that property owners and corporations who are victimized by the state warrant compensatory treatment. But what about a people like the Afro-Americans who are victimized by the state and, under cover of the state's example and authority, by its citizens? By any serious historical and moral measure, Afro-Americans are such a victimized people. Do they warrant the Affirmative Action policies and practices promulgated by the American government since the late 1960s, including a compensatory component? Here the reverse-discrimination spokesmen invent moral criteria. The moral valuations for property are different from those for people.

The moral confusion has several variations among the reverse-discrimination spokesmen. Wilson offers the following comment in an article in *The Washington Post* (March 4, 1981): "Affirmative Action is often thought of as a form of restitution, but those who pay (e.g., Alan Bakke) are not those responsible for the earlier evil." No serious conservative — by which I mean one of Burkean fiber — would apply this formulation to, e.g., indemnities imposed on the German state for expropriating property and destroying careers of millions of Jews during World War II. Nor would a serious conservative apply this formulation to a court's award of damages against a chemical corporation for contaminating the drinking water or housing sites of an American community. Successor politicians to the German state and citizens, too, while not having injured Jews themselves, are nevertheless liable to acknowledge restitution, and the same is true for the new executives and stockholders of the polluting firm, while not themselves polluters.

Thus in their restitution dimension Affirmative Action policies — enacted by our government and courts after findings of massive rights violation — are as valid as the indemnities from the German state or damages against a polluting firm. That the cost-allocation dynamic for restitution under Affirmative Action policies renders the individual citizen's role in restitution more visible and immediate is undeniable. It does not, however, negate the equivalence — actual and moral — of these several restitutionary processes. Only a special moral posture, one rather callous in its indifference to the wreckage racism has visited upon Afro-Americans, can allow one to assert such negation. And when combined with the relatively easy politicization of white Americans who understandably resent or are doubtful about the immediacy of their relationship to the restitutionary dynamic under Affirmative Action, the politics of Affirmative Action is inevitably precarious, highly contentious, and risk-laden for those politicians who brave the storms to support it.

AFFIRMATIVE ACTION AND MERIT

Owing to the immediacy for citizens of the cost-allocation in the restitutionary process under Affirmative Action, the issue of merit looms large. In the reverse-discrimination proponents' lexicon of values, merit is the other side of the coin of equality. If Affirmative Action negates equality it also negates merit.

In this ahistorical perspective — treating merit and equality as unqualified by time and space... and morality — Affirmative Action is an illegitimate encroachment upon merit, and as such distorts it. However, in the real world of American history — of unfair distribution of opportunity and leverage — there is no intrinsic reason why merit is more grossly distorted under the politics of Affirmative Action, than it was during the patronage-fueled politics of city, county, and state machines, North and South. Merit is, in these political realities, modified, fabricated, above all, *circumstantially defined.* It happens that American pragmatism — our greatest gift to the democratic process — has allowed us to interpret meritocratic criteria broadly, not rigidly; circumstantially, not ideally.

Thus, in the real historical world of American society, meritocratic criteria are in practice relative, not absolute. Any Brahmin WASP in early 20th century Boston could tell you of the variety in meritocratic criteria found at American universities of that era — the gaps between, e.g., Harvard Law School controlled by the WASP elite and the numerous professional schools founded as social mobility mills for working-class ethnics, schools such as Suffolk Law School or New England School of Law. The latter, for instance, did not require a Bachelor degree before pursuing legal training until the middle 1950s; the

Irish in Massachusetts had enough political clout to shape the laws and standards governing the establishment of professional schools and did not hesitate to circumstantially qualify merit. And, of course, one need hardly mention the plethora of correspondence-degree institutions that commenced in the early 20th century, the kind where my father, exercising pragmatic good sense, gained his first degree in religion in the 1920s, Chicago's Moody Seminary. Here, then, was a redefinition of meritocratic criteria as defined initially by upper-class institutions — a redefinition on behalf of equality of opportunity for social mobility for low-status Americans of one description or another.

Ignoring the intricately pragmatic fabric of American political culture, the "reverse-discrimination" critics of Affirmative Action are as wrong about the nature of merit as about the individual's perception of its circumstantial definition. Though there is some perception of a stigma of self-doubt on the part of beneficiaries of Affirmative Action, the critics of Affirmative Action forget that stigmas are usually relative, not absolute. Thus, in relation to the stigmas that Afro-Americans endured during the century-old era of deep-rooted American racism, the stigma associated with Affirmative Action policies are, to put it mildly, rather easier to shoulder. Affirmative Action critics also ignore the fact that, since the late 19th century, groups other than blacks carried stigmas associated with policies that disproportionately aided them with jobs, contracts, educational resources, etc. Brahmin WASPs in late 19th century Boston, and well into this century, despised the thousands-upon-thousands of Irish Americans who gained jobs and other benefits through the patronage of city machines and state bureaucracies. Italian construction contractors in Massachusetts who gained multimillion-dollar contracts through a combination of patronage and corruption are stigmatized in the eyes of some fellow citizens, and perhaps in their own eyes, too. But the Irish, Italian, Jewish, and alas, even the WASP patronage recipients who, remorseful over the stigma from affirmative assistance in jobs and social mobility, surrender the latter in atonement are few, if any.

Thus the point here is that the stigma associated with Affirmative Action is traded off, as it were, for the benefit produced by this policy. In a perfect political system these issues might not arise. But, though the reverse-discrimination proponents seem to believe otherwise despite their air of realism, ours is not a perfect or ideal system. In a pluralistic and intricately pragmatic democracy like ours, highly prized ideals such as equality-before-the-law and meritocracy cannot avoid circumstantial adjustment through politics.

CONCLUSION: AFFIRMATIVE ACTION AND THE POLITICAL SYSTEM

The politics of circumstantial adjustment of meritocracy are not likely to be system-threatening, though some opponents of Affirmative Action believe otherwise. The unlikelihood of non-system-threatening conflict related to Affirmative Action is due, in part, to the fact that many American ethnic and interest groups have, indeed, experienced some variant of affirmative advancement, even though middle-class status compels some not to admit as much.

Precisely how individual and collective memory among ethnic and interest groups reasserts itself across the life cycle of a political system, providing some semblance of cognate political identification in otherwise conflicting policy milieux, is not yet well understood. There is evidence suggestive of what I call cognate political identification in regard to black/white competition and conflict over Affirmative Action.

A Sindlinger Poll in 1980 sought to uncover whether whites' opposition to certain features of Affirmative Action (e.g., 84% opposed quotas) produced any lasting resentment toward persons who benefited from this policy. One question asked whether Affirmative Action assistance to blacks and women "affected your own attitude toward minority and women doctors and lawyers." Some 20% of whites said "yes," but 75% said "no." When asked whether "qualifications of a black lawyer or woman doctor are likely to be worth as much as those of other doctors and lawyers," some 17% said "no," but 67% said "yes." Finally, when asked "if you would, *yourself,* avoid dealing with a black doctor or a woman lawyer," some 11% said "yes," but 84% said "no."

What is the function of cognate political identification among politically warring groups? For one thing, it produces some degree of "political empathy" between the established and claimant groups. Political empathy, in its turn, reduces the tendency of zero-sum or foreclosure perceptions of the conflicting claims. To the extent that political empathy persists throughout the high-fission period of black/white conflict over Affirmative Action, the potential system-threatening outcomes remain manageable. At this point, the conflict is capable of changing its parameters, with fusionary pressures checkmating fissionary ones.

This scenario of conflict management in regard to Affirmative Action might well be off the mark, of course. The reverse-discrimination perspective, however, is probably not even near the mark when considering conflict management. It tends to intensify fissionary parameters of the conflict over Affirmative Action, owing its indifference to the role of political empathy, as well as its support of foreclosure perceptions among the competing groups — a tendency intrinsic to the "reverse-discrimination" analysis. This alone induces me to question the efficacy of the reverse-discrimination perspective, regardless of my theoretical differences with it.

REFERENCE

(1) Kilson, M. 1977. Generational change among black Americans. **In** Samuels, R. Political Generations and Political Development, p. 36. Lexington, MA: Lexington Books.

Minorities: Community and Identity, ed. C. Fried, pp. 375–384
Dahlem Konferenzen 1983. Berlin, Heidelberg, New York, Tokyo: Springer-Verlag.

THE CASE FOR A GROUP LIBEL LAW IN GREAT BRITAIN

J. Gewirtz
Board of Deputies of British Jews
London WC1H 0EP, England

INTRODUCTION

The term "group libel" in this paper refers to that genre of legal processes in which the right to maintain an action is delegated to a representative body empowered by statute to maintain an action for defamation on behalf of a particular racial, ethnic, national, or religious section of the general public. The nature of the action varies from country to country.

In France (and in Belgium, which patterned its law on the French model) an association whose "statutory aim (is) the defence of human rights or the combating of racial discrimination is (provided it has been in existence for at least five years prior to the actual defamation) entitled to initiate proceedings under the penal code in cases of incitement to discrimination, segregation, hatred or violence against ethnic or racial groups or members of such groups"(1,2).

In Italy, representatives of the Jewish community were enabled to recover damages following a successful criminal prosecution of neo-Nazis for "publicly praising" genocide (3). In Switzerland, by judicial extension of the Swiss civil code which affords protection of individuals from illegal encroachment on their personal interests, the Federation of Swiss Jewish Communities was able to obtain an order banning the distribution of copies of an anti-Semitic book on the grounds that it was insulting to Jews and advocated curtailment of their rights (4,5).

The United Kingdom has no group libel law, although it has enacted legislation which makes it an offense to incite racial hatred.

It is my contention that the British law, despite legislative attempts to strengthen it, has been largely ineffective in stemming the considerable increase in racist propaganda emanating in the main from neo-Nazi groups and directed against ethnic minorities and immigrant groups. I believe that the failure comes not from any lack of purpose or skill in draftsmanship, but because the present law has set itself the enormously difficult and impractical task of altering, through coercion, public and private attitudes to racism in a country in which the right to dissent or to express any opinion no matter how obnoxious, without let or hindrance, is considered the inalienable prerogative of every subject of Her Majesty's realm.

INCITEMENT TO RACIAL HATRED

In 1965, Parliament introduced Britain's first race relations law. Passage of the law reflected the changing nature and composition of society in the United Kingdom. The first decades after the Second World War saw a marked increase in immigration into Britain of former colonial peoples. Some, like the East African Indians, came because of difficulties in adjustment within the emergent independent African states. Others, notably the West Indians, were encouraged to come to meet the serious shortage of workers in the hospitals.

The tendency of the new immigrants to cluster in certain areas gave the impression that there were more of them than there really were. The fact that they were easily identifiable by color made them a convenient target for political agitation by extremist groups on the far Right.

The new law was "based on a clear recognition that the overwhelming majority of the colored population is here to stay, and that a substantial and increasing proportion of that population belongs to this country, and that the time has come for a determined effort by Government, by industry and unions and by ordinary men and women, to ensure fair and equal treatment for all our people, regardless of their race, color, or national origins" (6).

Racial discrimination was declared to be "morally unacceptable" and "socially wasteful." The main preoccupation of the legislation was to eliminate discrimination in the work place. Tacked onto the law, almost as an afterthought, was a provision which made it an offense to circulate written matter or use words in public which were threatening, abusive, or insulting and which were intended to and were likely to stir up racial hatred (7).

The inclusion of a provision which made the incitement to racial hatred a criminal offense constituted recognition by the Government that the elimination of discrimination could not be achieved in a climate of racial antagonism whipped up by politically motivated extremists.

The late 1950s and early 1960s had seen an alarming growth of activities of neo-Nazis groups, who were taking to the streets to demonstrate against the colored immigrants and with even greater virulence against the Jews, whom they blamed for conducting a conspiracy to "mongrelize" Britain's white population.

The use of "threatening, abusive or insulting words" at a public meeting to provoke a *breach of the peace* was already covered by the Public Order Act (1936). The new law extended the offense to written matter and applied to situations in which the likely result of the publication was to cause racial hatred.

The law against incitement has never been popular. Thirteen years after passage of the original act, *The Times* saw fit to attack the enlargement of the Public Order Act to take in *incitement to racial hatred,* which it saw "as a state of mind, not a mode of action."

"Englishmen," said *The Times*, "have a strong attachment to freedom of speech. The freedom was won...not just to enable people to say pleasant, fraternal and acceptable things ... but to say distasteful, unacceptable, provocative, antagonistic things .... Any criminal statute which is framed to circumscribe that freedom is likely to be given a bumpy ride, however desirable or even necessary its purpose may be" (8).

Indeed, in anticipation of such criticism the Government had hedged the law about with considerable obstacles, not the least of which was to require the Attorney General's consent before any prosecution could be brought. The police were thus discouraged from making arrests even where offensive literature was distributed outside schools or football grounds (9).

Between 1965 and 1976, when the law was amended to remove from the prosecution the burden of proving subjective intent, only 20 prosecutions were brought: four of them against blacks (successfully) and the rest against whites, with acquittals in over a third of the cases.

In fact, the 1976 amendment has not been the panacea hoped for by its proponents. Using as his criteria for prosecution the chances of securing a conviction, the Attorney General has reserved his consent only to "strong cases." In practice this has meant bringing before the bar of justice only those cases in which the language used or illustrations displayed were so shocking as to border on the obscene. Even here, however, the prosecution has come unstuck in the most flagrant cases of racial abuse either because of judicial obtuseness or a recalcitrant jury which saw each defendant in terms of "there but for the grace of God go I."

Thus, since 1976, despite the amendment of the law designed to improve the chances of conviction, there have been only sixteen prosecutions, scarcely an improvement over the preceding twelve years, even if the ratio of successful prosecutions to failures (13:3) has increased (10).

Among the more notorious failures to convict has been the acquittal of the leader of the racist Democratic National Party, Kingsley Read, who in a public speech in London's East End referred to "niggers, wogs and coons," and said of an Asian youth killed in a race riot, "One down, a million to go."

In his summation the judge ruled that the use of the words "niggers, wogs and coons" was not itself unlawful. "Britain," he added, "was still a free country and people should be able to say what they liked provided they did not incite to violence."

To the defendant he said after the jury's verdict, "By all means propagate your views. You have been rightly acquitted. But use moderate language. I wish you well"(11).

Kingsley Read was prosecuted and acquitted under the 1965 act. Cole and Jones were found not guilty of incitement to racial hatred in the very first case brought after amendment of the Act. At a public meeting they, too, referred to "wogs" and "coons," predicting civil war between whites and colored people and alleging that black people were responsible for bringing diseases into the country. Jones, however, was convicted by a majority verdict of 10–2 for using insulting and abusive words likely to cause a breach of the peace (12).

In the case of written material, the Attorney General has given his consent to prosecution only in the most extreme instances of racist abuse and provocation.

Relf, a racial activist with a long record of affiliation to fascist and neo-Nazi organizations, was jailed for 15 months for publishing leaflets containing slogans such as "You can take niggers out of the jungle, but you can't take the jungle out of niggers" and which juxtaposed photographs of Negroes and apes to indicate common descent.

Another of his leaflets threatened "A second Battle of Britain ... of black against white. The blood of innocent English people will flow in the gutters. White man awake, the Zionists and blacks are taking away your country"(13).

In a recent case (14) which received national publicity, Joe Pearce, the 20-year-old editor of the Young National Front newspaper *Bulldog*, was sentenced to six months' imprisonment after a retrial when the first jury failed to reach a verdict. The prosecution was based on articles in *Bulldog* warning of the formation of black terrorist gangs and accusing immigrant doctors of killing, injuring, and raping their patients.

In what came close to a defense of justification, defending counsel argued that the material was factual and came from reports in national and local newspapers.

The seriousness of the incitement was aggravated by its having been published at a time when summer race riots were taking place over wide areas of the country, a fact to which the judge alluded in his summation. Here again, there was a nexus between incitement to racial hatred and the danger of street violence.

A third case involved publication of a comic strip titled "The Stormer," patterned after the notorious Sturmer cartoons of Nazi Germany which contained material grossly obscene and defamatory of Jews and colored people. The cartoonist, Edwards, prominent in the National Front Constitutional Movement and the League of St. George, a British liaison group for international Nazis, was convicted for racial incitement and sentenced to twelve months' imprisonment (15).

An analysis of the cases studied — including several in which the incitement was combined with acts of physical assault or illegal collection of arms — tends to confirm the view that the danger of consequential physical violence is a major factor not only in the Attorney General's decision of whether or not to prosecute, but in judge and jury arriving at a conviction.

In thus acceding to the popular view supported by a preponderance of the press — that the law on incitement to racial hatred is an unwarranted intrusion on free speech — the Government has turned full circle, reverting to a threatened breach of the peace as the gravamen of the offense, the original point of departure.

The only practical difference is that the law now also affects written material.

It could be argued, too, that despite the considerable restraint shown by the Attorney General, a small gray area exists beyond that of breach of the peace in which incitement to racial hatred in certain exceptional circumstances will serve as the basis for an attempted prosecution.

It is significant in this respect that the Race Relations Act, as amended in 1976, instructed that the incitement provision was to be incorporated (as section 5a) within the Public Order Act (16).

The Government's dilemma in setting the boundary between incitement to racial hatred that needs to be curbed and the need to defend traditional free speech is expressed in the *Review of the Public Order Act 1936 and Related Legislation* as follows:

> "Given the existing provision for preserving public order, is it necessary or desirable to have a criminal offence that aims to prevent the use of spoken or written words which incite hatred against racial groups and thereby directly provoke disorder? Or should the objective be to prevent the use of language or behaviour which incites or is intended to incite hostility against racial groups, which could lead to subsequent, though not immediate, disorder or to criminal activity directed against members of racial groups? Should the offence be extended to words which could create racial disharmony, either with or without regard to the likelihood of subsequent actions; if so, should provision be made to safeguard fair and reasonable comment as well as, for example, the publication of the results of academic or medical research? It seems doubtful whether a provision could be drafted that would prevent damaging language and behaviour and would not encroach on the democratic right to express even unpalatable ideas nor present serious problems of enforcement for the police"(17).

It is difficult to see how this view of the Home Secretary as to the state of the law in this field can be interpreted in any other way than as a repudiation of the law itself.

Indeed, Bindman, legal adviser to the Commission for Racial Equality, is severely critical of the Government's view. Writing in the *New Law Journal,* he rounds on "the basic failure of the Government's review to recognize that racialism is not a legitimate political philosophy, fit to take its place with others in the market place of ideas, but is a disease which cannot be allowed to spread in a civilised society. While it may not be possible or desirable to prohibit the holding of racialist views, the expression of such views publicly is not required by any acceptable doctrine of freedom of speech"(18).

The fact remains, however, that considerable differences of opinion exist within the British body politic on whether it is right to limit free speech even for so laudable a purpose as preventing incitement to racial hatred.

TOWARDS A GROUP LIBEL LAW

In view of the confessed reluctance of the Government to prosecute under a law hedged about with obstacles to prosecutions which are made even more difficult by sustained opposition from judges, the press, and the general public, it would appear that the time has come to reconsider whether the present law serves any practical purpose.

The steady flow of racist material continues unabated, apparently undeterred by any fear of prosecution. If anything, the material now monitored by organizations such as the Commission for Racial Equality and the Board of Deputies of British Jews shows an increasing boldness and outspokenness in expression.

This is not an unexpected development given the paucity of prosecutions and the even rarer convictions. The matter is complicated further by the need to prove causation. How does one effectively show that the material, offensive as it is, is likely to affect recipients who are either already hopelessly prejudiced against the ethnic minorities or are so tolerant of them that they are likely not to be affected by appeals to bigotry?

In the area of incitement at public meetings or demonstrations the audience usually consists of racist fanatics, antiracist activists, and members of the ethnic minorities themselves: none of these are likely to alter their ideas in response to racist propaganda expressed in slogans and placards (19). It is no wonder, then, that the law has gradually been fudged by irrelevant questions pertaining to breach of the peace (19).

It is precisely this very conundrum, in which the Attorney General is required to make what is in many cases a political decision, which provides the most persuasive argument in favor of a group libel approach to the problem of racial abuse. By shifting responsibility for initiating an action to the victim of the abuse, the issue is removed from the political arena and placed where it belongs, both logically and morally, to be decided within an adversarial context.

The failure of successive Attorneys General to prosecute out of an unwillingness to curtail free speech or because of uncertainty as to the outcome would no longer serve as a license for demagogues and bigots to attack racial minorities with impunity.

Under the present administration of the law, minority groups are left without redress for defamations of the most offensive nature which, were they directed at individuals, would give to the victims a right to substantial damages.

To cite but one example: there is currently a sinister propaganda campaign of massive proportions extending over many countries in which the Jewish people are accused of having "invented the myth of the Holocaust" in order to fraudulently obtain reparations from the German Federal Republic. The campaign is centered in Britain and is organized and conducted by neo-Nazi groups.

One of the booklets, "Did Six Million Really Die?" written by a leading National Front activist under the pseudonym "Richard Harwood," has been translated into several languages and distributed in tens of thousands in England and abroad. But because the allegations are couched in pseudo-historical language the Attorney General has refused to consent to a prosecution. If similar charges involving fraud and deceit were aimed at an individual or a corporation, legal redress would be readily available. Ironically, where the victim of the libel is an entire community and the consequences of the defamation infinitely more serious and reprehensible, no right of redress is available.

While readily conceding that there is some merit in a system that prevents the state from placing its imprimatur on a historical fact — no matter how certain — it is an argument that would have no relevance in a proceeding brought, e.g., by a representative Jewish body on behalf of the Jewish community.

The defendant in such an action would be called upon to substantiate his allegations, failing which he would be required, as in a private action for libel, to face the risk of paying aggravated damages for invoking truth as a defense, either frivolously or out of malice.

The Common Law crime of seditious libel still exists, but in practice it has been allowed to pass into disuse because of its historical associations with political oppression. It is interesting that the Porter Committee, which in 1948 considered the possibility of Parliamentary legislation in this area, rejected proposals for a group libel law on the grounds that it would endanger "free and frank — albeit hot and hasty — political discussion and criticism."

While the effect of a group libel law would undoubtedly be to limit to some extent a "free exchange of ideas," it would likewise have the effect of elevating discussion to a higher moral plane. There is no virtue in a state of the law which allows the traducer of truth deliberately to malign sections of the community which are then denied access to the courts.

The right of free speech is not absolute, nor is it sacrosanct. It is sometimes forgotten that the most cogent exposition of the right of free speech was delivered by Justice Holmes in a case in which the United States Supreme Court sustained a judgement *curtailing* free speech (20).

That is not to say that the argument for preservation of free speech should be lightly regarded. But to allow it to override the right of a racial or ethnic minority to protect itself against vilification and abuse is to accord it a sanctity and a status to which no philosophical principle should be entitled.

NOTES

(1) See report of Faurisson case (Le Monde, 18 July 1981) for application of French statute which authorizes representative communal organizations to maintain an action for racial defamation.

(2) Law for the suppression of racism enacted 30 July 1981.

(3) Italian Law No. 962, Prevention and Punishment of the Crime of Genocide, 9 October 1967.

(4) Article 28 ....

(5) Federation of Swiss Jewish Communities v. Mathez (Canton of Vaud, 19 December 1968).

(6) Racial Discrimination [White Paper Command 6234, HMSO September 1975].

(7) Race Relations Act 1965 [Section 6].

(8) *The Times* (London), 9 January 1978.

(9) For a study of Nazi recruitment of schoolchildren and teenagers, see "Nazis in the playground," published by the Centre for Contemporary Studies, London, England, May 1981.

(10) Review of Public Order Act Command 7891, HMSO April 1980.

(11) Judge McKinnon, quoted in *Sunday Telegraph* (London), 8 January 1978.

(12) *Daily Telegraph* (London) report, 25 July 1978.

(13) *Daily Mail* (London), 30 January 1979.

(14) *Guardian* (London), 13 January 1982.

(15) *Guardian*, 29 September 1982.

(16) See Section 70, Race Relations Act of 1976.

(17) Command 7891 (as above).

(18) New Law Journal, 25 March 1982.

(19) Under Section 5A [subsection 6], publication and distribution to members of an association of which the publisher and distributor is himself a member is exempted from prosecution under the Act.

(20) Schenk v. US 249, US 47 (1919).

Standing, left to right:
Hugh Roberts, Shlomo Avineri, Marty Kilson, Michael Banton, Charles Fried.

Seated, left to right:
Jacob Gewirtz, Ann Dummett, Carl Cohen, Abigail Thernstrom, Mike Sandel.

Minorities: Community and Identity, ed. C. Fried, pp. 387–401
Dahlem Konferenzen 1983. Berlin, Heidelberg, New York, Tokyo: Springer-Verlag.

FORMS OF STATE AND RIGHTS
Group Report

A.M.A. Dummett, Rapporteur
S. Avineri, M.P. Banton, C. Cohen, C. Fried, J. Gewirtz, M.L. Kilson, H.J.R. Roberts, M. Sandel, A.M. Thernstrom

Our discussion is concerned with the rights of minorities within those states where majority rule is the basis of legitimate government. These include both unitary nation-states and federations. We exclude, however, states where the composition and policies of the government are not subject to periodic change by popular elections in which the votes of the majority prevail. These other states have, indeed, minority populations in the sense of ethnic and religious groups, but there are special problems for minorities who vote in countries with majority rule simply because, by definition, a minority cannot assume governing power by electoral means in such a state. Where there are substantial noncitizen minorities who cannot vote, they have no direct means whatever of asserting their rights through the electoral process. But whether or not minorities are enfranchised, a country in which the principle of majority rule is generally accepted faces intellectual, moral, and practical difficulties in determining how the rights of minorities can properly be established.

The concept of the state itself differs from one such democracy to another. In some, it is assumed that the state's sovereignty extends to jurisdiction over all the people within the territory; in others, the legitimacy of the state itself is challenged or its jurisdiction limited by treaty. The rights of minorities will arise from different kinds of claim in these two types of state.

The distinctions made above between types of government and concepts of state provide us with a working basis for discussion: they are broad descriptions rather than close definitions. On this basis, we consider the options open to minorities, to public authoriities, and to other institutions for securing minority rights.

GROUP RIGHTS AND INDIVIDUAL RIGHTS

We begin from the position that a right arises from a claim. We distinguish between a "mere" claim and a "legitimate" claim: the former is one that cannot give rise to assertion of a right, while the latter may do so.

Where a legitimate claim establishes a right, such as the right to free speech or to safety of the person, there may be a conflict between rights. Decisions must then be taken to determine which is a strong right and which a weak one. It is inevitable, however particular rights are defined, that in any estimation some will be "strong" and some "weak."

Those decisions are already difficult where two individual rights clash. The problem is more complex when the rights of minorities are at issue, because it is often hard to distinguish between the claims of an individual and those of a group member or of a group as such. For example, should we characterize the right of assembly for public worship as a right held by the individual members of the faith concerned or as a right belonging to the group as such?

We shall speak of the "aggregative" view of minority rights when we refer to the view that sees any minority as an aggregate of individuals, each of whom holds individual rights, and the "organic" view where there is a communitarian conception of the group. On the aggregate view, the whole group is equal to the sum of its individual parts and the group is an instrument of shared individual interests. On the organic or communitarian view, the whole is greater than the sum: the group is an expression of a common identity.

The ways in which members of any minority group may exercise their rights are determined by the recourse their society makes available. In the United States, for instance, an example of a liberal-democratic state where the individual citizen can claim rights by legal process, a minority member can claim redress as an individual, and his protection against majority oppression is achieved through his links as an individual to the body politic. His option is then an "immediate" one. But where a society recognizes that certain rights attach to groups (the organic view) and the individual's rights are in practice guaranteed by these group rights, his option is "mediated." Mediation may be performed by some other body, not necessarily by the minority group concerned; for instance, if a state subscribes to an international agreement which lays certain obligations on the state towards people under its jurisdiction. In such a case, a foreign worker who was being paid far below the legally-established rate for the job might be able to obtain a fair rate of pay not as a citizen (being foreign, this may not be open to him) or as a member of a national minority in a country where that minority was small and unorganized or not recognized as having any group rights, but through the mediation of an international instrument which would lay down rules on equal pay for all workers within a contracting state.

In practice, the options open to minorities for establishing rights are limited in innumerable ways. All constitutions pay lip service to freedom, equality, and toleration, but no state has a clean record. Where there is due process for the individual citizen and a tribunal supposedly impartial, there may also be lynch law for the minority and acquittal for the murderer who belongs to the dominant majority. Where a group, qua group, has preserved rights, the following questions arise: Who defines the group? How are its representatives recognized? Whatever the theoretical model, there will be disadvantages in practice. Even a well designed system operated conscientiously by noble human beings cannot remain effective over time, because societies change: people who at one stage counted as minority members are transformed into majority members; new minorities appear; the balance of power in society at large shifts about. But in the more common case of an imperfect system operated by fallible and corruptible human beings, the dice are always loaded against the less powerful — that is, against "minorities" in the present sense. The conflict is not simply between conflicting "strong" rights as defined above, but between the easily asserted right of the powerful and the practically insubstantial right of the powerless. If one takes a pragmatic view of how societies work and assumes that moral rights, and even legally established rights derived from legislation and judicial decisions, must compete on unequal terms against "old Fury"*, then one must look for options that involve maneuvers, allegiances, and entrenched constitutional safeguards.

The right to preserve a minority language is a useful example of a right which cannot be asserted by an individual merely on his own behalf. Nor only is language used socially, it cannot survive at all without interchange within a group and without some special facilities for its use: teaching in schools, use in the

* "I'll be judge; I'll be jury," Said cunning old Fury, "I'll try the whole cause And condemn you to death." —Lewis Carroll, "The Mouse's Tail"

media, publication of books, and so on. A right to a language (if it is a right) is not upheld simply if the use of the language is not forbidden, if majority pressure is so strong as to stifle counter-pressure. The preservation of a language may require some positive action by the state authorities. How can such positive action be obtained?

First, how can relatively powerless minority members use the system available to them? Although language is a matter of group rights, it may not be necessary to operate as a group to secure it; under some constitutions an individual might be able to go to court and establish a precedent. This would secure the right by making use of the society's "aggregative" approach to group questions. If the tribunal seemed likely to be biased against his particular minority language, he might seek alliance with more powerful or popular linguistic groups, one of which could try for the same type of individual redress, whose results would be transferred. If the state were of a type whose legitimate jurisdiction might be challenged where matters of language were concerned, the whole affair would become a political confrontation between the state authorities and the challenging group. If the state's legitimacy were not in question, and where state authorities recognized (taking the organic view) that certain groups within the state had a special communal identity which the state was bound to respect by means of specific arrangements, these arrangements could be utilized. (Yugoslavia provides a useful example of special arrangements concerning various linguistic groups. Other examples of special arrangements concerning matters other than language abound: many states have special provisions for religious schools, for varying marriage and inheritance laws, and so on, for specific groups.)

Between the liberal-individualist remedy and the specific group guarantee, there are many possible intermediate arrangements which have the effect of promoting minority rights. One of these is the type of structure found in a federal state, with popular representation chosen on at least two different levels. A simple example is the United States, where one House is elected on the basis of territorial divisions with roughly equal numbers of people residing in each area, and the other House on the basis of divisions between States which vary greatly in size and population but which are entities seen as deserving representation in their own right. One state in the German Federal Republic has a specialized arrangement for ensuring some representation to an ethnic minority which might not otherwise qualify under its general system (i.e., the Danish minority in Schleswig-Holstein). A more elaborate model was the plan drawn up by socialists in the Austro-Hungarian Empire before 1918 for a multiethnic commonwealth. This commonwealth was to have two kinds of representation: territorial and ethnic. The territorial divisions would be geographical; the ethnic groups, on the other hand (Czechs, Slovenes, etc.), would be scattered through different parts of the territory. They were to deal with matters supposed to relate particularly to ethnic minority needs such as schooling, religion, and culture, and to be empowered to raise taxes where necessary for these. Each person in the commonwealth was thus to have two sets of rights: one qua citizen and one qua group member. Unfortunately, this structure weakened those institutions which already cut across ethnic and territorial divisions. (The Austrian scheme was favored by left-wing groups in the United States, for at least twenty years after its promulgation, as a policy for democratization and the extension of power to black American groups.)

ACCESS TO POLITICAL POWER

In "majority-rule" societies of the general kind we are considering, should there be a right to electoral arrangements which *ensure* representation in the elected body to members of particular ethnic minorities? This question does not assume that an ethnic group is the same as a political interest group; the rationale of the representation would be group membership and not political affiliation or socioeconomic status. Nor can the answer assume that circumstances in all states containing minorities are the same: the historical relationships between groups will make a crucial difference. Thus, in the United States such electoral arrangements might make good practical sense for minorities whose territorial concentration was nowhere great enough to ensure the election of a representative, but the idea of an ethnic quota would be in inappropriate where members of a minority are regularly elected under existing arrangements. In the United Kingdom, Jews make up less than one percent of the general population but provide about five percent of Members of Parliament. They do not sit as representatives of British Jewry but as ordinary territorial constituency representatives.

It may also be argued that arrangements for reserved seats are liable to misuse and could be turned against the underprivileged instead of to their aid (as in some former British colonial territories). A reserved-seat system might also make ethnic divisions a permanent part of the political culture and undermine attempts at cooperation and integration on other bases. There are examples of a reserved-seat system for small minorities (Danes in north Germany, Germans in south Hungary) which work well in practice, but where minorities are large or scattered, alternative arrangements may be better. Bermuda is a society where two groups, black and white, compete for representation, and the solution adopted has been to use a voting system with two-member constituencies. By this method, each political party can put up two candidates, one black and one white, in each constituency, and voters have the opportunity to back, at the same time, a group member and a party member as they choose.

Such arrangements can deal more or less adequately with minority groups of citizens, but not with a situation where minorities are excluded from the electoral process on the ground that they are noncitizens. Some states such as Sweden already give voting rights to noncitizens, but in special-purpose elections and in inconsistent ways. The United Kingdom allows the vote to Irish citizens and Commonwealth country citizens but not to resident aliens.* Neither of these methods ensures actual representation by a member of the noncitizen group.

However, there are dangers in group voting, very evident from the experience of Northern Ireland. There, voters for representatives in the Westminster Parliament which controls their fate do not have the choice of voting for the mainland political parties, all of which have up to now refused to operate in Northern Ireland. They are left with the choice of voting for sectarian parties or for ineffectual, because purely provincial, non-sectarian parties. This has a double effect: the political primacy of sectarian divisions is perpetuated, since the common interests which transcend these divisions are denied effective political expression; and Northern Ireland electors are barred de facto from full participation in the British political system. They cannot influence the outcome of general elections, nor participate in the formulation of national policy (towards Northern Ireland itself, or in any other respect) because they have no opportunity to vote for a party capable of forming a government at Westminster.

If, for any reason, a system that *ensures* minority representation is unacceptable, there is the possibility of adopting a system which at least *facilitates* the election of candidates preferred by minority groups. Whereas the first option will commend itself to a society that takes the organic view of group identity, the second will find more favor on the aggregative view. The answer, in practice, will also depend on the nature of the demands put forward by minorities themselves. The re-drawing of electoral district boundaries — otherwise known as gerrymandering — is one possible means of facilitation. Another might be a voting system with multiple-member constituencies. A party-list

* Commonwealth and Irish citizens are not "aliens" according to British nationality law.

system could be devised which facilitates minority representation without distorting party-political preferences. Any facilitating system will produce borderline problems: the adverse effects such problems could create in a particular society must be weighed and every attempt be made to ensure a variety of options for voters.

Whatever electoral system a society adopts, the system will be open to misuse unless practical steps are taken to provide efficient and equitable means for its administration. For example, arrangements for the registration of voters vary widely from one country to another. Should illiterate voters be barred, and if not, should it be permissible for illiterate voters to be helped in the polling booth? Must registration be done or the vote be held at hours or on days when certain people because of their job hours or religious obligations are unable to attend? Is there a right to administrative arrangements which facilitate registration and voting, and if so, under what conditions does it apply? Each society must find an answer suited to its circumstances; these will vary far too widely for any general guidelines to be listed usefully here.

Associated with the electoral process are many factors outside the polling booth and the voting system, such as access to television and radio time during an election period, access to the press, rules on election expenses, and facilities for candidates to hold meetings. If minorities are to be assured of their rights, the public authorities may have an obligation to undertake affirmative action on their behalf. The allocation of state broadcasting time offers a particularly useful and flexible means for such action. Where public authorities fail to afford such help, and where minority resources are inadequate, there will be a role for "mediating" institutions to play. For instance, a trade union with more power, money, and organization than a given minority group may assist a group of immigrant workers and help to promote its interests; a church or mosque association may use its resources to help an ethnic group, many of whose members practice the religion concerned.

Even if arrangements for elections are as fair as they can be made, the importance of getting elected will vary according to the real power structure of the country concerned. If a minority has a right to access to political power, it may not be enough that members of the minority, or other persons ready to promote that minority's interests and wishes, get elected. Access to employment in senior posts in the government service may be almost as important: so may appointments to government boards, state broadcasting services, and commissions. Here, affirmative action in government employment is a possible answer. Where a minority group has a particular concern with one area of public policy such as education, the group's political power might be better advanced through appointments to the educational bureaucracy than by election of a representative to the national or local assembly, where power for the rank-and-file assembly member is not always very great. This sort of difficulty has to be considered first of all by the minority group itself, but also by the public authorities and by possible mediating groups.

CHALLENGES TO THE LEGITIMACY OF THE STATE ITSELF

In some societies, majority/minority politics is the outcome of a challenge to the legitimacy of the state. The demands made by minorities are no longer demands for shared power within an agreed and settled system: they go further.

There are several ways in which this situation can develop. Sometimes we see an inversion of the majority/minority relationship take place. In Cyprus, for example, the Turkish population used to enjoy political hegemony under the Ottoman Empire; then it became a minority living alongside a Greek majority under British colonial rule, and then a minority in an independent state where the Greek population was dominant. Both groups in Cyprus have had the historical experience of being minorities and both have been marked by this trauma. In Israel today, the Jews are a majority, but for two thousand years the Jewish people have had the experience in most parts of the world of constituting a minority. Indeed, they are a minority, in a slightly different sense, in the Arab world of the Middle East today. The Arab population of modern Israel can look back to having enjoyed majority status in the land (though under non-Arab rule) within living memory. Such historical inversions produce particularly difficult problems for a state. Sometimes such problems prevent from the outset the formation of a projected new state: British India never became the single independent Indian state that Congress had struggled for but instead was partitioned into India and Pakistan. Ireland did not become the unified, independent state which both Irish nationalist leaders and British governments had envisaged. In both the Indian subcontinent and Ireland, the claims of the main nationalist movement were contested by minorities (Muslim and Protestant, respectively). In such circumstances, partition may be the least unsatisfactory and most democratic solution available, but it is likely to leave behind it a legacy of conflict, where some elements of the different populations find themselves outside the states of their choice.

Conflicts of this sort often spill over into regional and international conflicts. The problems of the Greek and Turkish populations of Cyprus are complicated by the existence and activities of the Greek and Turkish states nearby. On the other hand, the Kurdish issue has never become a real challenge to the legitimacy of those states where Kurds live, because these countries have at least found a common interest in resisting the demands of the Kurds for independence. Nor has Basque separatism in Spain led to regional conflict across the state boundary with France; had there been a Basque nation north of the Pyrenees, it might have done so.

A country is more likely to split under such tensions where there are just two main groups involved. Where there are many, the desire for some common sovereignty may hold together the different groups. One can identify something like a dozen main groups in Lebanon, for instance, where the desire for national integrity remains strong despite the appalling strains of recent years.

A challenge to the legitimacy of the state itself is likely to arise once the claims of a minority as an aggregate become the claim to status as a community. It was thus that Zionism emerged. Jews in European countries after the French revolution had demanded emancipation or equal treatment within each country without questioning the legitimacy of the body politic, but Zionism made the new claim that Jews had a right to emancipate themselves. Similar claims to self-determination brought about decolonization of the former European overseas empires: Algeria did not challenge the right of France to be France, but it challenged the legitimacy of French rule over Algeria.

Violence has often erupted when state legitimacy is challenged or partition resisted. Nigeria is still in the process of change. From being an arbitrarily defined unit of Empire cut out of old kingdoms it became an independent state with a three-tier system which broke down in the civil war of 1967. Army coups over the last fourteen years have progressively re-elaborated the design for a framework within which competing ethnic units can coexist. The larger federation is still unstable. There are similar patterns of violence and instability in Chad.

What can we learn from these difficulties and from comparing them with the successful states, such as Switzerland, where different ethnic groups and languages coexist and the federation has enjoyed a long and stable history? There an individual can feel simultaneously a direct loyalty to the state as well as to the group, perhaps because loyalty to the state is mediated by loyalty to a particular group that is clearly part of the Swiss body politic. Where loyalty to the group but not to the state exists, there can be no mediation, and there is a risk of the challenge emerging and even of violent conflict breaking out. But immediate loyalty to the state alone is not usually desirable, either. It seems that there are real advantages for the minority group member as well as for the community at large in a "mediated" system where a variety of groups and institutions are interwoven in a flexible framework.

Mediating institutions include such bodies as trade unions, political parties, religious groups, and regional governments within a federal system, but this list is not exclusive. What makes a mediating body effective is a hard question we shall not attempt to answer for all possible circumstances, but examples from particular societies will readily come to mind.

ACTION BY INSTITUTIONS AND ASSOCIATIONS OTHER THAN THE STATE AUTHORITIES

The constitution of a state may be designed or amended so as to promote minority rights. Public policy, through legislation or government executive action, may attempt to promote these rights. But there is also a wide range of possible options that other bodies may undertake to the same end. We call these "non-state options."

The most obvious instrument for protecting or establishing minority rights is a minority itself, as a group. But some minorities are not organized as coherent groups. Sometimes there are several rival organizations, each claiming to be the true interpreter and champion; sometimes there are splits along party political lines, or between different socioeconomic groups within the minority, or following regional patterns derived from the country of origin, and so on. There are at least two kinds of difficulty in identifying and defining a minority group: How does the group identify itself? and How is it identified by other groups in society and by the public authorities? (For example, in the United Kingdom the Rastafari claim to be a religious body and so to have the right of religious freedom, but most public authorities do not admit this claim, and there are differences of opinion between other religious bodies on the question.)

Suppose, however, that we have to deal with a clearly identified group, organized and with leaders acknowledged both by group members and by the public authorities. This group may have either high or low viability: that is, it may have strong financial resources and a form of organization that can readily deploy manpower to do voluntary tasks, and it may include highly educated individuals who know their way around the society's power-structure; or, on the other hand, it may be poor, scattered, and completely lacking in powerful friends. In either case, for different reasons, there may be a tendency for the group to be coopted by mainstream agencies, but, again for different reasons, it may successfully maintain its integrity and independence.

Non-state options may be preferred to public policy options for various reasons. For instance, in the United States non-state options may be preferred by neo-conservatives on the principle that opposes in general governmental solutions to problems. But they may also be preferred by political radicals who mistrust the paternalistic character of government action. They may also be preferred by minorities because they can stimulate the self-confidence and self-respect of minorities and reduce internal tensions.

In many countries, certain minorities are well-known at least by name to the public at large and, though relatively powerless vis-a-vis the majority, are relatively better provided for than other minorities. These others, even if they are sizable groups with urgent needs, are somehow not counted when minority affairs are under discussion: they remain outside the debate and outside the bounds of positive action. For example, the Japanese Americans and Armenians in the United States belong to this category, as do the Chinese in the United Kingdom. Chinese migrants in most of the countries where they have settled in only small numbers have usually adopted a policy of keeping their heads down, running their own affairs in a self-sufficient way, and avoiding making demands upon the public authorities or upon society in general to be considered as a minority group. The special activities they undertake, such as Chinese-language teaching for their children, are generally carried out entirely with their own resources and without publicity. Some "outsider" groups, however, less well organized than the Chinese, remain underprivileged or lose their sense of communal identity. The former kind of "outsider" does not look for any mediating agency to assert his rights or link him with the state; the latter either becomes isolated and powerless or finds a mediator to help integrate him into the majority or at least some alternative association or group.

An option may be attractive or repellent to a minority for many reasons. Some acceptable kinds of mediation may be chosen because of the social structure of the country: in an industrialized society a trade union may offer better means for security to a minority group member than participation in an ethnic organization, for instance. If an Irish worker joins a trade union in a country where the Irish are a poor minority, the union will press for his rights as a worker, not his rights as an Irishman, but this sort of pressure may be of much more practical use to him than any other. (The case is not altered if the great majority of members of his union branch all happen to be Irish: the structure of the group is a workers' structure and not an ethnic one. But the union can still be instrumental in eventually promoting and safeguarding the rights of the Irish.) Attitudes as well as structure are important; someone from a family of small businessmen may be hostile to the idea of trade unions; someone who would enthusiastically adopt a group self-help project if it were advocated by a socialist may turn away from it if it is praised by someone with known right-wing views as a good way of saving public money and keeping undesirables off the streets.

Countries vary greatly in the range they permit in general, and not just to minority groups, to freedom of association and to initiation of non-state projects. The French system has many constraints: it has long been part of French political culture to bring the exercise of initiative under a centralized bureaucracy. (For instance, though the law has now been changed, it was for a long time the case that a parent could only register a child under a first name from an approved list, and Jewish names were among those excluded.) Group mobilization is thus far more difficult in France than in, for example, the United Kingdom, where voluntary associations of astonishing variety abound: anybody can start one without any need for registration or approval of any sort. Yet, very oddly, the United Kingdom, which has contained sizable, socially mixed, articulate, and highly publicized minorities from the Caribbean and the Indian subcontinent for over twenty years, has no nationwide representative minority organization drawn from either group. (The three rival Indian Workers' Associations come nearest to being such organizations.) This surprising lack may perhaps be due in part to the public policy of successive British governments which have succeeded in coopting potential leaders into a state-financed community relations structure, but further explanation is still needed. The conditions necessary for successful group mobilization are clearly an important determinant of the success of non-state options.

It is arguable that some non-state options are inconsistent with democratic values. All industrialized states take for granted, for example, that the state has a duty to assure universal education and to promote equality of opportunity in and through education. A minority which insists upon its right to special educational arrangements and which uses its own resources to press for these could perhaps hamper the opportunity of minority children to compete on equal terms with majority children (because, for example, of lower proficiency in the majority language, or inadequate time spent on subjects needed to qualify an applicant for higher education.)

Some have argued that preferential arrangements for minorities, such as job quotas, or affirmative action on behalf of minorities, such as practical assistance in taking up existing rights, may be undemocratic. The arguments on these questions are complicated and much-rehearsed, and we have not gone into them here. Are such options humane? Are they just? It may be helpful to consider the debates on "special provision" and "affirmative action" in the context of argument on other options. If one were forced into a choice of some option or another to achieve minority enjoyment of the right to equal opportunities in employment, which option would be admissable and which would be, come what might, totally unjustified: the use of force (an option to which the United States owes its national origin); nonviolent resistence and non-cooperation, as used by Gandhi and by Martin Luther King; civil disobedience; strikes, self-help, separatism, mediation through other institutions, organized lobbying of the public authorities? It is not unreasonable to ask a further question as well: which options are likely to be most effective and why?

SUMMARY OF MAIN THEMES

The description and analysis of our discussion has brought out a variety of approaches to the problem of how to assure minority rights. There is no consensus among us as to which theoretical approach is the best one, yet we have reached some agreed conclusions after travelling to them along different paths.

We were divided on the question as to whether a group, as a group, can possess rights. Some took the organic view, some the aggregative view; some were reluctant to accept even the aggregative view of the rights of members of a minority and preferred to see the matter as one of individual rights only. These differences were apparent in our discussion of group libel and of language preservation. Some rejected the concept of group libel altogether and denied the right of public authorities to limit freedom of speech by prohibiting defamatory remarks about ethnic or religious minorities. Others held that such defamation damaged all the individuals in a group, and so should be prohibited; still others thought that the group, as a group, had the right not to be defamed, whether or not damage could be shown to result to specific members of it. There was, however, rather wider agreement on the language question. It was generally agreed that it is consistent with the values of a majority-rule democracy to take special measures, including expenditure by public authorities and special schooling arrangements, to preserve a minority language.

Some of us believe that, in the long run, the best hope of minority group members is to rely on an immediate link between citizen and state, in the same circumstances as majority group members would so rely, to obtain rights, yet we are agreed that mediating institutions should help minority members attain their rights, at least in the short term. We do not all have the same reasons for coming to this conclusion, and we have differing views on the best mediating institution to use, but we all accept the principle of mediation.

There is a perennial problem about who decides where a group's best interests lie. How does one identify the group's own choice: from individual choices by group members or from some corporate decision? On what occasions should public policy considerations override a minority's preference? Though we have not resolved such difficulties, we have agreed that there are particular fields where minorities have a special claim to arrangements which facilitate their preservation of group identity and achievement: for example, in matters of religion, education, access to the media, and sport or recreation.

A minority can reasonably claim that its own situation justifies measures which affect public policy towards the entire community and not only towards the minority. For instance, though a minority could not reasonably claim to decide what languages the majority should be taught in school, a minority could justifiably ask that the training of all policemen should seek to eliminate race or religious prejudice.

We are agreed that policies toward minorities must safeguard the principles of free association and free exit.

We agreed that public policy and private action should seek to eliminate discrimination against members of minority groups, but there was a strong division of opinion on the use of affirmative action, some holding it necessary and others inadmissible. However, although there were undoubtedly differences between us on principles, we were able to agree on the desirability of some kinds of positive action to remedy the disadvantages of minority members.

Our discussion as a whole was based upon a structure which used the majority/minority distinction in preference to all others. The minority group members we considered were seen mainly as group members rather than as employed or employers, rich or poor. We probably did not pay sufficient attention to the different kinds of stratification that existed within the societies we mentioned, nor to the differences between states themselves. Nor did we pay adequate attention to the differences between types of minority, such as the one with a markedly different culture from the majority, on the one hand, and the one which shares in the majority culture but is distinguished by political powerlessness rather than internal cohesion, on the other.

We have attempted, however, to distinguish between the varied and interconnected problems faced by minority groups within majority-rule societies, and we hope that these distinctions will be useful starting points from which others can move forward.

LIST OF PARTICIPANTS

AVINERI, S.
Dept. of Political Science
Hebrew University
Jerusalem 91905, Israel

Field of research: Socialist thought, national conflicts in the Middle East, Marxism and the national problem

BANTON, M.P.
Dept. of Sociology
University of Bristol
Bristol BS8 1UQ, England

Field of research: Racial and ethnic competition

BARIC, L.F.
Dept. of Sociological and
Anthropological Sciences
University of Salford
Salford M5 4WT, England

Field of research: Ethnic minorities and language — communication, participation, control

BRAND, J.
Dept. of Politics
University of Strathclyde
Glasgow G1 1XQ, Scotland

Field of research: Nationalist movements in modern Europe (especially Spain and the U.K.)

BRUN-ROVET, J.
Ecole des Hautes Etudes
en Sciences Sociales
Centre d'Etudes Nord-Américaines
75005 Paris, France

Field of research: History of heterogeneity in the USA, with a special interest in the 18th century

COHEN, C.
Dept. of Philosophy
University of Michigan
Ann Arbor, MI 41809, USA

Field of research: Moral and political philosophy, ethical issues underlying political conflict

DUMMETT, A.M.A.
Joint Council for the Welfare
of Immigrants
London, WC1X 8SP, England

Field of research: British nationality and immigration law

EICHENER, V.
Institut für Arbeitssoziologie
und Arbeitspolitik
Ruhr-Universität Bochum
4630 Bochum 1, F.R. Germany

Field of research: Housing situation and integration of migrant workers

FRIED, C.
Harvard Law School
Cambridge, MA 02138, USA

Field of research: Philosophy of law — social justice, legal and moral rights

GEWIRTZ, J.
Board of Deputies of British Jews
London WC1H OEP, England

Field of research: Anti-Semitism and political extremism

GRAUMANN, C.F.
Psychologisches Institut der
Universität Heidelberg
6900 Heidelberg 1, F.R. Germany

Field of research: Social psychology

GUMPERZ, J.J.
Dept. of Anthropology
University of California
Berkeley, CA 94720, USA

Field of research: Ethnography of communication and discourse analysis in urban institutions

HECHTER, M.
Dept. of Sociology
University of Washington
Seattle, WA 98195, USA

Field of research: Sociology (sociological theory)

HECKMANN, F.
Hochschule für Wirtschaft
und Politik
2000 Hamburg 13, F.R. Germany

Field of research: Minority relations, history of sociology

HEIBERG, M.
International Peace Research
Institute
Oslo 1, Norway

Field of research: Ethnic nationalism and state formation with special reference to the countries of the Mediterranean basin

HOPF, D.
Max-Planck-Institut für
Bildungsforschung
1000 Berlin 33, F.R. Germany

Field of research: School problems of children of migrant workers

HÜTTEROTH, W.
Institut für Geographie der
Universität Erlangen-Nürnberg
8520 Erlangen, F.R. Germany

Field of research: Historical geography of Near Eastern countries

HUTH, L.
Communications Consultants
6301 Linden-Leihgestern, F.R. Germany

Field of research: Mass communications — mediation between social groups and roles

KILSON, M.L.
Dept. of Government
Harvard University
Cambridge, MA 02138, USA

Field of research: Studies in American political culture, political sociology of Afro-Americans

KITROMILIDES, P.M.
University of Athens
Athens, Greece

Field of research: Theories of the state, historical change and national identity

MARDIN, S.A.
Dept. of Politics
Bogazici Universitesi
Istanbul, Turkey

Field of research: Fundamentalism and religious change in Turkey and the Middle East

MARX, R.
Amnesty International
6050 Offenbach, F.R. Germany

Field of research: Right of asylum, international law

MORIN, F.
Institut des Sciences Sociales
Université Toulouse Le Mirail
31058 Toulouse Cedex, France

Field of research: Minorities, ethnicity, and new nationalism in Latin America, USA, and France

MURRAY, A.D.
Further Education Unit
Dept. of Education and Science
London SEI 7PH, England

Field of research: Public services and the multiracial community

PATTERSON, O.
Dept. of Sociology
Harvard University
Cambridge, MA 02138, USA

Field of research: Slavery, underdevelopment in the Caribbean

PATTERSON, S.
New Community
Commission for Racial Equality
London SW1E 5EH, England

Field of research: Comparative study of ethnic minorities in Britain (USA and Canada)

PETTIGREW, T.F.
Stevenson College
University of California
Santa Cruz, CA 95064, USA

Field of research: Intergroup relations in many forms (especially black-white relations in the USA, intergroup education, etc.)

PRADELLES DE LATOUR, M.-L.
E.R.A. du C.N.R.S. no. 974
Université Louis Pasteur
67000 Strasbourg, France

Field of research: The effects of frontier movements on identity and linguistics allegiance — a case study in the mining area of Lorraine

QUIGLEY, D.
Dept. of Philosophy
London School of Economics
London WC2A 2AE, England

Field of research: The Newar caste system (Nepal), cohesion and mobility in urban caste society

RAU, H.
Max-Planck-Institut für Ausländisches und Internationales Privatrecht
2000 Hamburg 13, F.R. Germany

Field of research: Spanish, Portuguese, and Latin American private law

REX, J.
Research Unit on Ethnic Relations
St. Peter's College, University of Aston
Birmingham B8 3TE, England

Field of research: Race relations in Britain

ROBERTS, H.J.R.
School of Development Studies
University of East Anglia
Norwich NR4 7TJ, England

Field of research: The character and evolution of the Kabyle question in Algeria from the precolonial period to the present day

SAIFULLAH KHAN, V.
Linguistic Minorities Project
Institute of Education
University of London
London WC1H ONS, England

Field of research: Language, identity, and ethnic relations (especially among South Asians) in Britain

SANDEL, M.
Dept. of Government
Harvard University
Cambridge, MA 02138, USA

Field of research: Political philosophy

SCHIFFAUER, W.
Boppstrasse 1
1000 Berlin 61, F.R. Germany

Field of research: Change of conception of world and self-image with the process of migration from Turkish rural areas

SHUE, H.
Center for Philosophy and
Public Policy
University of Maryland
College Park, MD 20742, USA

Field of research: Philosophical theories of human rights

SMITH, M.G.
Dept. of Anthropology
Yale University
New Haven, CT 06520, USA

Field of research: Social and cultural anthropology, pluralism, ethnic and race relations, Africa and the Caribbean

SONDHI, R.
Asian Resource Center, Handsworth
Birmingham B19 1NH, England

Field of research: Research connected with practical community work with Asian minorities in Birmingham, England

THERNSTROM, A.M.
The Twentieth Century Fund
New York, NY 10021, USA

Field of research: The electoral rights of blacks and Hispanics in the USA

THERNSTROM, S.
Dept. of History
Harvard University
Cambridge, MA 02138, USA

Field of research: History of American ethnic groups

URAN, S.
Laboratoire d'Ethnologie et de
Sociologie Comparative
Université de Paris X-Nanterre
92001 Nanterre Cedex, France

Field of research: Comparative study of nationalism, ethnic relations, and ethnic identity. Current research on the Jews and pieds noirs in French Algeria and on the Afrikaners in South Africa

WALLMAN, S.
London School of Economics
London WC2A 2AE, England

Field of research: Resource options for economy and identity in the inner city

WALZER, M.
Institute for Advanced Study
Princeton, NJ 08540, USA

Field of research: Political theory — distributive justice, equality and inequality

WILPERT, C.
Institut für Soziologie
Technische Universität Berlin
1000 Berlin 10, F.R. Germany

Field of research: Migration and ethnicity, intergroup relations, future orientations of minority families and youth

YALCIN, L.A.
Dept. of Anthropology
London School of Economics
London WC2A 2AE, England

Field of research: Ethnicity, tribalism, and local politics in S.E. Turkey with reference to the Kurds

ZOLBERG, A.R.
Dept. of Political Science
Graduate Faculty
The New School for Social Research
New York, NY 10011, USA

Field of research: International migrations and related policies, politics of multilingual countries

SUBJECT INDEX

AUTHOR INDEX

MIX
Papier aus verantwortungsvollen Quellen
Paper from responsible sources
FSC® C105338

If you have any concerns about our products,
you can contact us on
ProductSafety@springernature.com

In case Publisher is established outside the EU,
the EU authorized representative is:
Springer Nature Customer Service Center GmbH
Europaplatz 3, 69115 Heidelberg, Germany

Printed by Libri Plureos GmbH
in Hamburg, Germany